AF768401

Smart Grid Redefined

Transformation of the Electric Utility

This book discusses how smart grid will transform the way society will produce, transport, and utilize electricity. Dr. Vadari delivers an in-depth analysis of how business processes, regulatory regimes, and government policies interact and how these factors can transform the utility industry differently in all regions of the world.

Anjan Bose, Regents Professor, Washington State University

Dr. Vadari expands from my book by sharing new use cases and methods on how smart grid and DERs have become an integral part of the power grid of the future. His focus on the need for a utility to transform itself to meet this new challenge makes it a must read for those wanting to build it.

Andres Carvallo, CEO & Founder, CMG Consulting &
Author of "The Advanced Smart Grid"

Dr. Vadari's book is the first to address the existential threat faced by utilities' business model because of the coming transformation of smart grid products, services, and value. Only someone with his vast experience across technology, operations, and strategy could deftly capture the complexities. This book's insight will empower our industry to solve tomorrow's problems.

John (JD) Hammerly, CEO, The Glarus Group

A must read for anyone getting involved with smart grid and utility transformation. This book looks across not only the technology side but brings awareness to the need to look at business processes critically to meet ever changing intelligent customer requirements!

John A Ryan. Drector, Strategic Projects, Exelon Utilities

Dr. Vadari's book offers one of the most comprehensive reviews of state of the art and best practices for successful business and digital transformation considering a quasi-exhaustive list of available technology options. A must read for actors at the edge of that next #DigitalGrid transformation!

Laurent Schmitt, Secretary General ENTSO-E

A truly thoughtful and provocative read that reveals the role smart grid will continue to play in the transformation of electrification in the 21st century. Moreover, Dr. Vadari shares his insights about smart grid enabling technologies and their impact on current utility business models that must change in the new transformational paradigm.

Paul Wyman, General Manager, Smart Grid Solutions, Lockheed Martin Energy

For titles in the *Artech House Power Engineering Series,* turn to the back of this book.

Smart Grid Redefined

Transformation of the Electric Utility

Dr. Mani Vadari

ARTECH
HOUSE

BOSTON | LONDON
artechhouse.com

Library of Congress Cataloging-in-Publication Data
A catalog record for this book is available from the U.S. Library of Congress.

British Library Cataloguing in Publication Data
A catalogue record for this book is available from the British Library.

Cover design by John Gomes

ISBN 13: 978-1-63081-476-2

© 2018 Mani Vadari

All rights reserved. Printed and bound in the United States of America. No part of this book may be reproduced or utilized in any form or by any means, electronic or mechanical, including photocopying, recording, or by any information storage and retrieval system, without permission in writing from the publisher.

All terms mentioned in this book that are known to be trademarks or service marks have been appropriately capitalized. Artech House cannot attest to the accuracy of this information. Use of a term in this book should not be regarded as affecting the validity of any trademark or service mark.

10 9 8 7 6 5 4 3 2

I dedicate this book to my wife, Anu, who has patiently supported me in everything I have accomplished in my life, and without whose support this book could not have become a reality. She has been my inspiration.
This book is also dedicated to my children Mayukha and Akhi without whose love and affection, I would be nothing.

—M. V.

Where there is righteousness in the heart, there is beauty in the character
When there is beauty in the character, there is harmony in the home
When there is harmony in the home, there is order in the nation
When there is order in the nation, there is peace in the world.

—His Excellency Dr. A. P. J. Abdul Kalam
Former President of India
Career Scientist and Statesman

Contents

Foreword

Grid technology disruption is inevitable, and has been accelerating over the past several decades. A new electric generation, particularly distributed energy resources (DERs), coming onto the grid offers fundamentally different characteristics and capabilities from our past generation mix. The increasing electrification and digitalization of our economy is also growing the demand for a more resilient and reliable grid. New communication technologies are being connected to the grid and allowing operators to have a significantly improved awareness of current grid conditions. Customers have a better understanding of their own energy use and are developing new ways of lowering their costs through more sophisticated and interactive approaches to energy consumption. And, in the middle of it all, our grid is facing a rising rate of infrastructure reaching the end of its useful life.

These changes present significant challenges but also present tremendous opportunity. Industry and U.S. federal government have played a major role in this transition through the American Reinvestment and Recovery Act of 2009 that leveraged $7.9 billion in public-private capital to install 16 million smart meters, 1,400 networked synchrophasors, and 82,000 automated devices along the distribution system. This investment has had immediate impact on the affordability, reliability, resilience, and security of the grid and demonstrates the promise of new technology and the massive scale of the opportunity.

New grid technology must be thoroughly studied, vetted, and understood before it gets connected to the system. It must meet stringent regulatory and policy requirements on top of any technical ones, and must have a robust marketplace. Only when it meets these stringent criteria, can innovation can reach critical scales of commercialization to move the sector forward.

Information communication technologies and DERs are emerging through this gauntlet and quickly advancing toward mass adoption. Traditional system networks—information and energy—are now intertwined. As this connection becomes more synergistic, new value streams (that capture location and time) are emerging to feed a virtuous cycle. This vision has been around for decades, but we are starting to see the convergence accelerate. New partnerships are being developed, new businesses formed, and new ideas are taking shape challenging us to understand the new complexity necessary to maintain a healthy electricity grid.

The industry is at an inflection point, poised to enter a new era of operations, generation, and consumption where a defined vision is crucial to coordinate this transition. DERs are creating a new paradigm for where our energy comes from and how we pay for it. Regulatory, technological, and cost barriers are coming down, and the pace of DER adoption is speeding up. Utility customers are poised and ready, but is our electricity delivery system ready?

In my 30 years in the industry, I've learned that, in any complex network, you cannot change one thing and expect everything else to remain the same. One must pay attention to the technical details while keeping an eye on the big picture implications that a change has on the complex ecosystem of the business of electric power grid. In the equally long time that I have known Dr. Mani Vadari, he has always challenged the status quo of the utility business as it grapples with injection of new technologies, hardware, and software. He has focused on transformation, whether in the form of control center consolidation or smart grid implementations. Together, our belief has always been that technology implementations on their own result in less-than-adequate return on the utility's investments.

Dr. Mani Vadari's unique proficiency in this area is an essential element to grasp and explain the evolutionary shift to a new era of grid operations and management. This aspect makes *Smart Grid Redefined: Transformation of the Electric Utility* essential reading to understand a vision for how utilities can respond and even thrive in this new environment.

Dr. Michael Pesin
Deputy Assistant Secretary, U.S. Department of Energy
Office of Electricity Delivery & Energy Reliability
Advanced Grid Research & Development

Preface

At one of my keynote speeches, I started with these words: "The industrial revolution spanned the years from the mid-1700s to the late 1800s. The automobile industry and its revolution occurred from the mid-1800s to the late 1900s. Starting in the early 1900s, industry in general was all about planes, trains, and automobiles. Then came computers, leading to the internet, cloud computing, and the information age. The term smart grid was coined in the early 2000s and gave birth to an era of change in the electric utility industry. The next fifty years will drive the transformation of the energy and utility industry."

This book is about the next fifty years. This book is about the technologies that will impact the electric utility industry, the lifeblood of any country's economy. The electric utility industry is in the throes of change, the change of a magnitude never seen before. This is not a North American situation or even a Western situation. The impact will be worldwide and will change the future of the electric utility forever.

The lessons I learned working on smart grid started in 2005 when my team built the world's first smart grid roadmap for a major utility in the United States. Since then, through the American Recovery and Reinvestment Act (ARRA) funding period of 2009 to 2013, I have seen the smart grid grow from small pilot projects to full implementations, delivering benefits to the utilities and their rate base. There were successes and failures along the way—some companies succeeded, while others failed.

During my years with Accenture, I was also fortunate to work on major transformation projects with some of the largest utilities in the United States. During these experiences, I learned that transformation was not a destination; it was a journey. A utility must always be transforming, because the next challenge could come from any direction. These experiences led to my interest in writing this book.

This book is intended for people who either work at a utility or support the industry through solutions and services. It is also intended for engineering and M.B.A. programs at universities. The book is designed to introduce the reader to all the dimensions of the smart grid, in detail, and help readers to understand how each dimension of the smart grid can impact utilities worldwide. This book discusses the complex dimensions of the smart grid and provides a deep exploration of the technologies that comprise those dimensions. It also investigates how technologies change to deliver greater value to a utility and its customers. In addition, every chapter ends with a section on the impact of a specific technology's ability to transform the utility in a viable and meaningful manner.

Books such as this are a labor of love. I have been thinking of this topic for over 10 years and working in the field throughout this time. I have always been amazed at the impact of technologies to the electric utility marketplace. There are several books on smart grid. I still refer to the book written by my friends Andres Carvallo and John Cooper, *The Advanced Smart Grid.* Their work has become a classic in our industry. I would be remiss if I did not mention another book from someone I greatly admire, *Smart Power* by Peter Fox-Penner. His book moved the dial from technologies to business models. Given these excellent works, was another book in smart grid needed? Yes.

The industry needed a book that focused on the imperative to transform, not just the how of the industry, but the what, why and when. This requirement to transform the industry makes the book unique. It also draws on my experiences working on transformation projects worldwide, from Dubai to China and Europe to North America.

The book is organized to expose the reader to utility transformation by opening one door at a time. Chapter 2 starts with redefining smart grid. Given the title of the book, it was important to ensure that each reader started with a common framework to set the stage for the rest of the book. The next seven chapters present the key dimensions of the smart grid, namely: distribution automation, energy storage, distributed energy resources, microgrids, data analytics, electric transportation, and smart homes and buildings. Each chapter defines the breadth and depth of that dimension, ending with the section on how the dimension contributes to utility transformation. Chapter 10 is the lynchpin to bring it all together, revealing the importance for utility transformation. The last chapter explores a new dimension around North America and the world—the transformed utility as a springboard to the smart city and beyond. Every chapter also includes a minimum of two case studies from the industry.

I end this preface with the words of George Santayana. "Those who cannot remember the past are condemned to repeat it." It is important to learn from history. History teaches us to learn from the fate of revered companies such as Xerox, AT&T, "Ma Bell," and Kodak, in addition to companies such as Comcast, which have survived and flourished by transforming. If the utility industry transforms itself, its future can be bright.

Acknowledgments

I am deeply indebted to a large number of colleagues, friends, and generous strangers who have helped me think about and write this book. The inspiration for this book came from my experiences with working on transformation projects and delivering grid modernization courses at utilities worldwide.

No book can be written without support from specific people whose contributions are critical to the quality of the book, form, content, and editing.

Significant thanks are due to Mrudhula Balasubramanyan, who was instrumental in doing much of the research. In addition, material from the training delivered by Modern Grid Solutions formed the basis for the core content of this book. Her contribution has made this book that much better in both form and content.

It is with a deep sense of gratitude that I thank the following people. John (JD) Hammerly, who has been like a brother to me as well as coconspirator in many of my projects; JD reviewed every chapter and made several significant additions. Mike Harrison, whose review comments specifically on Chapters 1 (Introduction), 10 (Electric Utility Transformation), and 11 (Transformed Utility: Springboard to a Smart City) were invaluable. He brought his deep knowledge of utility transformation and literally transformed these chapters. In addition, I would like to thank Anne Cleary for her review of the chapter on storage, Soorya Kuloor for his contributions to the review of the chapter on data analytics, Ben Rushwald and Kate Garman for their support with the case study on smart cities, and Gerry Stokes for the joint work we did in defining the next generation microgrid. I would be remiss if I did not mention the Smart Cities Council, and especially Phillip Bane, for removing all obstacles and allowing me to use content from the Smart Cities Readiness Guide™.

There were a few people who went out of their way to remove obstacles, taking steps such as getting copyright permissions—David Luedtke, Marc Keyser, Julie Munsell, Tracey Bertilson, Rahul Chopra, Karyn Houston, Michael Stadler, Murali Baggu, Aimee Mills, Annie Haas, and many others.

The editing quality of the book goes entirely to Elena Hartwell and Gayle Wooster. Their changes made the book that much more readable and insightful. I also wish to thank my editors (Aileen Storry and Soraya Nair) who were very supportive of the process and worked with me throughout.

Thanks are due to several companies (Siemens, GE Grid Solutions, Lawrence Berkeley National Laboratories, Mid-Continental Independent System Operator, National Renewable Energy Laboratories, Electric Power Research Institute, and the U.S. Department of Energy) for providing me with pictures of their offerings which made this book much richer than it otherwise would have been. I must add here that while I am extremely thankful for their assistance, this book does not in any way or form endorse their products, offerings, or services.

1

Introduction

A friend once made a comment about the electric system, "You are in the only industry in which you send us a product and take it right back." He was alluding to the closed loop, which needs to exist for electricity to flow.

While being somewhat humorous, it is a very profound statement. The grid still depends on electrons flowing from a generating source to the consuming source. It still depends on physics for the flow to occur with a product that flows at the speed of light.

The key takeaway from this is: Don't forget Kirchhoff's laws, Maxwell's equations, and other laws of physics that make the electric grid work, but also do not let it stop you from changing how service is provided to the customer.

Background

The industrial revolution spanned the years from the mid-1700s to the late 1800s. The automobile industry and its revolution occurred from about the mid-1800s to the late 1900s, even though one could argue that, with the advent of electric car, we are now in another revolution in the automobile industry. Starting with the early 1900s, industry in general was all about planes, trains, and automobiles. World War I and World War II created such an impetus to industry, driving U.S. dominance in this space, something that continues even today. The development of the Electronic Numerical Integrator and Computer (ENIAC) computer at the University of Pennsylvania in 1946 led to the introduction of computers. The internet, which took off from the Defense Advanced Research Projects Agency (DARPA) project in the early 1960s, became a mainstay around the late 1990s and is as common around the world as apple pie is in the United States. With computers and the internet becoming so familiar, this led to the information age that has been upon us from about the early 1990s and continues into the new future with Internet of Things (IoT) where sensors and the internet are coming together.

Throughout this entire time, one key aspect was either completely ignored, taken for granted, or both: the electric grid and its evolution. Steam could only take the industrial revolution so far; it was electricity that created a level of efficiency that steam could never deliver. The electric grid is an amazing integrated system of machines spanning an entire continent. The National Academy of Engineering has called it one of the greatest engineering achievements of the twentieth century. In fact, in poll after poll, electricity and the electric grid has been called the greatest invention ever (http://news.gallup.com/poll/17881/Electricity-Retains-Power-Greatest-Invention.aspx). It is because none of the great inventions listed on this Website would have been possible without electricity and the electric grid.

According to Wikipedia, the birth of the Pearl Street Station in Manhattan was the beginning of what we would eventually come to recognize as the modern electrical grid by providing power to 85 customers [1].

Over time, this grid evolved to a system that includes (mostly) centralized generation, transmission, distribution, and the end-use customer. For the longest time, this electric grid evolved with much of the progress happening in generation and transmission. So much had evolved there that in the mid-1990s, the U.S. Federal Energy Regulatory Commission (FERC) decided to set up wholesale energy markets across the country. These markets are still successfully operating today.

Progress at the distribution level was sparse and sporadic until the term smart grid was coined sometime in the early 2000s. Then it exploded. Smart grid gave birth to new terms and technologies such as distribution automation, energy storage, distributed energy, microgrids, smart homes, and data analytics. Over the last 10 to 15 years, as these technologies were being rolled out and maturing, the utility industry has stayed the same. Utilities are still, for most intents and purposes, regulated monopolies, designated as the franchise entity that will deliver power to the customer.

For the electric utility, this is a tough place to stand because everything under them is shifting. Distributed energy allows the customer to produce energy and either use it for themselves or sell it back to the utility. Storage allows anyone to store energy when it is plentiful and cheap and to release it when it is not. Microgrids allow a group of homes, offices, or industries to ring-fence themselves and manage their own electric needs whether still connected to the electric grid or not. Electric vehicles are becoming more relevant, bringing a completely new type of load to the electric grid; a life-saver to the declining power consumption faced by most utilities. Homes and buildings are becoming smarter, thereby allowing for the control of their consumption and generation if available. Lastly, a new genre of automaton is becoming available, enabling improved sensing and control of everything in the grid and beyond the meter. These topics are also chapter titles in this book.

The book is about these changes and how they are affecting the transformation of the utility of yesterday and influencing their journey to the utility of tomorrow. Before embarking, let us look at the breakdown of each chapter and how they flow through the book.

Sequence of Chapters and Their Description

The book starts with redefining what the smart grid is all about with a specific focus on how the original definition has morphed from its humble origins in the early 2000s to now. We then discuss the various dimensions of the smart grid, namely, distribution automation, energy storage, distributed energy resources, microgrids, electric transportation, smart homes and buildings, and finally to the underlying value lynchpin, data analytics.

The topic of demand response was considered as a separate chapter and then combined into the Smart Homes and Buildings chapter because the similarities warranted merging the two topics. Domestic and commercial demand response (DR) programs are integral to smart homes and/or buildings being able to shift their loads during peak and off-peak periods. Controlling energy is a key aspect of the smart home or building, but not all of it.

Each chapter covers the full dimension of the themes with a focus on topics such as definition, drivers, technological components, ongoing challenges, and its future. Every chapter also offers case studies describing real-world implementation at a utility. Some of these cases are in the United States and some are in other countries. Every chapter reviews the impact of the technology (the topic of that chapter) on utility transformation; some reactive, some disruptive but all focused on business transformation.

Chapter 2—Smart Grid Redefined

This chapter, in many ways, is the foundational chapter for the book itself. Much has been said about the smart grid since it came to the forefront of the utility/power system engineering consciousness. However, I still believe that if one asks four people to define the smart grid, you may get five or more different answers. Defining the foundation upon which this book is built is an important starting point. The smart grid has evolved significantly since its early days and several newer terms such as distributed energy resources (DERs) and others have come into the lexicon.

The chapter starts with updating the definition of smart grid and all its dimensions[1]. The dimensions covered include: distributed energy, energy storage, transmission and distribution automation, advanced operational and decision-support systems, microgrids, big data and analytics, electric transportation, smart meters, smart homes and buildings, demand response and energy efficiency, communications, and cybersecurity.

Where appropriate, each dimension also points to the specific chapter in the book that explores it in more detail. The chapter also delves into the ongoing challenges that have been faced and what they mean to the practitioner in this area. The two case studies presented in this chapter are the gridSMART program at AEP-Ohio (United States) and the Grid4EU program (European Union).

This chapter brings forward a specific discussion on utility transformation using a hypothetical case study about how improvements in smart grid have supported a transformed utility to function in a manner very different from the utility of the past as it moves into the utility of the future.

The key takeaway is that the smart grid will change everything related to the grid. It will change what the grid looks like, it will change how the grid works, and it will even change everything we know about the utility business. This is a journey of continually improving the operations of the grid through increasing levels of intelligence and the ability to use this intelligence to improve the operational efficiencies of the grid and a better service to the customer.

The next set of chapters goes into detail on some of these technologies that have the potential to transform the electric utility industry.

Chapter 3—Distribution Automation: Path to the Self-Healing Grid

The core vision of the smart grid was always about the self-healing aspect of the grid. Right from the beginning, this aspect was always highlighted to the lay person as the main reason for moving toward the smart grid. Hence, this chapter starts with the key characteristics of distribution automation (DA) that make the grid self-healing. It is all about sensing, analyzing, controlling, and communicating.

Chapter 3 then covers the core components of DA, such as sensing and measurement devices, control methods and devices, and advanced components. The description of control methods explains centralized, decentralized, and other decision-support systems, in effect offering a refresher to some of the systems covered in my first book, *Electric System Operations: Evolving to the Modern Grid*. The last segment on advanced components also presents information on

1. A version of this definition was also presented in my first book, *Electric System Operations: Evolving to the Modern Grid*. Chapter 2 also evolves from that point to consider the developments that have happened since then.

some of the newer advances in devices such as smart inverters that support the connection of photovoltaic and other DERs that generate in DC but must supply their power to an AC network. The two case studies presented in this chapter focus on the work done with DA at Duke Energy (United States) and Stedin (Netherlands).

The key takeaways of this chapter are the examination of how DA can transform the utility industry. It is not about implementing technology for the sake of technology. It is about optimization of costs, of reducing time to restoration, of grid performance, and of integrating renewables in the grid. The ability to optimize allows the utility to deliver more cost-effective services to its customers and even support their interests with going green while still providing for a reliable and resilient grid.

Chapter 4—Energy Storage: Electric Value Chain Disruptor

Like Chapter 3, which introduced the concepts of the self-healing grid, this chapter discusses electric energy storage and its disrupting role in the energy value chain[2]. Until storage started showing up at utility scale, power flowed one way from generator to consumption and in real time; every ounce of energy generated needed to be instantly consumed.

This chapter starts with a definition of key terms and concepts associated with storage devices and then lists the different types of storage technologies and their core characteristics. I point out that all types of storage are not the same and provide a figure in this chapter comparing most of the technologies against each other based on a comparison of their storage capacity versus discharge speed, characteristics that define how they can be used in the grid. This aspect is further expounded on by looking at the various applications of storage on the grid before making the business case for storage.

The business case for storage is important because of the different players' perspectives that may exist. As a result, the author offers the business cases from the perspective of the utility, investor, customer, and regulator. The four entities presented here are all important because their stakeholders are very different and come with completely different expectations of cost and return on investment (ROI).

The key takeaway of this chapter is in the analysis of the transformational impact of storage on the electric utility industry. The present design of the grid and its associated components are all based on peak power. Storage, with the

2. The term *electric value chain* is used to describe the flow of electricity from the source generators to the customer who is typically the consumer. It was important because electric energy could not be stored in an efficient manner and hence this value chain was managed in real time with generation and consumption always balanced.

right characteristics, changes that dynamic by its ability to store and deliver energy for use at a later point in time.

Chapter 5—Distributed Energy Resources: The Challenge of Integrating Supply and Demand Diversity

Distributed energy resources (DERs) are another of those transformative technologies impacting the flow of energy through the value chain from generation to consumption. Depending on the scale of implementation, DERs can generate energy at any point across the value chain all the way to the customer level.

The chapter starts with a discussion about the various types of DERs with the realization that, unlike common perception, not all of them are from renewable sources. In addition, it also highlights that while technologies such as energy storage (Chapter 4), demand response, and others are considered as DERs in many jurisdictions, some, such as hydropowered generators, are not considered renewable. The chapter also assesses the technical and business challenges of DERs supplying energy to the grid. The assessment looks at the challenges from the perspective of the utility, investor, customer, and regulator, a recurring theme due to their importance as stakeholders in the modern grid. The next step is to focus on the building blocks of integrating DERs into the grid along with a key set of integration do's and don'ts.

The key takeaway in this chapter is that, like storage, DERs have the potential to change everything. With DERs (and sometimes supported by storage), the role of the customer changes from a consumer to a prosumer (energy consumer and producer). While this appears to be a small change from the outside, it changes the utility-customer relationship, and along with it, how the utility handles reliability, procures energy and may have to deal with the advent of retail energy markets.

Chapter 6—Microgrids: Fragmentation of the Grid

Microgrids represent the first time the set of technologies presented in this book come together. It represents a business (and technical) model in which segments of the utility's customer base have access to technologies to disconnect themselves from the grid and in turn can cease to be the utility's customer, if they desire.

The chapter starts with defining the different types of microgrids and their key drivers. A section on the features of microgrids raises attention to key technologies making the modern microgrid a potent force in the industry. A unique concept presented in this chapter is a write-up on the next generation microgrid, a new construct in which I propose the next generation of the distribution grid that is essentially designed as a stitching together of many microgrids of different sizes. I then move on to assessing storm response under this new construct by highlighting the potential for improvements.

The key takeaway from this chapter is that microgrids are not new, but could pose a genuine risk to utilities if not thoughtfully managed. They have been in existence for several decades, but are mostly found in remote areas such as Alaska, Africa, and other places where it is either difficult or considered too expensive to bring in power from the infrastructure of a centralized utility's system. The change to today's microgrid comes from the availability of new technologies and control mechanisms allowing the emergence of the microgrid within an existing utility's territory. So, if not managed properly, the microgrid could represent the first real threat of the utility losing customers.

Chapter 7—Data Analytics: Bringing Intelligence to the Grid

Data analytics is all about mining existing data to gather intelligence and enhance operations within a utility. The advent of the smart grid has resulted in the generation of a significant increase in new data at a utility. While all this data is still stored, utilities are barely scratching the surface in terms of drawing intelligence or insights from them.

To establish the right framework, I start this chapter with an analysis of how data analytics is used in other industries and the benefits extracted by them. Then the chapter moves to defining data analytics and the main sources of data followed by some of the key data analytics drivers at a utility. The chapter then presents a conceptual view of an architecture for implementing data analytics at a utility with the introduction of a new term, *master data management,* focusing on bringing together a diverse dataset and extracting intelligence across the various data siloes. I then present a section on enabling technologies as used in other industries and could be used within utilities as well. The chapter ends with the presentation of a roadmap for a utility to implement a strong flexible and scalable solution.

The key takeaway from this chapter is that data analytics at a utility is still in its infancy. The utility industry can learn a lot from other industries and how they have not only survived but profited from the innovative use of newer technologies such as the cloud and artificial intelligence to take their business and technical analytics to a new level thereby differentiated from their customers.

Chapter 8—Electric Transportation: First Mover to a Mobile Carbon-Free Future

While electric cars are still new to the industry, they have evolved from hybrid cars to plug-in hybrid cars to all electric cars. However, in other forms, they have already been in use in other locations such as airports, railroads, delivery fleets, and warehouses. At the core, they share a lot of similarities in their design. However, regarding the impact on the grid, the electric car represents both new challenges as well as new opportunities.

The chapter starts with electric transportation drivers and then moves on to defining the different kinds of electric vehicles, some on-road (e.g., cars) and some off-road (e.g., diesel-electric railway engines). I then move on to the infrastructure needs of electric transportation from the onboard components such as the battery, connecting power electronics and the electric motor to the charging infrastructure. The chapter then looks at the grid impacts of these modes of electric transportation when they connect both from a capacity and charging to voltage and system imbalance. A still new and developing concept called Vehicle-to-Grid (V2G) is explained in detail along with an initial view of its applications and their benefits. The chapter wraps up with the impacts and options available to the utility in the future when electric vehicle (EV) adoption moves from today's less than 1% to numbers in the ranges of 30% to 50% penetration.

The key takeaway from this chapter is that, while EVs are still in their infancy, they are gathering an incredible amount of momentum worldwide with several companies and countries all discussing the move to a carbon-free and all-electric fleet in the not-too-distant future. This is both a challenge and an opportunity with a new device that can both consume and generate energy, and even more importantly, do so in a mobile manner.

Chapter 9—Smart Homes and Buildings: The Final Frontier

Smart homes and buildings are that final frontier in the transformation of the electric utility of the past to the transformed utility of the future.

The chapter starts with defining smart homes and buildings, their characteristics, and a comparison between them and then moves on to key drivers and their importance from the perspective of the utility, owner/operator, business, and regulator. The chapter then moves to the elements of smart homes and buildings focusing on the architectures that drive their implementations. Given that much of the utility's demand exists in homes and buildings, I introduce DR and its impact.

The key takeaway of this chapter is that technologies such as DERs, smart cars, energy storage, and/or microgrids will impact the utility through their implementation at smart homes and buildings. This is where they all come together. These technologies have the potential to permanently impact the relationship between the utility and its customer, moving from a one-way interaction to a two-way collaboration, or, if the wrong move is made, then losing the customer altogether. The last section on how smart buildings will transform the utility industry is an important read. The impact of these changes to the utility's core business and in the regulatory landscape is discussed to ensure the utility can survive and thrive though these changes.

Chapter 10—Electric Utility Transformation

This is the chapter where all the parts converge. In this chapter, I make the overall case for utility transformation. The chapter starts with the transformational journey undertaken by utilities over time starting with the FERC orders 888/889 and how they changed the wholesale energy space.

The chapter makes its point through two major case studies: ARRA and its impact to the technology landscape and the lessons learned from the transformation (or lack thereof) in other industries, some that survived and some that did not. The examples of companies in other industries are AT&T (Ma Bell), cable/internet companies, Xerox, and Kodak. While the parallels are not perfect, there are a lot of experiences and lessons to be learned from the successes and failures. A key area of difference between the utility industry and most other industries listed above is there are many utilities around the world: over 3,000 of them in North America alone. In addition, most of them are tightly regulated either at the state level, the federal level, or both.

The chapter then presents a conceptual roadmap for consideration by utilities worldwide. While realizing there are serious differences between utilities, the number of customers that they serve, the regulatory mandate under which they serve, and public versus private ownership, the roadmap sets up some broad guidelines for utilities to follow as each of them react to the changes in their landscape.

Chapter 11—Transformed Utility: Springboard to a Smart City

This chapter sets the stage for the next phase of evolution for the transformed utility: enabling the smart city.

A typical city provides several services to its citizens. Key among them include power/energy, water, wastewater, sewage, transportation, emergency services, and others. The smart city of the future cannot exist with each service functioning in a siloed environment regardless of whether the service is provided by the city itself or by another entity, from the private or public sector.

The smart grid is allowing the traditional utility to move to a much better place. Traditional utilities had the monopolistic franchise that provided a degree of certainty around investments and earnings and return thereon. Over a period of decades, however, the downside was that regulators and customers have been wary of the utility putting its self-interest over everything else. The utility will stay in this pseudo penalty box and will not be allowed to come out of it.

The transformed utility can develop a lean-in message: become transformed into a services and solutions marketplace around energy. Unless the utility shows itself to be fully equipped and responsive, the city will not open the doors to bring in the utility as a partner.

This, in turn, opens the door to a fundamental repositioning of the utility in the future landscape that will enable it to become a full partner in the smart cities game across industry verticals and in partnership with public entities as opposed to just having them as customers. The utility can partner with the city and bring that vision to life.

The chapter puts forward a hypothesis that if the utility can transform and streamline its operations and be customer-friendly, then its value improves in the customer's eyes. The customer will perceive the utility as driving value and being operationally sophisticated. When this state is achieved, the customer and the city will open the doors to the utility to become a partner on the smart city landscape.

Conclusions

Information and operations technologies (IT and OT) are being implemented at utilities across all areas of their enterprise and even coming together in some systems like the Advanced Distribution Management System (ADMS). Some of the changes are coming from their customers who are installing technologies such as photovoltaic (PV) on their roof, storage in their garage, smart inverters to interface with the grid, and so on. Other changes are being implemented by the utility either in response to the normal course of action necessary to keep the grid functioning in a reliable and resilient manner, or in response to the customer's installations. Regardless of where the changes are coming from, much of this is resulting in disruption to the business model and is already being felt by the utility. In addition, as utilities and grids join with cities to become smart cities, newer technologies such as IoT are making an appearance and becoming mainstream.

Utilities are known for their 5, 10, and 20-year capital and maintenance plans. Their processes are unique, extremely well laid out and followed carefully. This approach worked when the changes were slow and predictable.

This is not so anymore. Everything around them is changing. State and federal mandates are influencing DER penetrations that are accelerating on a year-over-year basis. The classic electromechanical meter installed at the home is now being replaced by an electronic smart meter that has a lifetime of about 5 to 10 years. More electric cars are coming to the marketplace. The solar roof is coming, as are other new technologies such as nanogrids, which are integrating all the devices at the premises.

There are many questions that are getting asked. For example, how should the modern utility respond when these conditions exist:

- DERs become more viable in terms of cost, performance, and reliability and come with increased levels of dispatchability.
- The price of storage comes down to a level that, when combined with DERs, allows customers, aggregators, and/or the incumbent utility to deliver energy where needed and when needed along different parts of the energy value chain.
- Distribution automation supported by adequate communications and decision support mechanisms allows various stakeholders both within the utility and outside a full suite of situational awareness tools enabling the entire value chain to work as one entity.
- Microgrids are formed when parts of the utility spin off into their own semi-independent entities and interact with the incumbent utility either during steady-state or emergency situations.
- Electric transportation takes on critical mass in terms of percentage penetration requiring an extensive network of charging stations paralleling the existing gas station network in urban, suburban, and rural territories.
- Homes and buildings become truly smart with nanogrid-like controls performed in a cost-effective manner with increased level of automation, connecting to the home over the cloud and reducing the need for direct customer involvement.

Moving on from technical changes to the business model changes, how should the modern utility respond when these circumstances exist:

- A New York Reform the Energy Vision (REV)-based (or similar) retail market is required to be established in their franchise territory, which in turn could force the distribution services portion of the utility to unbundle.
- Aggregators aggressively enter the marketplace and take customers away from the utility.

While this is not the entire set of scenarios that may appear on the utility's horizon, this is a good enough subset that should be on the mind of every utility executive and personnel and also on the minds of every person who is involved in delivering products and services to the utility industry. What is the role of the utility in the future? Does the utility go by the wayside and cease to exist? Or does the utility succeed, and, if so, in what form?

These and other similar questions will be answered in this book.

Reference

[1] "The Electricity Grid: A History," *BURN: An Energy Journal*, http://burnanenergyjournal.com/the-electricity-grid-a-history/.

2

Smart Grid Redefined

Intelligent electric grids require increased network awareness and real-time response, leading to increased operational effectiveness for the utility and an improved experience for the end customer. In short, it represents a complete transformation of today's electric grid and the entities in the businesses of managing and running it.

Introduction

More than anything else impacting the electric utility industry in the last 40 years, the smart grid has made the biggest impression by far. Was the grid dumb before and is now on its way to becoming smarter? Yes and no.

If you ask four people to define what a smart grid is, you will get five answers. This chapter starts by looking at what the smart grid can and should deliver and uses that as a baseline to form a broader smart grid definition. It also presents a construct to identify various components of the smart grid followed by their descriptions. Understanding these components is important because as technology progresses, one or more of them may be replaced by another. This chapter concludes with a description of two prominent smart grid implementations, one in the United States and one in Europe. It concludes with some projections about what the future may hold for this area and the challenges it may pose.

The smart grid is not simply about implementing technology. It is really a complete business transformation for electric utilities challenging the status quo and requiring changes to people and processes [1]. If these changes are considered in totality, the implementation will be successful. If not, the benefits will be lower than expected.

Intelligence in the grid is generally associated with the ability to:

- Sense and understand the state of the network;
- Control devices in the field to alter the state of the network if necessary;
- Use decision-support tools allowing the sensed information to be converted into controls.

Mainly because of cost, the first two elements until now were sparse. There was no single, pivotal event that triggered the onset of the smart grid. Rather, it was a series of somewhat disconnected events and expectations that led to this revolution.

- *Modern customer expectations:* The customer who is more used to the iPhone era is expecting its utility to provide quick feedback on status of outages, more choice on power use, and the ability to interact via smartphone apps.
- *Modern customer actions:* Electric utility customers are also enacting changes impacting the grid. They are installing wind farms and solar photovoltaic (PV), buying electric cars, and changing the delivery landscape and the traditional utility/customer relationship.
- *More affordable information technologies:* For newer information technology (IT) that is smaller and consumes less power, it follows that it should be more affordable. Sensors and controls are being designed that, thanks to cheaper access to ubiquitous communications, allow the utility operator to better control the flow of power at a lower cost of installation.
- *IT and architecture advances:* Cloud computing, mobile computing, machine learning, big data analytics, and artificial intelligence are enabling companies to implement advanced solutions more easily and at a lower cost.
- *More options for solving the same problem:* Newer operations technology (OT) alternatives under a broad grouping called distributed energy resources are providing increased possibilities for generation, transmission, and consumption of power.

Much progress is being made is all the areas. Newer technologies are being developed, costs falling, and more capabilities are available. These are all leading to more opportunities to support increased sensing, controls, and intelligence in the network. This perfect storm being influenced by the changes identified above is called smart grid and is altering the entire utility power system landscape.

Defining the Smart Grid

The smart grid can be defined as a modernized electrical grid, a reliable and secure transmission and distribution (T&D) infrastructure that can meet demand growth in the future, while intelligently responding to the behavior and actions of all the electric power users connected to it, delivering power in a reliable, efficient, economic, and sustainable manner [2].

The greater focus of smart grid is on distribution. The transmission system already has extensive sensing and control that are increasing with newer technologies such as synchrophasors, associated controls, and decision-support systems. The long overdue emphasis on distribution is due to several factors, such as increasing customer expectations for greater reliability; increasing complexity in types of consumption devices being added to the grid, such as newer TVs, smartphone and electric vehicle (EV) charging, and EVs leading to an increasing need for better power quality; increasing the need to reduce the dependence on fossil fuels and reduce greenhouse gas emissions resulting in increasing penetration of distributed renewables; an increasing need to reduce costs to the customer; and several others [3].

The attributes as laid out in Figure 2.1 illustrate the smart grid:

- Is intelligent in sensing system overloads and taking corrective action;
- Can accommodate renewables and distributed energy;
- Is sustainable by reducing dependence on fossil fuels and decreasing carbon emissions;

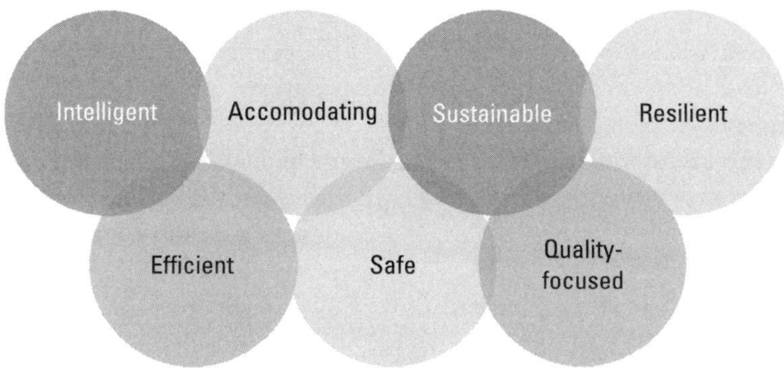

Figure 2.1 Key qualities of a smart grid. (© Modern Grid Solutions.)

- Is resilient to natural disasters and attacks;
- Is efficient in meeting increased consumer demand without adding infrastructure;
- Provides a safe environment for utility workforce and consumers;
- Can deliver the power quality needed for a digital economy.

This means there are several dimensional components to a smart grid.

Dimensions of the Smart Grid

Figure 2.2 shows the smart grid and its various dimensions. The dimensions, aligned against the distribution/smart grid value chain are distributed energy, energy storage, transmission and distribution automation, advanced operational and decision-support systems, big data and analytics, microgrids, demand response and energy efficiency, smart meters, electric transportation, smart homes and buildings, communications, and cybersecurity.

Figure 2.2 demonstrates the smart grid is more than just the electric grid becoming smart. Each segment of the electric value chain becomes smarter and more connected with deployment of modern technologies: some OT and some IT. These dimensions of the smart grid are critical to understanding how the new infrastructure is coming together in the age of the smarter grid. In addition, depending on the circumstances, one or more of these technologies could be used instead of others to deliver similar outcomes.

A short description of the 12 smart grid dimensions listed above is provided here.

Distributed Energy Resource

DERs are defined as the set of small-scale power generation technologies located close to the load being served. DERs are characterized by small and modular construction with outputs typically in the kilowatt range that can be based on renewable or nonrenewable sources of energy and are quieter and less polluting than large power plants [4–6].

There are various kinds of DERs, some renewable and some not, with differences in characteristics and services that they provide [7]. New York State's Reform the Energy Vision (REV) report classifies the following as DERs:

- Biofuels (including biogas), converted or burned;
- Cogeneration;
- Demand response (DR);

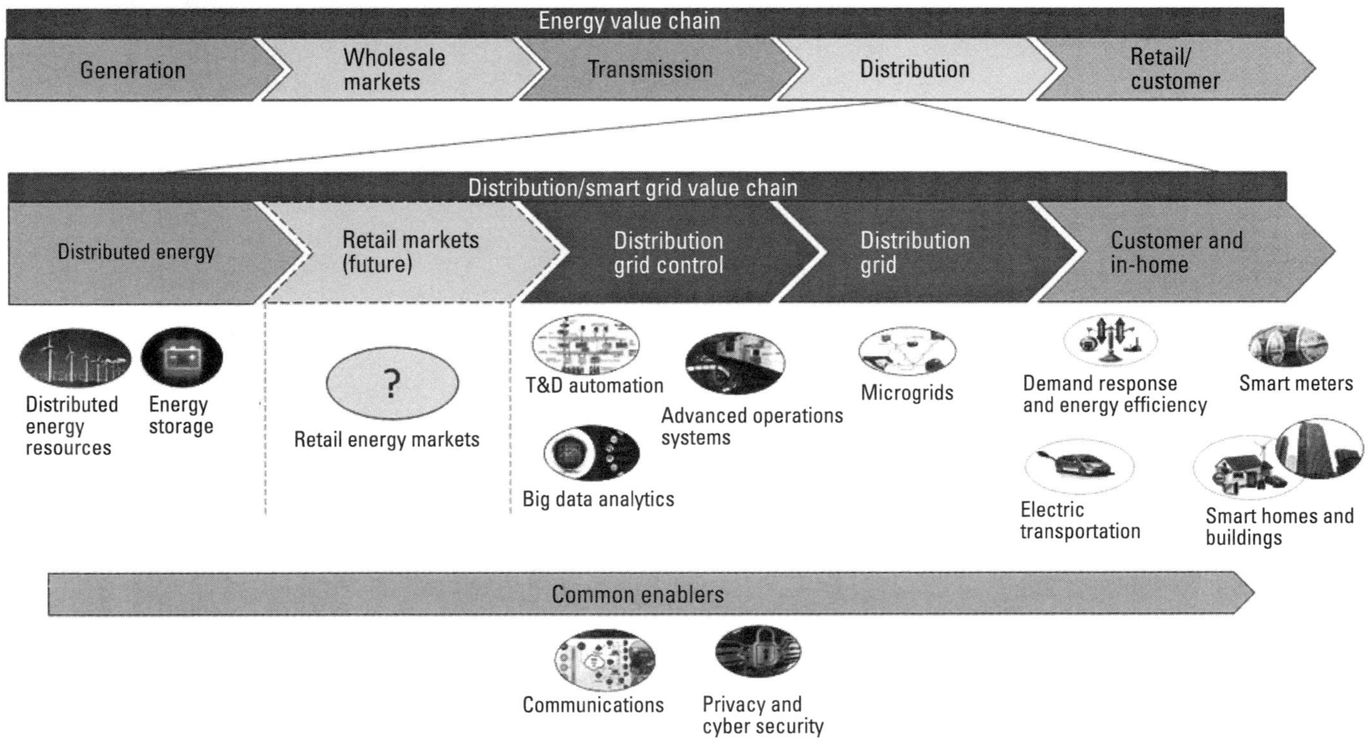

Figure 2.2 Key dimensions of a smart grid, aligned against the value chain. (© Modern Grid Solutions.)

- Energy efficiency (EE);
- Energy storage (including batteries, fuel cells, flywheels, thermal);
- Hydroelectric generation;
- Solar/PV;
- Wind.

Figure 2.3 presents a prototypical utility substation connect to a diverse set of DERs.

Some of the DER technologies that are commercially available, as well as those still undergoing development, are microturbines, combustion turbines, internal combustion engines, Stirling engines, fuel cells, uninterruptible power supply (UPS), photovoltaic systems, wind systems, and hybrid systems. Typically, DERs provide the end user with higher reliability, sufficient power quality, and the possibility to participate in competitive electric power markets. They also have the potential to mitigate overloaded transmission lines, control price fluctuations, strengthen energy security, and provide greater stability to the electricity grid [8].

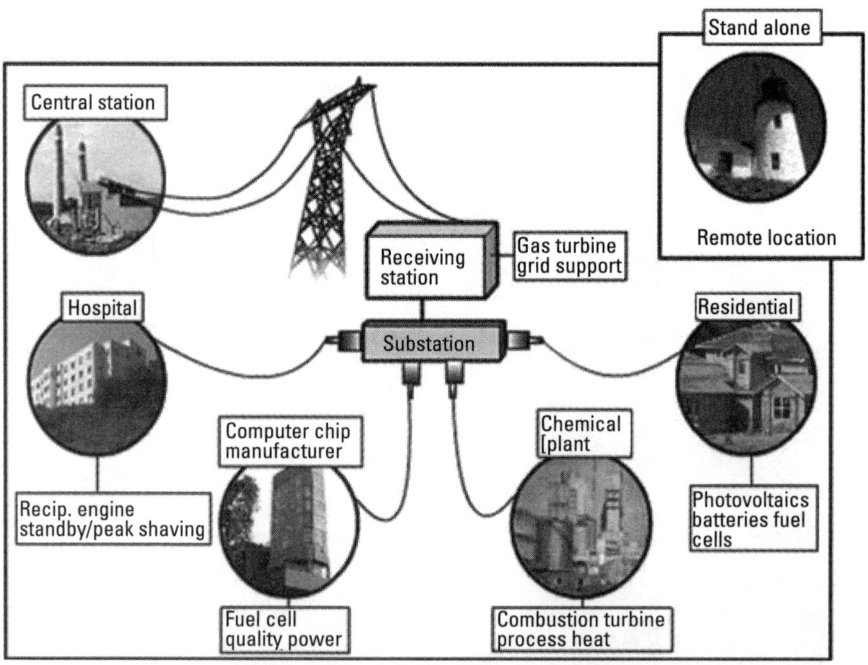

Figure 2.3 Distributed energy resources. (Source: California Energy Commission. https://www.wbdg.org/resources/distributed-energy-resources-der.)

The key to managing and controlling DERs is heavily enabled by the advances in IT using a combination of integrated centralized and distributed computing, cloud computing and automation all enabled by ubiquitous communication.

For more information on DERs, see Chapter 5.

Energy Storage

Electric energy storage is defined as a set of technologies capable of storing electrical energy previously generated and can be released later. It can use chemical, kinetic, thermal, and potential energy forms to store energy so it can be converted to electricity later. Figure 2.4 presents a picture of a typical utility-scale storage installation.

Energy storage has applications along the entire electric value chain. It can be used to store energy from large-scale renewable generation for use later. It can also be deployed at substations to provide power supply to substation equipment and computers. It can be installed at commercial, industrial, and residential customer premises to provide grid-independent or emergency power. There are various storage technologies in different states of development and their usage depends on their energy and power characteristics.

Energy storage is a game changer for the electricity grid. It has the potential to solve the age-old problem of having to generate and consume electricity in real time and to lead to a more reliable and flexible grid. Energy storage systems provide a wide array of technological methods to manage the power supply in order to create a more resilient energy infrastructure and bring cost savings to utilities and consumers [9].

Figure 2.4 Electric energy storage. (Source: Sandia Report. https://www.smartgrid.gov/files/155242_ARRA_Storage_Lessons_Learned.pdf.)

Three major categories for energy storage are:

1. *Chemical:* Examples are solid state batteries, lithium-ion batteries, lead-acid batteries, and flow batteries.
2. *Mechanical:* Examples are flywheels, compressed air energy storage, and pumped hydropower.
3. *Thermal:* Examples arc ice energy storage, combined heat and power (CHP) storage, and molten-salt technology.

Storage technologies not only improve the quality of power through frequency regulation and allow companies to produce power when it is cheapest and most efficient, but they also provide an uninterruptible source of power for critical infrastructure and services.

For more information on energy storage, see Chapter 4.

T&D Automation

T&D automation is the process of monitoring and controlling the grid via the use of intelligent devices, instruments, and advanced components. It is enabled by integrating the devices and components in the field with the analytical tools in the control center via two-way communication networks.

Automation technologies enable smarter sensing of grid issues such as faults, congestion, overloads, and real-time control to alleviate these issues and lead to a more optimal utilization of existing assets. In addition, automation is also enabled by advances in computing such as mobile/distributed computing, cellular, and other forms of wireless communication all connected to deliver real-time responses.

For more information on T&D automation, see Chapter 3.

Advanced Operational and Decision Support Systems

Advanced operational systems (see Figure 2.5) are the set of applications, algorithms, and technologies enabling analysis, diagnosis, and prediction of conditions in the modern grid. They help to determine and take appropriate corrective actions to eliminate, mitigate, and prevent various conditions such as outages, power quality disturbances, and so on. These systems provide control at the transmission, distribution, and consumer levels. The monitoring and control action can be at the control center level or the local power system level.

Examples of operational systems include:

- Supervisory control and data acquisition (SCADA);
- Energy management system (EMS);

Figure 2.5 Advanced operational and decision-support systems. (Source: Mid Continental ISO. Used with permission.)

- Outage management system (OMS);
- Distribution management system (DMS);
- Distributed energy resource management system (DERMS), an emerging system to manage distributed energy resources.

Getting data into these complex systems and getting controls into the field requires distribution automation which includes sensors and controls supported by real-time communications.

For more information on advanced operational and decision-support systems, see the book *Electric System Operations: Evolving to the Modern Grid.*

Microgrid

A microgrid is defined as a group of interconnected loads and generators, very often also supported by DERs within an electrical region allowing it to act as an independent entity that can be controlled independent of the larger grid. Some microgrids can connect and disconnect from the grid to enable it to operate in both grid connected and island modes [10].

Microgrids (see Figure 2.6) can be designed to meet the needs of the users it serves and can be replicated in any system where the power infrastructure is locally owned and managed. They can be deployed at university campuses, commercial and industrial locations, military bases, islands, and in communities. Because microgrids are able to operate while the main grid is down, they

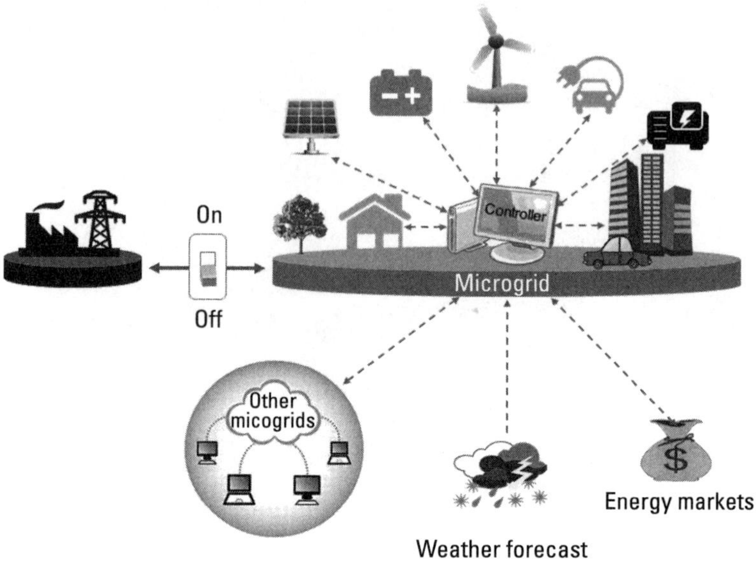

Figure 2.6 Microgrids. (Source: Lawrence Berkeley National Laboratories. https://building-microgrid.lbl.gov/about-microgrids. Used with permission.)

can strengthen grid resilience and help mitigate grid disturbances, functioning as a resource for faster system response and recovery. Not only do microgrids enable the integration of growing deployments of distributed energy resources such as renewables like solar, they also enhance the use of local sources of energy to serve local loads thereby reducing energy losses in transmission and distribution.

There are several smart grid technologies enabling microgrids. They are distributed generation, energy storage, automated demand response, islanding and bidirectional smart inverters, and microgrid control systems.

As microgrids increase in number and scale (size), solution providers are taking advantage of cloud computing to provide customized controls at a lower-cost profile. For more information on microgrids, see Chapter 6.

Data Analytics

Data analytics (see Figure 2.7) are the techniques and processes of converting data collected from meters, sensors, switches, and other devices deployed in the field into actionable intelligence, for use by the utility. Data analytics provide the utility with insights into grid performance, consumer energy use, peak demand, and business risks.

Utilities can benefit significantly from smart grid data analytics [11]. From an operational perspective, they provide the following benefits and

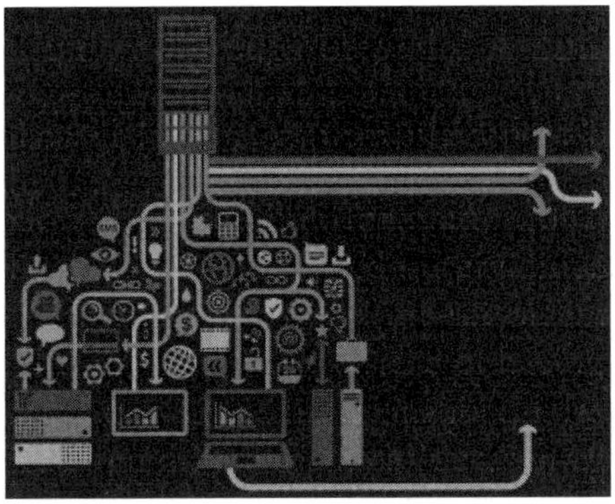

Figure 2.7 Data analytics. (Source: Rocio von Jungenfeld. http://datablog.is.ed.ac.uk/2013/12/. Licensed under Creative Commons Attribution 4.0 International License.)

more:UImprove outage management by aggregating thousands of outage alerts to a common upstream node on the distribution grid and target crew dispatch to appropriate outage zones.

- Enhance the accuracy of load forecasting using more granular point-of-consumption data from smart meters, as well as using consumption data aggregated at the distribution level.
- Optimize the grid through aggregated smart meter voltage data correlated to upstream feeder voltage to optimize voltage regulation implementation of conservation voltage reduction for power delivery efficiencies.

Data analytics are the critical element to tie data from T&D automation to advanced operator decision making in real time.

For more information on data analytics, see Chapter 7.

Electric Transportation

Electrification of transportation (see Figure 2.8) is the use of hybrid electric or all-EVs instead of pure petroleum-based vehicles. It also includes the infrastructure to charge the EVs.

An EV's propulsion system contains one or more electric motors that contribute, partly or entirely, toward providing the motive force to drive the vehicle. Innovation in the areas of power electronics and communications is also paving the path for vehicle-to-grid (V2G) in which a fully charged EV could provide temporary power to a residence or the grid during a power outage. Electric motors react more quickly than internal combustion engines making them highly responsive resulting in very good torque. They are more digitally

Figure 2.8 Smart transportation. (Source: Energy.Gov. https://energy.gov/eere/electricvehicles/electric-vehicle-benefits.)

connected than conventional vehicles with some charging stations providing the option to control charging even from a smartphone app [7].

For more information on electric transportation, see Chapter 8.

Smart Meters

Advanced metering infrastructure (AMI) is defined as a system that collects, measures, and analyses energy usage data via a two-way communications network connecting advanced meters (called smart meters) and the utility's back-office systems. The smart meter (see Figure 2.9) measures, collects, and stores end-user energy consumption data. Smart meters provide greater granularity of usage data enabling accurate billing and other services. AMI enables remote meter reading for billing, remote connect/disconnect capabilities, outage detection and management, and tamper and theft detection, all of which lead to a more reliable and smarter grid. The customer systems in the two-way commu-

Figure 2.9 Smart meters. (Source: https://commons.wikimedia.org/wiki/File:Landis%2BGyr_FocusAXR-SD_Hydro-Quebec.JPG Licensed under the Creative Commons Attribution-Share Alike 4.0 International.)

nication between utilities and customers include in-home displays, home area networks, energy management systems, and other customer-side-of-the-meter equipment enabling smart grid functions in residential, commercial, and industrial facilities.

Smart Homes and Buildings

Smart homes and buildings present the integration of building energy systems with information and communication technologies. Empowered by its automation system, buildings provide actionable information enabling the owner or facility manager to optimize energy usage, space, and services provided to occupants. A key objective is to improve home energy performance by developing and demonstrating advanced energy efficiency technologies and practices making homes more efficient, affordable and comfortable. Figure 2.10 is the depiction of a typical smart home with a home energy manager (HEM) controlling various appliances (smart or otherwise) in the house.

Characteristics of a smart home or building include:

- Tools and technologies for energy conservation and environmental sustainability;
- Proactively monitoring premise energy usage;
- Providing actionable information regarding the performance of building systems and facilities;
- Integrating with systems for real-time reporting and management of energy;

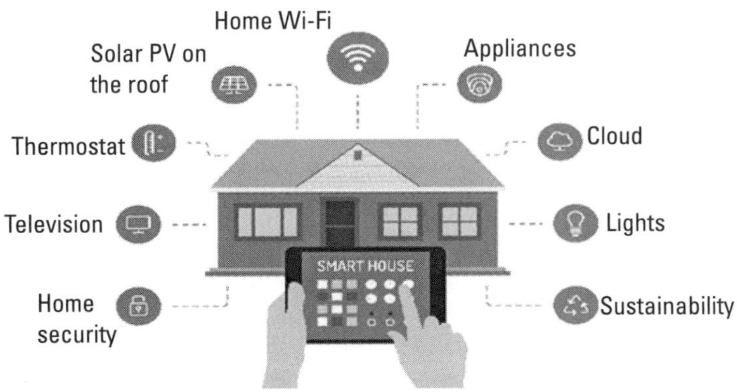

Figure 2.10 Smart homes and buildings. (After: http://engineering.electrical-equipment.org/energy-efficiency-building/smart-home-energy-efficiency.html. Licensed under the Creative Commons cc logo Attribution 2.0 France [CC BY 2.0 FR]).

- Operations for occupant comfort [12, 13];
- Secure communication network.

For more information on smart homes and buildings, see Chapter 9.

Demand Response and Energy Efficiency

The U.S. Department of Energy defines demand response (DR) as "changes in electric usage by end-use customers from their normal consumption patterns (see Figure 2.11) in response to changes in the price of electricity over time, or to incentive payments designed to induce lower electricity use at times of high wholesale market prices or when system reliability is jeopardized." Consumer actions changing any part of the load profile of a utility or region may be part of DR. Engaging customers have applied some new methods for DR including offering a retail electricity rate reflecting the time-varying nature of electricity costs or programs providing incentives to reduce load at critical times [14]. Another example is Internet-controlled switches on residential air conditioners or electric water heaters. Similarly, energy efficiency (EE) is the overall reduction of energy consumption through the incorporation of specific changes while still providing the same or improved level of service to the end user [15]. EE

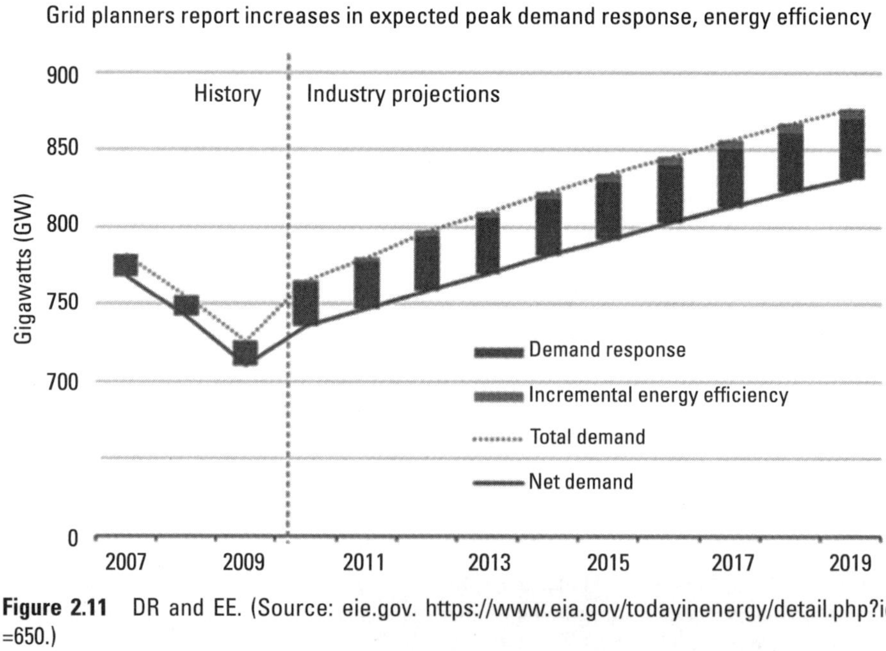

Figure 2.11 DR and EE. (Source: eie.gov. https://www.eia.gov/todayinenergy/detail.php?id=650.)

offers one of the easier and more cost-effective ways to combat climate change and improve the competitiveness of businesses while reducing energy costs for consumers.

DR and EE are critical pieces of the smart grid puzzle [16]. The normal mode of operation in an electric grid is load changing based on customer use and supply following the load. For the first time, DR and EE present mechanisms for controlling load thereby allowing both utility and customer to become partners in the supply-demand equation. This dimension is expounded in several chapters.

Communications

Communications is the backbone of smart grid architecture. It is the key enabler of information transferring from devices and applications to where it is needed for the right decisions to be made at the right time. It is the foundational medium integrating the smart grid dimensions and their applications.

Communications exist at different levels (inside the customer's premise, in the grid, inside the utility), all to carry various types of data and controls from source to destination (see Figure 2.12). Communications tie utility-side and consumer-side technologies to enable a smarter grid. This dimension is expounded in several chapters.

Smart grid applications are supported by several communications and networking technologies. These include traditional twisted-copper phone lines, cable lines, fiber optic cable, cellular, satellite, microwave, WiMAX, power line carrier, and broadband over power line, as well as short-range in-home technologies such as Wi-Fi, Z-Wave, and ZigBee [17]. Some applications built on such communication technologies, for example, are home area networks, networks for wide area situational awareness, enhanced substation supervisory control and data acquisition systems, and distributed generation monitoring and control, to list a few.

Cybersecurity

The electricity grid is facing an ever-increasing amount of data and controls being generated from an ever-increasing number of sensors. Much of the communications is via IP-based protocols thereby opening a communications network that until now was very much closed. These changes have also increased grid vulnerabilities due to the increased potential of external points of entry either authorized or unauthorized [18–20].

Smart grid cybersecurity supports both grid reliability and confidentiality or privacy of the information transmitted. The National Institute of Standards and Technology Interagency Reports (NISTIR) guidelines for smart grid

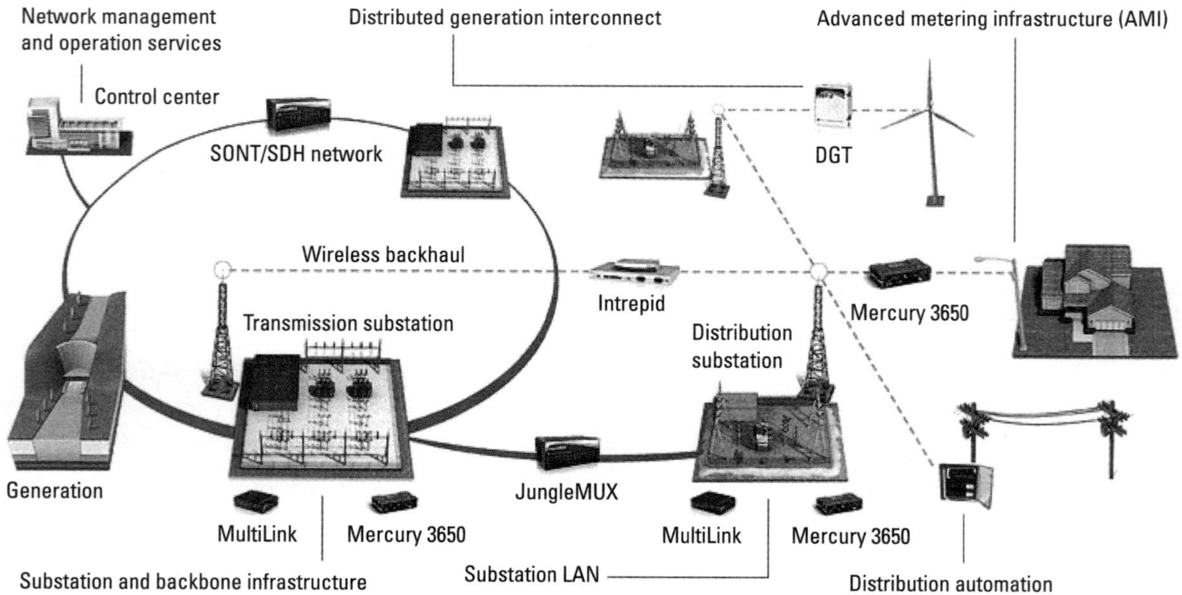

Figure 2.12 Communications. (Source: GE Grid Solutions. http://www.gegridsolutions.com/Communications/smartgrid.htm. Used with permission.)

cybersecurity [20] emphasize how a continuing transformation of the electric power infrastructure would lead to new risks and evolving threats. There is a significant need to be vigilant to ensure energy efficiency, reliability, and security while transitioning to renewable sources of energy and reducing greenhouse gas emissions. Cybersecurity is the set of collective measures and processes setup to address the security concerns of IT and communications infrastructure of the smart grid (see Figure 2.13). it must address vulnerabilities from deliberate attacks as well as inadvertent compromises, to enable reliable and secure operation of the electric grid.

Advances in Technical Architectures and Computing

The architectures followed for the implementation of both automation and decision-support systems are being designed in very different ways than before. These technical architectures are changing from the older monolithic designs to the more distributed designs [21]. Now, instead of having a primary/secondary pair, the specific pieces of software or functionality can be deployed in multiple pieces of hardware located in different places, pieces integrated with the sensor itself and other parts combined in a central location. There is also a slow and steady move to the cloud. This move is allowing functionality to be shared with others, located remotely and sharing in costs.

Utilities are also bringing in data in extremely large quantities and working on converting them into meaningful information and extracting intelligence from them using normal and artificial intelligence methods. Systems are being designed and implemented that take data from different sources from

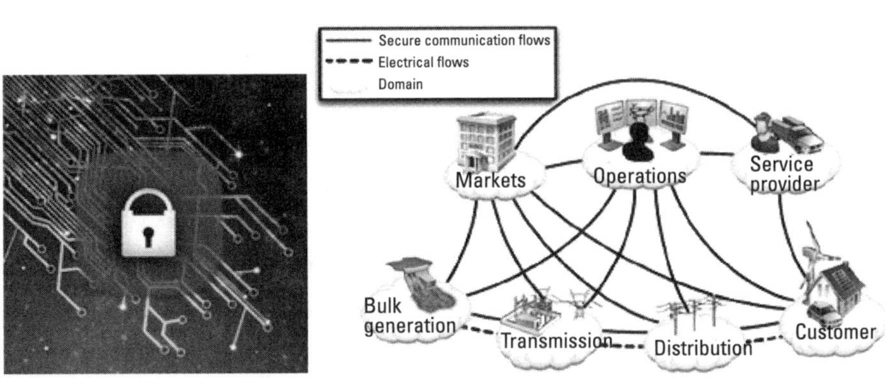

Figure 2.13 Interaction of actors in different smart grid domains through secure communication flows. (Source: NIST Framework and Roadmap for Smart Grid Interoperability Standards, Release 1.0 [NIST SP 1108].)

within the utility and delivering them to customers, field staff and other stakeholders through a set of mobile applications designed to enhance productivity.

Ongoing Challenges and the Future

Making the grid smart a reality is not easy. Grid modifications are expensive and difficult. Investments needs to be considered based on their business benefits. In addition, innovation is happening at many levels in every dimension such that an implementation performed today may become obsolete within a few years. Some examples of innovation causing challenges are provided here:

- *Residential meters:* Just in the last 10 years, technology has moved from electromechanical meters to automated meter reading (AMR) to AMI.

  Utilities that invested in AMR are now facing stranded investment because AMI can provide much more capabilities than AMR or the old electromechanical meters. Moreover, business cases that expected electromechanical meters to last 30 to 40 years now must plan for smart meters that last only about 10 years.

  AMI technologies are moving to second and third generations in which they can perform many more actions and sense more quantities in the field. Smart meters and technologies within them are rapidly moving forward with the inclusion of advanced sensors and other sophisticated electronic components such as microprocessor control and are capable of supporting different wireless technology mechanisms. They also include mechanisms capable of sensing energy, voltage, power quality, and other quantities. Additionally, they perform control actions such as remote connect/disconnect.

  AMI meters are allowing utilities to perform actions at a level unforeseen in the past, so much so that utilities have not even scratched the surface in terms of what they can do with the data [22].
- *Telecommunications:* The smart grid is heavily dependent on a ubiquitous telecom network. The network is necessary to take data from where it is created to where it is needed and processed. It is also necessary to allow the various control actions both at the residence and at the grid.

  Information and communication technology (ICT) includes voice, data, and all forms of communications in an integrated form, along with the software, hardware and all the systems that go with them. ICT technologies have come a long way since the days of wired fixed-line communications that were prevalent many years ago. These days, it includes wired, wireless (cellular, mesh, and others), and other forms that can talk with each other.

Telecommunications at utilities have moved from completely utility-owned private networks to public networks for some implementations, to cases of some major utilities even considering cellular-based networks. Just like AMI investments, telecom investments can be expensive to implement and are further complicated by the existence of legacy installations that cannot be easily replaced.

Any smart grid implementation needs to be ready to take advantage of improvements so investments can be protected.

- *Automation (sensing and control):* There is a significant dependence upon sensors providing information on the state of the grid and the ability to react to adverse conditions by changing that state through the use of controls. Advanced power electronics has demonstrated the ability to dramatically reduce costs of these devices allowing the grid to be smart. These sensors and controls sometimes have the intelligence to make decisions locally and sometimes centrally for implementation remotely.

  Technological advances have allowed these sensors and controls to become smaller, cheaper, and more rugged. Their compute and connectivity capabilities are also improving. A tremendous amount of innovation is going into it in both small and large companies alike.

  Given these rapid innovations, and existing investments in automation, utilities should be careful in spending on automation. In all practicality, solving priority problems capable of delivering the most benefit makes more sense than taking on a system-wide project.

- *DR versus energy storage:* The advent of American Recovery and Reinvestment Act of 2009 (ARRA) [23] efforts in the United States led to the identification of DR as low-hanging fruit for smart grid implementations focused on peak shaving. DR, which started as load management or demand side management (DSM) several years ago, has continued to be the leading and most used mechanism for peak shaving and/or peak shifting.

  The initiation of electric energy storage has created a new opportunity to store energy when it is cheap and discharge let it out when energy is expensive. Storage also can be placed in places and activated with no customer involvement or action, something that DR cannot claim. DR can only work if the customer acts by reducing load at the right time. It also can support DERs by smoothing out the load profile. However, energy storage is still new and expensive to install at utility scale.

As utilities consider implementing DR across their territories, they also need to consider the potential for storage options that can deliver the same outcomes.

- *Privacy:* The inception of smart meters (AMI) has allowed utilities access to large amounts of customer-specific data. The industry is moving from collecting one piece of data per month per residence, to 8 to 10 pieces of data every 15 minutes per residence. This data represents an unprecedented level of access to customer energy consumption, which, in turn, leads to an unparalleled level of access into customer behavior inside the home.

  Analyzing this data with the right set of analytic mechanisms can allow a person or company with questionable intentions to: (1) understand the energy consumption habits (and cycles) of people living in the house, (2) deduce when they are in the house or not, (3) deduce whether they are on vacation or not, and (4) and also get enough insight to determine whether they are a candidate for a targeted marketing campaign for some products. As a result, these large amounts of very personal data stored over a period of time within the utility's archives are causing an increased concern for privacy.

  The treasure trove of data that the utility now has through the AMI system can also assist the utility in making good decisions on a broad set of value-added functions such as asset management, grid operations, DER penetration, and so on. Utilities will need to manage the data in such a way that the privacy aspect is not violated.

- *Cybersecurity:* The multitude of sensors in the field, such as smart meters, distribution automation, and others, has resulted in a large number of devices connected to the power grid across communication networks running on industry-standard networks. This is a significant departure from the sensors and networks of the past, where most of them were operating in closed and proprietary networks.

  The move to greater number of end-points and the use of industry-standard communication mechanisms are immediately resulting in many points in a network that can now be hacked. The impact of hacking can range all the way from modifying the behavior of specific power system components to causing widespread blackout or disconnecting all the customers through the smart meters.

  The more obvious is in internal utility (IT) people not wanting to connect anything to the internet or network because of cybersecurity risk. Utilities need to weigh the risk/reward scenarios from the enhanced amount of data received and the greater amounts of control capability. As a result, much of the cybersecurity effort across the value chain is focused on: (1) making the network more secure (2) making the endpoints more capable of thwarting cyberattacks, and (3) (equally important) setting up schemes to detect abnormal behavior in the grid.

Utilities are faced with cybersecurity challenges every day and need to make tough decisions. A set of key considerations critical for utilities to consider on their journey to the smarter grid are included below:

- Always focus on the business benefits and not technology for the sake of technology.
- Invest in technology but based on business need and where benefits justify the cost. Anticipate the technology, the costs, and the benefits to change over time.
- Anticipate and expect new technologies to declare existing solutions obsolete, for example, the example of energy storage against DR.
- Train your personnel as well as your customers. Employees need to learn about new technologies and how they will impact operations. Customers need to understand how these changes will impact services provided to them.
- Anticipate new players to threaten the utility business model and traditional sources of revenue.

It is important to note while the challenges of the smart grid are many, the benefits are even more. The smart grid has delivered an electric utility future in which distributed and centralized renewables can play a significant part leading to reduced greenhouse gas emissions, a more reliable grid with fewer outages, and more choice for the customer in terms of service providers, all with the increased potential for lower costs of electricity at the home/premise [24, 25].

Case Studies

Case Study #1: United States—AEP Ohio gridSMART Demonstration Project

The AEP gridSMART project (see Figure 2.14) introduced multiple technologies to AEP Ohio's project service territory. These technologies include AMI, which enables:

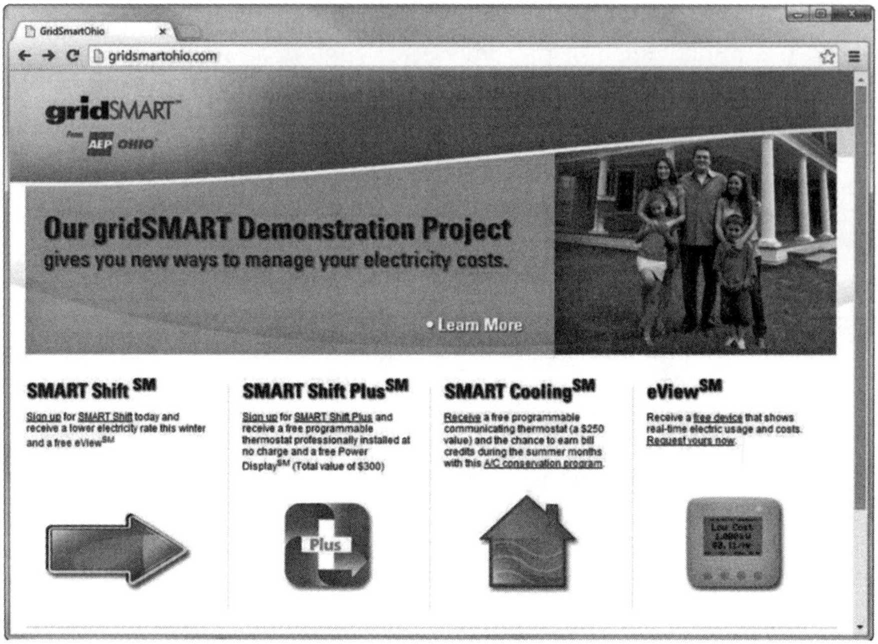

Figure 2.14 AEP-Ohio demonstration Website. (Source: Smartgrid.gov https://www.smart-grid.gov/files/AEP_Ohio_DE-OE-0000193_Final_Technical_Report_06-23-2014.pdf [26].)

- Two-way communication between smart meters and the utility control center;
- Distribution automation circuit reconfiguration which is the automation of distribution assets;
- Volt/volt-ampere reactive (VAR) optimization which involves voltage control and optimization;
- Consumer programs which provide cost-saving opportunities to customers through enhanced communication.

The deployment of these technologies enabled two-way communication with consumers and allowed for tailoring of consumer-related programs and products. It provided significant cost, reliability, and environmental benefits for the utility and its consumers. The success of this holistic approach to smart grid implementation enabled AEP Ohio to move forward with the gridSMART Phase 2 filing [27].

The project resulted in enhanced reliability in service delivered along with achieving important practical outcomes of saving money through reduction of energy consumption. The users participating in the consumer programs were also able to receive near-real-time information about their electricity usage. AEP Ohio also benefited on multiple fronts including reduction in costs through the elimination of meter reading routes and reduced field visits, improved employee safety, system reliability, customer satisfaction and reduced peak demand. In addition, they performed tasks remotely such as reading meters while taking up proactive maintenance and repair.

Case Study #2: Grid4EU

Grid4EU is a large-scale demonstration project of advanced smart grid solutions in Europe. It is a consortium of six European energy distributors: ERDF, Enel Distribuzione, Iberdrola, CEZ Distribuce, Vattenfall Eldistribution, and RWE. With an overall cost of €54 million, with €25 million financed by the European Commission, it is the biggest smart grid project to be funded by the European Union. It is a European initiative at developing a smart grid that distributes electricity safely and efficiently anywhere anytime to all consumers regardless of how they use energy and their source of generation.

The main objectives of the project [28] are to develop and test innovative technologies, define standards through the setup of demonstrators, guarantee the scalability of these new technologies, guarantee the replicability over Europe, and analyze smart grid cost-benefits. Figure 2.15 shows the overall picture of the how this demonstration project came along.

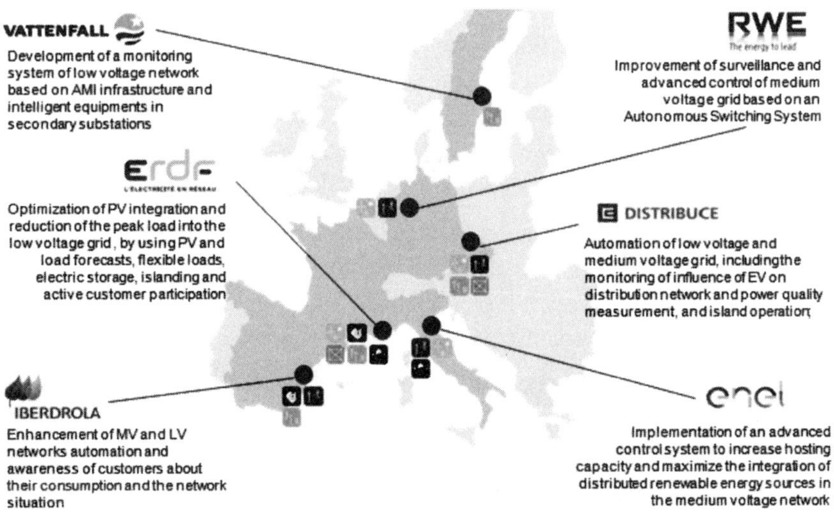

Figure 2.15 Grid4EU project partners. (Source: Grid4EU http://www. grid4eu.blob.core.windows.net/media-prod/29375/grid4eu-final-report_normal-res.pdf [29].)

The projects focused on using more DERs connected to distribution networks, implementing active and more efficient participation of customers to electricity markets, securing energy supply and increasing network reliability, improved supervision and automation of the medium voltage (MV)/low voltage (LV) network, improving peak load management through increased interactions between network operation and electricity customers, and incorporating DR, energy storage, and microgrids. In short, it was intended to be as complete as possible demonstration of the smart grid incorporating almost all the dimensions. Within specific boundary conditions, they implemented new applications like Automatic Grid Recovery, carried out controlled islandings lasting more than 4 hours to test grid resiliency, and set up new architectures for grid monitoring and control to better serve customers.

The project was completed in 2016 and identified the following benefits as outcomes.

1. The advanced control of On Load Tap Changer at MV/LV is a major resource for increasing the Hosting Capacity by up to more than 50%.
2. Grid storage tested at MV/LV level is able to contribute effectively to voltage regulation and grid power flow control and thus increase network hosting capacity.
3. In addition to decrease in the bill, the opportunities to act in favor of the environment and to contribute to improving the security of supply turned out to be major drivers in joining DR programs. Around 80% of the participants in the Customer Engagement program changed their consumption habits.
4. Fault localization and restoration time can be further reduced through more automation at MV and LV levels. Very significant improvements on the identification of outages have a potential positive effect on SAIDI (from 5% to 12%).
5. Automation at LV and MV levels also expand further levers to decrease energy losses. Solutions like the grid reconfiguration tend to very strongly decrease losses. It shows that the planning phase is a very important stage for the reduction of network losses.
6. For higher grid resiliency, it is technically feasible to operate the grid in islanding mode for more than 4 hours, with and also without rotating machines, while complying with strong requirements in terms of continuity of supply.
7. Given that smart grids systems rely very much on telecommunication systems, it is important and necessary to foster convergence between electric distribution and communication infrastructures.

8. Smart grids are not just about technologies. The human component is of paramount importance. The active involvement of the users of the smart grids solution, whether they are partners, workers, or customers, is key to enable the appropriate execution of smart grids solutions.

How Smart Grid Will Transform the Utility Industry

Let us start with a hypothetical case study[1].

"I just returned from Houston and my friend got a message on her cell phone that the power was out at their house, but that it would be back on in 2 hours, so we kept playing tennis. When she checked the app, she also showed me her car was only charged 80% but it was okay, because she was using her solar cells to charge it and it would be completed in 3 hours. She smiled and said she sold $75 worth of power last month back to her retailer and it paid for lunch today. She said her electricity bill now only includes a connection charge unless she does her clothes washing and baking on the same day. I am calling my utility to see what they can provide."

A deeper analysis of the situation provides the following context into what really needed to be in place for the conversation above to be real. Let us follow the conversation:

1. *Direct communication from the utility:* Through an app on her smartphone, this friend in Houston could get an exact status of the outage in her home, an update on the charging status of her electric car, and other specific billing information.

   So what? To provide this level of information to the customer, the utility needed to transform all aspects of both their back-office systems and processes and people. The result is all the information was in one place and the customer had easy access to it. The information was also presented to the customer in a way allowing them to make decisions and plan their life around it instead of being constantly surprised.
2. *Command of severe weather implications:* Whether the outage was planned or unplanned, the utility could take full advantage of the automation and sensors in the field to derive specific outage times, which are always based on severity of the outage, availability of crews, and availability of spare parts.

   So what? Outage and outage-related disruptions are some of the most serious events causing disruptions to customers' lives by taking away

1. Case credit to Charles Filewych, CEO, Smart Grid Interconnect. Used here with permission.

their electricity, which in more ways than one is their lifeline to many things. Making this information available to their fingertips in a timely and accurate manner allows for customers to be more satisfied. A satisfied customer results in a satisfied regulator.

3. *Reducing outage times and a reliable partner to restore on time:* Her confidence on the outage period resulting in playing tennis longer is an important observation because she trusted that at the end of that period, power would be back in her house and she could go back.

   So what? This means that the utility has transformed itself around the customer, who is now a part of the equation instead of being an afterthought. As a result, the customer can plan his or her schedule around utility plans.

4. *Monitor and control energy usage:* The utility had a full-featured AMI system in place with net metering capability and could determine exactly how much energy this person was consuming and how much energy she had delivered back into the grid.

   So what? The transformation in this case resulted in the customer information being pushed back to the customer in near real time, allowing the customer to take advantage of the energy tools available.

5. *Personal acquisition of renewable energy:* This person had installed solar cells and possibly other equipment (storage) in her home and the utility was able to both monitor and take advantage of it when needed under outage situations to provide for local access to power.

   So what? The customer has now transformed into a "prosumer" (meaning both a consumer and a producer) and is now working closely with the utility to take advantage of nonutility options such as solar cells on her roof. The customer and utility jointly decide on the best use of excess energy being sent back to the grid.

6. *Customer choice:* Net metering (the ability to track both production and consumption independent of each other) and retail competition were already in place so she could then purchase her power from a third-party retailer.

   So what? The utility is the intermediary in providing options to the customer to meet their energy needs by working with an entity that is different from their incumbent utility. For this customer, the utility is primarily a wires-only entity which still has the most important role of ensuring the energy reaches the customer in a reliable and resilient manner.

7. *Rewards for reducing consumption:* The fact she could reduce her energy consumption so much and almost completely depend on the renewables and other equipment installed in her house was significant to her, resulting in a $75 rebate from the utility and paying for her lunch.

   So what? The transformation of the utility continues with the customer no longer being a source of a bill (and resulting payment) for the utility. The prosumer now delivers energy to the utility and in turn, gets a payment back. A utility that plans these disparate sources of energy well can delay system enhancements resulting in further savings to the customer.
8. *Regulatory directive for energy efficiency:* The connection to time-of-use (TOU) rates identifying that she only paid money to the utility when she did her washing and baking at the same time is significant.

   So what? The transformation of the utility now includes the education of the customer to an extent that she also understands the concept of utility peak power and its impact on peaking resources, congestion, and cost. As a result, the customer is now a willing partner with the utility as they work together to ensure the best kind of energy (base load power) is used and/or available in plenty at the lowest cost.
9. *Plug-in electric vehicles:* The full integration of her EV into the grid both from a storage availability perspective, and using solar cells to charge the car are significant connections.

   So what? In addition to being a new source of energy in the utility equation, an EV is also a mobile source of energy and storage that can be moved where needed having the potential to create an unprecedented level of support for the grid operator. The customer and utility are now part of the same energy equation.

The utility front-office, back-office, and mid-office systems are all integrated and, most importantly, related to the customer. Much of the fundamentals come from automation and supported business architecture being defined here and need to be in place prior to deliver on the customer mandate. If the architecture is not in place, the utility will need to implement them on a piecemeal basis leading to increased cost to the customer and inefficient operations at the utility.

It is important to note the expectations of the typical customer in other industries are much higher than they expect from a utility. However, it is believed the electric utility will also get there in time. This means that in the next 10 years, utilities need to get their automation (and other) systems in place to deliver on this expectation. In addition, as New York (REV) and California

(Better than Smart) initiatives go operational, other states will start looking at their initiatives and push their utilities to move forward in the same direction.

This Is Transformation: The Entire Utility Needs to Change for the Conversation to Be Real

This transformation will affect system (T&D) operations, outage management, customer service, asset management, and corporate investment and planning.

The smart grid will change everything related to the grid. It will change what the grid looks like, how the grid works, and everything we know about the utility business [30]. Some of the major changes we can expect in the future will include:

- The entity known today as the consumer (or customer) will both consume and generate power (become a prosumer) by installing one or more DERs within their premise.
- The utility of today could morph into becoming the wires provider for transmission and/or distribution.
- Service providers such as telecom, security systems, or internet service could take on the role of providing electricity as a service and bundle it along with the other services being provided by them.
- Microgrids could form and, depending on their business need, could partially or completely separate themselves from the utility grid.

The smart grid is a journey of continually improving the operations of the grid through increasing levels of intelligence and the ability to use this intelligence to improve the operational efficiencies of the grid and a better service to the customer.

References

[1] Vadari, M., et al., "Disruption Becomes Evolution: Creating the Value-Based Utility," White paper, CMG, July 31, 2014, http://www.512cmg.com/services/white-papers/.

[2] Carvallo, A., and J. Cooper, *The Advanced Smart Grid: Edge Power Driving Sustainability*, 2nd ed., Norwood, MA: Artech House, 2015.

[3] Vadari, M., "Demystifying Intelligent Networks: Why the Next Wave of Transformation Is Already upon Us," *Public Utilities Fortnightly*, November 2006.

[4] Vadari, M., "A New Architecture for Distributed Energy Management," *IEEE Smart Grid Newsletter*, June 2014.

[5] Vadari, M., "Integrating Renewables: Opportunity for Advancement or Exercise in Futility?" *Public Utilities Fortnightly*, March 2010.

[6] Kramer, W., "NREL," March 2008, http://www.nrel.gov/docs/fy08osti/42672.pdf.

[7] New York State, "Reform the Energy Vision (REV)," March 2016, https://www.ny.gov/sites/ny.gov/files/atoms/files/WhitePaperREVMarch2016.pdf.

[8] Capehart, B. L., "*Distributed Energy Resources (DER),*" October 20, 2016, https://www.wbdg.org/resources/distributed-energy-resources-der.

[9] Energy Storage Association, "Energy Storage Technologies," http://energystorage.org/.

[10] Vadari, M., and G. Stokes, "Utility 2.0 and the Dynamic Microgrids," *Public Utilities Fortnightly*, November 2013.

[11] Vadari, M., "Smart Grid, System Operations and the Management of Big Data to Drive Utility Transformation," *CIO Review Magazine*, November 2013.

[12] Sinopoli, J., *Advanced Technology for Smart Buildings*, Norwood, MA: Artech House, 2016.

[13] Matiko, J., and S. Beeby, *Applications of Energy Harvesting Technologies in Buildings*, Norwood, MA: Artech House, 2017.

[14] Office of Electricity Delivery and Energy Reliability, "Demand Response: Policy," https://energy.gov/oe/services/electricity-policy-coordination-and-implementation/state-and-regional-policy-assistanc-4.

[15] Vadari, M., "Active Demand Management: A System Approach to Managing Customer Demand," *Public Utilities Fortnightly*, November 2009.

[16] Rose, A., M. Vadari, and L. Wigle, "How the Internet of Things Will Transform Energy Efficiency and Energy Services," *Proc. of the 2014 ACEEE Summer Study on Energy Efficiency in Buildings*, 2014.

[17] U.S. Department of Energy, "Communications Requirements of Smart Grid Technologies," October 5, 2010, https://energy.gov/sites/prod/files/gcprod/documents/Smart_Grid_Communications_Requirements_Report_10-05-2010.pdf.

[18] NETL, "Electric Power System Asset Optimization," March 2011, http://www.netl.doe.gov/energy-analyses/pubs/ElecAssetOptRep.pdf.

[19] NETL, "Framework for Improving Critical Infrastructure Cyber-Security," February 2014, http://www.nist.gov/cyberframework/upload/cybersecurity-framework-021214-final.pdf.

[20] NIST, "NIST Framework and Roadmap for Smart Grid Interoperability Standards, Release 3.0," NIST Special Publication 1108r3, September 2014, http://www.nist.gov/smartgrid/upload/NIST-SP-1108r3.pdf.

[20] NIST, *Introduction to NISTIR 7628 Guidelines for Smart Grid Cyber Security*, September 2010, https://www.nist.gov/sites/default/files/documents/smartgrid/nistir-7628_total.pdf.

[21] Koutitas, G., and S. McClellan, *The Smart Grid as an Application Development Platform*, Norwood, MA: Artech House, 2017.

[22] Green Tech Media Staff, "Data Analytics and Smart Grid: The Rising Tide for Power Utilities," December 20, 2012, https://www.greentechmedia.com/articles/read/data-analytics-and-smart-grid.

[23] Smart Grid Information Clearinghouse, 2013, http://www.sgiclearinghouse.org/Technologies.

[24] Market Design and Platform Technology Working Group, *Report of the Market Design and Platform Technology Working Group*, August, 17, 2015, http://nyssmartgrid.com/wp-content/uploads/MDPT-Report_150817_Final.pdf.

[25] GridWise Architecture Council, "GridWise Transactive Energy Framework Version 1.0," January 2015, http://www.gridwiseac.org/pdfs/te_framework_report_pnnl-22946.pdf.

[26] AEP Ohio, "gridSMART Demonstration Project – Final Report, based upon work supported by the Department of Energy under Award Number DE-OE0000193," March 29, 2013.

[27] AEP Ohio, *Final Technical Report, AEP Ohio gridSmart Demonstration Project*, June 2014, https://www.smartgrid.gov/files/AEP_Ohio_DE-OE-0000193_Final_Technical_Report_06-23-2014.pdf.

[28] GRID4EU, "GRID4EU," August 2015, http://grid4eu.eu/overview.aspx.

[29] Grid4EU, "Final Report," March 2016, http://grid4eu.blob.core.windows.net/media-prod/29375/grid4eu-final-report_normal-res.pdf.

[30] U.S. DoE, *2014 Smart Grid System Report*, August 2014, http://energy.gov/sites/prod/files/2014/08/f18/SmartGrid-SystemReport2014.pdf.

3

Distribution Automation: Path to a Self-Healing Grid

Transmission and distribution automation (DA) are the process of monitoring and controlling the grid via use of intelligent devices, instruments, and advanced components. They are enabled by integrating devices and components in the field with analytical tools in the control center via two-way communications networks.

DA holds the rightful position as the smart grid's foundation. They are components of a larger system that cover sensing, control, and healing. The technologies discussed will form the underlying driver for smart grid for the foreseeable future. As controlling the grid becomes more complex, DA will provide the most important data and control mechanisms for the utility and other operators to build a reliable and resilient grid.

This chapter defines DA and its components. Smart grid has a variety of components, understanding their implementation will clarify its capabilities. The dos and don'ts section explores key considerations for utilities around the world to consider when implementing DA in their grids. Finally, this chapter explores the ongoing challenges and potential future of DA.

Introduction

A self-healing grid is the holy grail of the smart grid. So what does this mean?

Today's grid suffers several disturbances. Some are weather-related, such as hurricanes, which are also called Acts of God, and some are man-made, such as when a car hits an electric pole or a cyberattack. These events generally impact the grid in the form of outages, which are a loss of power. This loss of power can be localized, such as solely one house or a block of houses, medium-sized, impacting a neighborhood or major parts of a city, or widespread with a regional reach, such as the 2003 blackout in the Northeast United States. Utility executives believe that there will be more of these outages and the grid's resiliency—the ability to recover readily from major attacks such as the ones identified above—will become more and more of an issue in the future [1].

Innovations in the smart grid arena also result in the grid getting new distributed energy resource (DER) components, including storage, distributed renewables like solar and wind, and microgrids. DERs bring new challenges to distribution grid operators. For example:

- Supply from sources within the distribution grid moves the grid from a one-way flow to a two-way flow.
- Supply from customer-owned instead of utility-owned sources requires decisions about how to utilize/transmit this power and how to compensate customers.
- Increased levels of intermittency from DERs, which put an additional burden on an operator's ability to manage and operate the grid in a reliable, efficient, and optimal manner.

The three challenging grid conditions identified above could easily morph into several offshoots that could drive the distribution grid into a variety of directions, each more complex than the other. With each change, the management and operation of the distribution grid still need to deliver a reliable and efficient grid, one that is capable of handling both the disturbances identified earlier but also and the intermittency and variability from newer forms of energy. For all this to happen, the distribution grid and, by definition, the grid operator needs access to several key technological components. These components allow the operator awareness information regarding of what is happening on their grid, and understand options they have for response. This process will improve an operator's ability and ultimately to select and implement the right response.

New network configurations such as microgrids[1] also pose inherent complexities. Parts of the grid, such as commercial applications/buildings and college campuses, are interested in creating microgrids to function in either grid-connected or grid-isolated conditions, based on their desire to save money

1. IEEE standard P2030.7 defines the microgrid as a system that can manage itself and operate autonomously or grid connected and seamlessly connect to and disconnect from the main distribution grid for the exchange of power and the supply of ancillary services.

and go green. This requires grid operators to collaborate with external entities within their distribution systems to ensure reliability and resiliency of their own systems as well as their interconnection with others.

In transmission, these sets of technologies are called transmission automation (TA). In distribution, these technologies are called DA. Before we delve further into DA, it is helpful to fully understand the difference between the two.

TA versus DA

The electric grid is generally distinguished between transmission and distribution systems based on voltage class. Equipment above 132 kV is generally considered transmission and equipment below 33 kV is considered distribution. Depending on location and/or the specific utility, equipment can be considered transmission or distribution. In North America, distribution is generally characterized as equipment that is 33 kV and below, whereas in Europe and other countries, distribution is 69 to 70 kV and below.

The primary goal of TA is to increase grid reliability through better overall grid situational awareness, monitoring, and control. Electric utilities typically have extensive control over their transmission-level equipment through SCADA and other wide-area monitoring systems. Since the implementation of the smart grid, these networks are increasingly being transformed from analog to digital systems through the widespread deployment of intelligent electronic devices (IEDs). IEDs are the foundation of TA and DA and are used to control equipment such as circuit breakers, relays, transformers, and capacitor banks. Examples of other TA technologies include phasor measurement units (PMUs), digital protective relays, and digital substations. Of these, PMUs are the newest and have the potential to change the entire landscape of TA in all forms.

DA and Its Connection to the Self-Healing Grid

DA is the process of monitoring and controlling the distribution network via the use of intelligent devices, instruments, and advanced components. Integrating devices and components in the field with analytical tools in the control center via two-way communications networks enables DA.

A fundamental capability of the smart grid is to self-heal. The U.S. Department of Energy's National Energy Technology Laboratory (NETL) defined one of the characteristics of the smart grid as "it will heal itself." NETL touts the smart grid as being able to perform continuous self-assessments to detect, analyze, respond to, and, as needed, restore grid components or network sections. Acting as the grid's immune system, self-healing will help to maintain grid reliability, security, affordability, power quality, and efficiency [2].

In other words, a self-healing grid is designed to sense and automatically respond to issues. It is designed to avert, confine, and reduce harm by protecting the power infrastructure and the human beings working on the infrastructure, as well as provide information to allow operators to make the right decisions at the right time to address those issues [3].

DA increases control over distribution equipment, however, to realize a truly self-healing grid, DA must extend to even smaller network entities such as DERs, industrial, commercial, and residential loads. By doing so, DA is considered one of the fundamental dimensions of the entire smart grid.

At a high level, the electric grid needs to sense, analyze, control, and communicate. Let's look at each of these functions and the impact of automation.

- *Sense:* A deep understanding of the state of the network is critical for automation to work efficiently. Different kinds of sensors are installed in the network to measure one or more quantities associated with a power system. Some of the quantities that are measured include voltage at a point, current through a component (such as a line, transformer), power consumed by a grid component at an instant of time and/or energy consumed by a component over a period, and other quantities, such as circuit breaker status and transformer oil temperature.

  Sensing occurs from a variety of sources with varying degrees of resolution and speed. Example sensors include smart meters, SCADA sensors, and/or PMUs.
- *Analyze:* Automation takes data collected by sensors, examines that data, and comes to conclusions about what steps need to be taken, if any. Some analysis may result in no action needing to be taken. The analysis supports any one of the goals of the grid, including reliability, optimization, and efficiency. It can also support new technologies such as storage and other DERs. Analysis can be centralized in a control center environment or distributed by taking place in the field at the substation or device level.
- *Control:* The result of analysis is often to take some action. The type of action could be one of the following: open or close a remote-controlled switch or circuit breaker, change a transformer tap up or down, or an alternative solution based on data considerations.

  The action could be information passed on to an operator to take manual action, where the operator performs a remote operation of a switch movement, or automatic. This resulting action, if automatic, is the self-healing aspect. The grid senses and analyses somewhat automatically, so it is the action automated in the smart grid, which makes it self-healing.

- *Communicate:* The combination of sensing, analysis, and control requires adequate communication for the sensing data to be sent to where it is needed and the control signal can be sent to operate the correct device in the appropriate manner. Additionally, information regarding the success of the operation must be sent back [4]. Even more importantly, the data needs to get to its destination at the right time and in the right frequency regardless of whether it is sensing or control. This requires a combination of bandwidth and latency to be considered in its design.

Smart Grid Dimensions That Make DA Self-Healing

In Chapter 2, the smart grid was defined, its dimensions were introduced, and DA was highlighted as a critical and fundamental dimension. To be truly effective, however, it also depends on other technological components. This section will elaborate on these technologies, showing how they augment DA in defining and delivering on the grid's self-healing capabilities.

Smart Meters

A smart meter is installed at the point where the utility's power line enters the premises. The premises could be residential, commercial, or industrial. The meter measures, collects, and stores end-user energy consumption and other data. It then takes this data and sends it back to the utility systems where it is used for a variety of purposes, such as billing. smart meters also sense or measure abnormal events, such as momentary outages on a per-customer basis, which are often a precursor to grid failure. Consequently, the smart meter is the utility's primary interaction point with customers' premises.

From an automation perspective, the smart meter is a point of contact and control installed at the grid edge[2]. In the event of an outage, the first step of a self-healing grid is to detect and identify outage locations. The last-gasp capability of smart meters provides outage alerts to the utility, thus pinpointing outage locations in real time. Outage prediction algorithms can then quickly aggregate this data to the nearest distribution-switching asset and perform

2. Grid edge is a recently coined term that means devices near the end-use customers such as homes or businesses or at the distribution system. These include devices such as solar panels, advanced metering infrastructure, smart inverters, energy storage systems, smart thermostats, smart appliances, and building controls.

restoration tasks. This can dramatically cut down outage durations. Depending on the scheme, this isolation and/or restoration process can be performed manually or through automation.

Data from smart meters are also used to support several other DA mechanisms such as voltage-VAR support. Newer mechanisms such as integrated volt-VAR control (IVVC), volt-VAR optimization (VVO), and conservation voltage reduction (CVR) are being implemented. These are augmented by access to data from smart meters on the grid edge. With the advent of generation at the grid edge from DERs, voltage deviations or power quality is becoming a bigger issue than ever before.

Big Data and Analytics

Big data and analytics can be thought of as the layer between raw data from the field and advanced operational systems. It is the layer that converts the plethora of data collected from meters, sensors, switches, and other devices deployed in the field into actionable information for use by the utility. It provides the utility with intelligence into the performance of the power grid, consumer energy use, peak demand, and business risks.

Techniques employed to perform data analytics at utilities range from pure power system algorithmic systems to advanced machine learning.

DA functions, such as fault location, isolation, and service restoration (FLISR), VVO, and CVR, gather a tremendous amount of data from automated feeder switches, reclosers, line monitors, and other equipment. Given the cost of designing and deploying these schemes, however, utilities must prioritize the feeders, which benefit the most from these functions. Data analytics highlight the worst performing feeders that would see a larger relative DA impact over feeders with less room for improvement. This provides the utility a substantial return on their investment.

Another area of application of big data analytics to T&D automation is grid asset diagnostics. Equipment health monitoring enables preventative maintenance scheduling. Utility personnel can use data to assess historical behavior of equipment and set priorities for maintenance and inspection activities. Data analysis can prediagnose equipment suspected of failures by reviewing operational parameters such as pressure, temperature, and oil levels prior to inspections to determine the best approach for proactive repair and/or replacement. Data analytics software can evaluate failure risks for specific devices and set priorities for maintenance and inspection activities accordingly.

The analytic results are fed into a broad array of operational and decision-support systems making the information processing capabilities of the control center more accurate and timely. In turn, this makes DA more powerful and accurate. Another aspect of analytics is in the predictive ability to provide future insight into grid conditions, thereby enabling operators to enhance, fine-tune, and test automatic restoration strategies ahead of grid disturbances.

Communications

Communication is the backbone of the self-healing grid. Coordination between sensors and advanced components in the field requires the transfer of data among them. A combination of bandwidth and latency driven by specific needs of the problem being solved creates the parameters for communication. The following list presents examples of the broad variation of data transfers that require communications mechanisms:

- Devices such as smart meters send data approximately once every 15 minutes.
- Automation devices that work through SCADA-based systems tend to end data approximately every 2, 4, or 6 seconds.
- PMU-based systems, in either transmission or the newly emerging distribution PMU, communicate geosynchronous data approximately 30 to 60 times a second.

To enable automation, utilities are either retrofitting existing grid devices or replacing with devices that come integrated with communication packages. Examples include radio-controlled smart switches, reclosers retrofitted with control boxes containing communications, and relay packages. Engineers can then program these devices as part of automatic restoration logic to implement self-healing on the distribution grid [5].

In order to enable DA, utilities typically expand their communications networks for distribution systems to acquire large volumes of new data from sensors, process the data, and send control signals with low-latency to operate equipment. Communications networks allow utilities to connect devices to each other and to SCADA, distribution management system (DMS), and other information and control systems, thereby greatly improving the capabilities of grid operators to manage power flows and address reliability issues.

Most utilities use at least a two-layer communications system to communicate between field devices and information and control systems. Typically, the first layer of the network connects substations and DMS at the control center. Some utilities use existing SCADA communications systems for this layer. Many utilities choose high-speed, fiber-optic, or microwave communications systems, while some contract with third-party telecommunications vendors to use their high-speed cellular networks.

The second layer of the network typically connects substations with field devices. Many utilities choose some form of wireless network for this layer, including radio frequency mesh or Wi-Fi.

DA projects typically include field device integration with the operational systems such as DMSs. Communications networks are the lynchpin connecting it all together.

Privacy and Cybersecurity

A self-healing grid should be designed to avert, confine, and reduce harm to the grid by protecting both its power infrastructure and data. The self-healing grid should be resistant to attacks both physical and virtual. For the energy system to be truly resistant to attacks, attention to privacy and cybersecurity is paramount. For cybersecurity, all aspects of the system—from sending to analysis to control—need to be designed to handle intrusions from any direction. Similarly, stored customer data requires adequate privacy measures [6].

Cybersecurity, specifically interoperability, remains an important technical challenge for modernizing electric distribution systems. Secure and interoperable technologies and systems require standards, protocols, tools, and techniques. These are currently being designed by standards bodies and vendors are in the process of accepting and incorporating the standards into their products.

Success in these areas involves ongoing activities for government and industry, including changes in regulations, business practices, and consumer data privacy protections, most of which require coordinated utility effort.

Core Components of DA

Figure 3.1 identifies the foundational components of DA. These components focus on distribution-level equipment with a goal of optimizing utility operations and improving reliability of the grid. These technologies are intended to do several things including reducing outage frequency and duration, and improve power quality and control demand. Let us discuss each of these further, first in general, and then in more detail.

Advanced Sensing and Measurement

The purpose of advanced sensing and measurement technologies is to acquire data from the field and send it where it is needed. This data would be sourced from a variety of locations in the distribution grid or customer premises and could include information related to fault location, transformer and line loading, feeder voltages, equipment health, power factor, outage notification, and

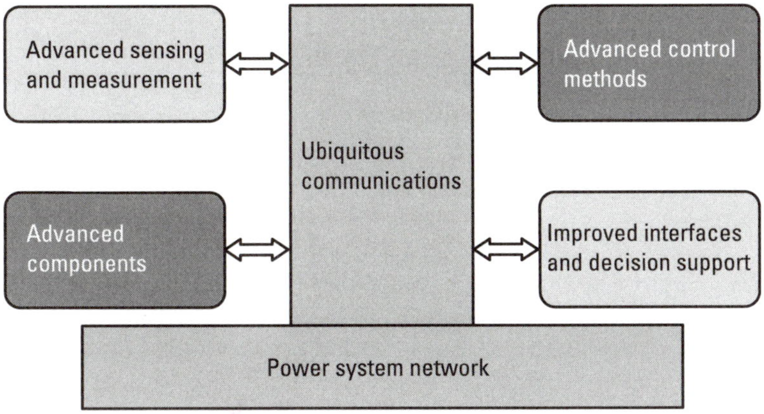

Figure 3.1 Key components of T&D automation. (Source: NETL.)

energy consumption. Some of these technologies are presented in Table 3.1 along with their corresponding applications [7, 8].

Advanced Control Methods

Advanced control methods are algorithms embedded in the devices that function in a distributed manner or are executed centrally in the control center environment. These methods, collectively called advanced operational and decision support systems, take actions to correct grid behavior along the electric value chain, all focusing on analysis, diagnosis, or prediction-level efforts on the grid [9]. These systems take data collected by sensors in the field and perform analysis resulting in decisions to manage and control grid behavior.

There are three specific aspects of advanced control methods to focus on centralized systems, decentralized systems, and advanced decision support systems. These three segments of advanced operational systems form the underpinning of the implementation arm of DA and provide the basic steps toward the self-healing grid.

Centralized Systems

Table 3.2 lists examples of centralized operational systems. Many of these systems are in use by utilities' control centers, while some are still evolving.

Decentralized Systems

These systems are installed in the field and designed to react autonomously and automatically. Their data inputs may come from the same sensors feeding the SCADA systems in the control centers, but they have the advantage of not requiring operator input or attention. Table 3.3 lists examples of decentralized operational systems.

Table 3.1
Advanced Sensing and Measurement Technologies by Application

Grid Application	Advanced Sensing and Measurement Technology
SCADA measurements	SCADA measurement devices such as RTUs connect to sensors and actuators in the field and are networked to the supervisory computer system. These devices typically have intelligent I/O and often have embedded control capabilities such as ladder logic to accomplish Boolean logic operations. SCADA systems are used to monitor and control a plant or equipment in industries such as telecommunications, water and waste control, energy, oil and gas refining, and transportation. Electric utilities use them to monitor and control the grid. SCADA measurements are fed back to the utility control center, where they are used by the energy management system and/or the distribution management system to assess the health of the grid. The system operator uses SCADA to issue controls to devices in the field. The system operator enables distribution automation through the remote control of devices such as breakers and switches. More recently, newer sensors such as PMUs, also known as synchrophasors, are also being deployed in the field. These devices deliver geo-time-synchronized data to the centralized systems enabling better analysis and assessment capabilities. Key components of SCADA in the field include current transformer (CT), potential transformer (PT), remote terminal unit (RTU), intelligent electronic devices (IED), and other sensors such as temperature, and transformer and oil pressure.
Metering	Meters are electronic devices that record consumption of electric energy in intervals of an hour or less and communicate that information at regular intervals back to the utility for monitoring and billing. Meters are installed on customer premises, which can be residential, commercial, or industrial in nature. They can also be installed on the grid in locations such as substations. Smart meters are enabled by two-way communication between the meter and the control center systems. This real-time data alerts the utility of power outages at customer locations, and as well the quality of power. Distribution automation and other schemes can use this information to automatically identify faults and restore power to parts of the affected grid. Key types of advanced metering in the field include smart meter, power quality meter, and revenue meter.
Monitoring and control	Monitoring and control devices are typically equipped with a mechanism that can automatically close or open the device after the occurrence of a fault has opened or closed it, as in the case of a smart recloser. They can also control the injection of reactive power into the grid as in the case of the capacitor bank. In addition, they can be used on overhead or underground distribution systems to detect and interrupt momentary faults, provide reactive power compensation, and assess the quality of power. The smarts incorporated into these controls enable a distribution automation scheme that includes these devices to automatically operate based on fault conditions. Key components of monitoring and control in the field include smart reclosers, recloser control, voltage regulator control, capacitor bank control, PQ monitors, and smart switches.
Asset condition monitoring	These sensors monitor the state of grid assets by tracking their health and providing early warning signs of problems such as overheating, insulation degradation, or mechanical movement within the asset. They are installed on valuable transmission and distribution assets such as transformers, and circuit breakers. The data provided by these sensors helps the utility take proactive steps in maintenance and repair and prevent asset failures that can not only prove expensive to the utility, but also cause widespread power outages. Key components of asset condition monitoring include dissolved gas monitors, moisture monitors, circuit breaker condition monitors, and load tap changer monitors.
Fault indicators and sensors	Fault indicators are devices that provide visual or remote indication warnings and locations of faults on the electric power system. These indicators can be installed on overhead or underground distribution lines. These devices provide a means of automatically detecting and identifying faults to reduce outage time. Key components of fault indicators and sensors include overhead fault indicators and underground fault indicators.

Table 3.2
Examples of Centralized Operational Systems

Systems	Descriptions
Supervisory control and data acquisition (SCADA)	SCADA is a centralized, real-time control system that interacts with devices in the field. This interaction includes getting data from sensors in the field and passing on that data to other applications, which need it to perform their analysis. The interaction also includes the ability to perform control actions on devices in the field such as circuit breakers and transformers. SCADA is the eyes and ears of the EMS and provides real-time data to centralized operations. SCADA is more often installed on the transmission, but over the last 10 years, it is also installed in the distribution system. From an architecture perspective, SCADA is shifting from a monolithic architecture into an integrated information and communications architecture, which enables data sharing between centralized services at the control center with decentralized systems in the field. This, in turn, enables smarter utility operations. More information on this topic is contained in [10].
Energy management system (EMS)	The EMS is a system used by system operators to take the data from SCADA and other sources to optimize the performance of the transmission system. The EMS can be configured with or without the inclusion of SCADA. Applications in an EMS use the data from SCADA and other transmission automation (TA) devices in the field to perform their analysis. This allows transmission operators to perform their core tasks. Typically, all actions needed to control devices in the field are communicated through SCADA. The EMS enables monitoring of power system dynamics in real time, identifies system stability conditions on the larger network, and enables design and implementation of measures to mitigate them. EMSs are primarily used for the transmission system, but they occasionally incorporate components from the distribution system, such as distribution automation functions. An EMS integrated with phasor measurement units (PMUs) enables wide-area situational awareness-prevention of system instability and stress. enhanced system reliability-prevention of cascading disturbances, outage reduction, faster restoration, enhanced system operations, planning, modeling and post-outage analysis, and market operations-advanced congestion management via visibility into congested transmission corridors. More information on this topic is contained in [10].
Outage management system (OMS)	OMSs are used by a broad range of utility personnel including system operators to track and locate outages and work with crews in the field with restoration of power. Information comes into the OMS from a broad variety of sources, including customer calls, smart meters, and SCADA. The OMS is the ultimate record of all outages. The data is used for outage reporting. The OMS is fundamentally a business system. It does not perform controls on the system, but provides the system operator with real-time data obtained from smart meters via last gasp or pinging. Integrated with AMI for automated outage alerts. The OMS enables the operator to assess the extent of power outages. Utilities typically deploy OMS on the distribution system and integrate it with D-SCADA for remote control of switches and circuit breakers, providing faster restoration. The OMS is dependent on the DMS for creating and executing switching orders for outage restoration. More information on this topic is contained in [10].

Table 3.2 (continued)

Systems	Descriptions
Distribution management system (DMS)	The DMS is a system used by operators to take the data from SCADA and optimize the performance of the distribution system. The DMS can be configured with or without SCADA. When configured with SCADA, it is sometimes called D-SCADA. DMS is deployed in the control center and used to monitor and control the distribution system. The DMS is very often integrated with an OMS and the systems feed each other to provide the distribution operator with a complete set of tools to manage and operate the distribution grid. The DMS integrated with AMI, D-SCADA, and GIS continually monitors the state of the distribution system and works to improve its efficiency and performance by proactively managing the network. Advanced distribution and substation automation functions such as fault location isolation and service restoration (FLISR), integrated volt-VAR control (IVVC), feeder load balancing, and equipment condition monitoring are incorporated into the DMS to enable smarter utility operations. A combination of D-SCADA, OMS, and advanced power-flow applications is also called an ADMS or advanced DMS. More information on this topic is contained in [10].
Distributed energy management system (DEMS or DERMS)	The DEMS or the DERMS as it is very often called, is a new and still evolving system used for active management of distributed energy resources (DERs) such as DG, energy storage, DR, and EVs from a single unified platform. Managing all DERs in a single platform as a portfolio of products allows the utility to take advantage of the variability in their characteristics in dispatching resources. The DERMS is a unified system that manages various distributed energy sources such as distributed generation and renewables, demand response, storage, and PHEV/PEV. This system focuses on customer operations and provides a comprehensive set of operational services to the customer while allowing the utility to manage and operate the grid in a safe and effective manner. A DEMS is deployed on the distribution system that has DERs deployed. DERMS is a less complex system than a DMS, and serves as a central repository to integrate DER data, programs, and core interfaces to DEMS integrates on a broader scale with distribution management and its automation functions. More information on this topic is contained in [10].

Advanced Decision Support Systems

Data gathered from sensors or created from various applications and algorithms can overwhelm the operator. The operator with a broader view of the grid could keep the system reliable and resilient to attacks. Advanced decision support systems are therefore a key piece of the automation puzzle, providing visualization, decision support, and system operator training, to utility control center personnel. Figure 3.2 presents examples of some of these technologies [11].

Color Contouring

Color contouring uses gradient changes in color on one-line displays to highlight areas of the power system that pose potential problems or require operator attention. Color contouring facilitates an operator's ability to detect and locate problems in the power grid with less effort while monitoring the system in real time. For example, knowing that red represents a very severe voltage violation

Table 3.3
Examples of Decentralized Control Mechanisms

Systems	Description
Protection and control (P&C)	These devices in the field can be either analog or digital and react autonomously and fast to save the power system during emergencies. They do this by mitigating and localizing the impact of faults to prevent cascading failures. Protection systems monitor current, voltage, and other power system variables to detect anomalies and open switches to automatically isolate the problem area. Protection systems operate in a generally decentralized manner. They are used for distributed control of the power grid for fault detection, mitigation, and clearance. Decentralized control mechanism has critical application during emergency operations such as blackouts and cascading failures. Rapid changes in T&D network topology during faults and impending blackouts require fast-acting decentralized adaptive protective mechanisms to mitigate and localize the faults to prevent cascading failures. This function is deployed in both transmission and distribution systems. An example of protection and control is automated feeder reconfiguration, currently a major feature of advanced distribution systems. This has led to the deployment of advanced coordinated decentralized protection, which can adapt to the rapidly changing system configurations. Sometimes these schemes are aided by distribution PMUs that provide system-wide visibility based on measurements gathered from phasor data concentrators (PDCs). They also detect disturbances and events at the system level by bridging the gap between the protection schemes in the field and advanced protection functions in the control room. Key components of protection and control devices in the field include line protection relays, transformer protection relays, bus protection relays, circuit breaker protection relays, and feeder protection relays.
Real and reactive power flow control (RRPFC)	These devices provide dynamic power flow control in power systems, based on real-time system conditions. They provide rapid responses to system stability issues via dynamic power flow control and improve the utilization of underutilized lines by leveraging information from IEDs on lines in a specific neighborhood. These DPFC devices and other IEDs located on transmission lines enable local measurement and control of line currents. It also allows them to communicate and interact with each other, which enables decentralized intelligent control. Local and/or remote voltage and current measurements typically report/monitor the status of the system. These measurements, in turn allow the devices to provide rapid response to line overloading by adjusting the impedance of the affected line and other lines in the neighborhood. System/local events triggers the automated response, which and takes only a few seconds. These generally low-cost, mass-produced, distributed power-flow devices are installed on each phase of a transmission line to provide instantaneous control of real and reactive power flow. Key components of real and reactive power flow control devices include devices installed in the transmission system that use advanced thyristor-based devices, such as flexible AC transmission systems (FACTS), distributed power flow controller (DPFC), and others to perform instantaneous control of real/reactive power in the grid allow the control of the flow of power along the transmission lines.

Table 3.3 (continued)

Systems	Description
Dynamic line rating (DLR)	Dynamic line rating (DLR) technology uses advanced sensors that measure the line current carrying capacity, conductor surface temperature, and ambient temperature of transmission lines in real time, to determine the safe and accurate dynamic thermal ratings of transmission lines. These devices allow the movement of the line ratings from a static number that rarely changes to one that changes rapidly to reflect system conditions, thereby allowing the operator to get maximum benefit from the grid. DLR devices monitor changing weather conditions in real time to improve system reliability. These devices optimize the usage of transmission lines over and beyond static line ratings (SLRs), which otherwise assumes constant weather conditions, over an extended period. This optimization enables integration of renewables, such as wind generation, via enhanced transfer capability thereby helping defer construction of new transmission lines. DLRs are typically deployed on transmission lines, especially in congested zones and lines serving as entry and exits to renewable generation. The various sensors communicate and interact with each other to determine the dynamic line ratings. Dynamic thermal line ratings are calculated automatically depending on system and local weather conditions. New voltage and current reference signals are communicated to DPFC controllers, to control the real and reactive power flows on lines, according to the dynamic line ratings.
Distribution integrated volt-VAR control (IVVC) and fault location isolation and service restoration (FLISR)	These are a newer set of DA controls that can either work in a centralized—unified with DMS—or decentralized and autonomous mode. They are generally designed to use peer-to-peer communication for sensors with smart meters and fault locators depending on the function to be implemented. Key functions include volt-VAR control and fault location, isolation, and service restoration. IVVC is a set of sophisticated volt-VAR control mechanisms that provide the benefits of power factor correction, voltage optimization, and condition-based maintenance in a single, optimized package. IVVC also enables conservation voltage reduction (CVR) on a utility's system. IVVC does this by functioning in a closed loop with sensing and controls at both ends of the feeder. FLISR networks groups of switches together on a feeder to vastly improve the reliability of utility-delivered power, by localizing outages. FLISR senses trips or faults in switches monitored and controlled by SCADA. FLISR identifies the faulted section, isolates the fault, and restores power to customers by automatically switching them to nonfaulted sections of the line. Both IVVC and FLISR are advanced distribution management components that enable automation in the field.
Grid Friendly Appliance Controllers (GFAC)	This is an evolving approach to consumer-level autonomous demand management using GFAC to autonomously control consumer loads based on the real-time pricing and transactive control. GFACs sense grid conditions by monitoring the frequency or voltage of the system and enables autonomous control of the load. GFACs can be installed in household appliances such as refrigerators, air conditioners, and water heaters, to monitor the power grid and turn appliances off from a few seconds to a few minutes in response to power grid overload. By triggering appliances to turn on and off at different times, these devices help control power oscillations, which occur in different parts of the grid. GFACs provide rapid, autonomous, and automatic response to grid emergencies. They enable mitigation of system under-frequency issues by providing decentralized control at the consumer level. This prevents the need to shed bulk load at the substation level. GFACs provide automatic demand response as needed by the grid operator.

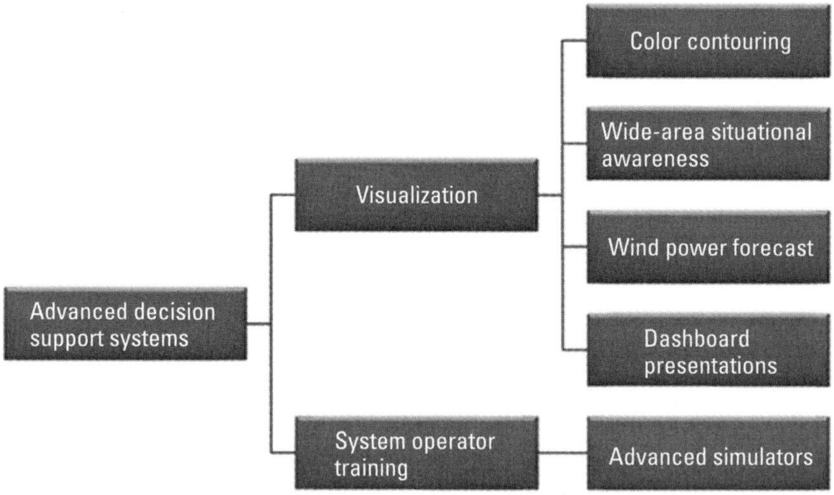

Figure 3.2 Examples of advanced decision support systems. (Source: Modern Grid Solutions.)

may aid operators in determining which buses have the worst violations faster than with a non-color contoured display.

Wide Area Situational Awareness

Comprehending complex system statuses to make informed decisions in a time critical manner and positively affect the system state is called situational awareness.

PMU sensors gather geosynchronized and time-synchronized data, gathered via WAMS, which can help grid operators and engineers to make better and more timely decisions. To improve situational awareness, PMU information can be used in many ways, including:

- Real time—direct visualization. This is essentially in a raw form, for example, contours of bus angles.
- Real time—embedded in other applications, such as state estimator (SE) for assessing system dynamics.
- Real time—embedded in data-mining applications to provide historical info.
- Offline—To develop better system models and understanding, for example, post-event analysis.

Wind Power Forecast

The need for wind power forecasting comes down to the inherent nature of electricity. In the electric grid balance must be constantly maintained between electricity consumption and generation, otherwise disturbances in power quality or supply may occur. Wind generation is a direct function of wind speed and thus not easily dispatchable. Electric utilities and transmission system operators must predict the output of wind generation to take into consideration for planning and operations. Depending on specific market structures, for utilities, inefficient commitment and dispatch caused by wind uncertainty increases fuel costs. Wind energy forecasting helps mitigate these costs.

Wind energy variability, especially rapid changes in wind energy production called ramps events, is both a cost and a reliability concern. A rapid ramp-up event requires conventional generators to turn down with comparable speed and vice versa for ramp-down events. Frequently, there is little to no notification of significant ramp events due to imprecise modeling of the underlying meteorological conditions. Large-scale events—such as cold fronts—may be predicted by weather models, but timing can be incorrect by tens of minutes to hours. Smaller-scale events, such as strong outflows associated with convective activity, are frequently not forecasted at all. All of these can lead to power outages or rolling blackouts. Accurate wind forecasting can serve as a valuable input into operational systems to better optimize and dispatch resources.

Wind power forecasting is an estimation of the expected production of wind turbines for a future time scale. Forecasting can be performed over different time scales from very short-term, such as a few hours, to long-term such as several days, depending on the application. There are several forecasting methods, but the most common ones are based on averages of past production values. Statistical prediction methods are based on one of several models that establish the relationship between historical values of power, as well as historical and forecast values of meteorological variables, and wind power measurements.

Solar forecasts are also emerging as a key decision-support mechanism.

Dashboard Presentations

Dashboards present a quick snapshot of key information such as outages and outage locations, substation load, and weather that assists grid operators to rapidly detect the status of key variables related to grid operations. Dashboards present a unified at-a-glance view of key system parameters via integration with disparate operational systems such as SCADA, OMS, AMI, GIS, and CIS. This dynamic operations data is presented to operators in easy-to-understand graphs, dials, and images.

Dashboards complement a utility's existing automation investments and allow more employees a concise overview of the present state of operations.

Advanced Simulators

Advanced Operator Training Simulators (OTS) provide dynamic graphical simulations by providing operators with a real-time view of the power system and its various parameters quickly, accurately, and in a format designed to increase situational awareness.

These simulators are especially advanced in the degree of realism allowing the operator to observe and practice procedures for abnormal conditions, such as voltage degradation, line overloads, cascading outages, system islanding with large frequency deviations, load interruption, and system restoration.

The development and use of these real-time dynamic simulators combined with industry-wide certification programs significantly improves the skill sets and performance of system operators. Advancements in system operations due to integration of new and enhanced smart grid technologies requires advanced training simulators to educate/prepare the grid operators to handle the complexities of the system competently.

Advanced Components

Examples of advanced components as shown in Table 3.4 are another critical group of technologies that enable automation [12]. Some of these are still in the research or pilot/prototype stage, but they extend out across a broad set of areas such as power electronics, super-conductivity, materials, chemistry, and microelectronics. These components can drive grid behavior and can also be applied in standalone applications or interconnected to create complex systems [13].

Power Electronics

FACTS devices such as UPFC, DVAR, and SVC, are examples of advanced components based on power electronic technologies. FACTS have been suc-

Table 3.4
Advanced Components by Category

Component Category	Examples
Power electronics	Flexible AC transmission system (FACTS) devices include: unified power flow controller (UPFC), dynamic volt-VAR controller (DVAR), and static VAR compensator (SVC)
	HVDC: Line commutated converter (LCC), voltage source converter (VSC)
	Smart AC/DC inverter
	Solid state transformer (SST)
Superconductors	First generation (1G) wire
	High temperature superconducting (HTS) cable
Composite conductors	Aluminum conductor composite core cable (ACCC)
	Aluminum conductor composite reinforced cable (ACCR)

cessfully implemented in many T&D applications, such as stability problems with energy transfer over long distances, voltage control at various load conditions, power quality enhancement, and reactive power balance. Other power electronics devices used in utilities include the following:

- SVCs improve transmission line performance by resolving dynamic voltage problems. SVCs provide high performance steady-state and transient voltage control by dampening power swings, improving transient stability, and reducing system losses by optimizing reactive power control.
- The high-voltage direct current (HVDC) system is a mature technology that relies on power electronics to convert AC power to DC and back. With this conversion, HVDC resolves many issues involving the power grid, such as coupling of asynchronous systems, stability problems with energy transfer over long distances, and the increase of short-circuit currents in meshed systems.
- Smart AC/DC inverters provide the grid interface with a variety of distributed generation sources. Future improvements in high-power semiconductors may make it economically viable to convert large areas of the grid to DC operation.
- SSTs are an ongoing research initiative to replace existing transformers with high-voltage power electronics. SSTs have transistors and diodes and other semiconductor-based devices that, unlike the transistors used in computer chips, are engineered to handle high power levels and fast switching. When mature, they will replace today's transformers with solid-state devices that will do much more than just transform AC voltages from one level to another.

  In response to signals from a utility or a home, SSTs can change the voltage and other characteristics of the power they produce. They can put out either AC or DC power or take in AC and DC power from wind turbines and solar panels and change the frequency and voltage to what the grid requires. They have processors and communications hardware built in, allowing them to communicate with utility operators, other smart transformers, and consumers.

  The functionality of SSTs could go beyond voltage transformation to other potential features surrounding voltage regulation, such as CVR, programmable output voltage, OV protection for PV-centric circuits, power quality—reactive power compensation and harmonic filtering, management and telemetry—integrated remote management, real-time load management, active line analysis, and other functions, such as the option for integrated storage.

Superconductors

As the name implies, superconductors have large power carrying capabilities because they have little resistance. There are several versions that have evolved over time. These include:

- *First generation (1G) wire:* Used in short line segments as exits from congested substations, in urban areas, and as fault current limiters. 1G wire is applied as a very low-impedance path to help control power flows on congested parallel lines.
- *High-temperature superconducting (HTS) cable:* Transmits large quantities of power at reduced voltages and high currents. Lower voltages reduce HVDC terminal costs by 25% to 50%. This cable may be competitive with underground cables employing large quantities of high-priced copper.
- *Second generation (2G) superconducting wire:* Provides lower-cost control of flicker, voltage, and transient stability. Prices could be 3 to 10 times lower than the 1G wire and have 10 times lower losses. This wire is expected to penetrate the replacement market for large industrial motors, power plant auxiliary motor drives, and power plant generators. Long-distance, low-impedance underground transmission of power is the goal.
- *2G wire fault current limiters (FCLs):* Under development to have 10 times lower losses, limit currents by a factor of 3 to 10, and have small footprints. These will significantly improve the cost and performance of HTS devices.

Superconductors are still in their infancy and have not yet delivered on their promise of moving to mainstream use.

Composite Conductors

Advancement in new materials have led to composite conductors, which enable increased utilization of right of way (ROW), allowing a doubling of amperage limits with little change to the line support or towers. Aluminum conductor composite core (ACCC) cable is superior to existing T&D cable in many key performance areas. It offers double the current-carrying capacity when compared to most standard conductors. It can dramatically increase system reliability by virtually eliminating problematic high-temperature cable sag. It is easy to install as conventional utility cable. Aluminum conductor composite reinforced (ACCR) cable can increase transmission thermal capacity by 150% to 300%. Composite core is also used to support the need for greater spans between poles to enable long distance crossings such as highways and rivers.

Dos and Don'ts of DA

As utilities look to implement DA in their jurisdictions, here is a set of items of consideration for the implementation.

- Don't ignore the business case. Implementing DA is expensive. Between the costs of the technology, implementation/maintenance costs, communications, and others, implementation can be quite expensive. All costs should be weighed against the value DA implementation provides.

  For example, progressive utilities assess their 50 worst performing circuits when looking at implementing key DA technologies such as IVVC and FLISR.
- Do expect technology to change. Any person who is tracking the dynamic changes in PCs, smartphones, or even in the electric power industry also realizes that technology changes rapidly in terms of capabilities and pricing.

  This means that any utility implementing DA technologies should be planning for changes and improvements. The planning process should include defining architectures that anticipate technology changes and plan the financial process in such a way that the cost for the implementation goes down over time.
- Do understand the need for geographical information systems (GIS). Automation works when it takes into consideration the connectivity of the power system it works with. The static model of this, often called the as-designed or as-built model, is generally held in a GIS or similar system. The dynamic model or the as-operated state is maintained in one or more DMS, SCADA, and/or OMS.

  Automation also works best when it is able to protect the grid under different connectivity configurations.
- Do design for cybersecurity. DA implementations are now moving away from private utility specific networks using proprietary protocols toward industry-standard Internet Protocol (IP)-based protocols and public networks. All such implementations are vulnerable to cyberattacks.

Cyberattacks can force the grid to perform and behave unpredictably, thereby causing interruption in services to customers. The design process needs to take cybersecurity into consideration from the beginning [14].

- Do plan for the safety of line and other personnel. Unlike communications and other networks where automatic routing of information can be performed, power networks are designed to transport energy. In addition, most distribution networks are also designed for one-way power flow. When a switch is opened, power flow is stopped, but when a switch is closed, then power flows in the direction of consumption.

  This means that specific safety steps need to be in place to ensure that any utility or other personnel working downstream is not fatally impacted if actions such as switch closings are not planned out.

Ongoing Challenges and the Future

The industry has just scratched the surface of DA; only two main applications are becoming prevalent. Those are volt-VAR control, called by other names such as IVVC and conservation voltage reduction and FLISR [15]. Some of these are managed centrally and some function in a decentralized manner.

It is important to note, however, that several other applications are in the pipeline. Some are in the pilot stages and some are still evolving [16]. The restrictions on many of these are not from a pure technology perspective. These restrictions stem from the need for more sensing, analysis, and control mechanisms in the field and a more accurate representation of the location and connectivity model of the distribution grid, which is generally implemented in systems such as the GIS. Some of these new capabilities are:

- Remote fuse condition monitoring;
- Feeder load balancing;
- Load survey monitoring;
- Equipment condition monitoring;
- Faulted circuit indication monitoring.

Visibility and controllability of the system will improve the accuracy of system representation. When this happens, sophisticated capabilities will move the grid towards the self-healing future that power system engineers are looking for.

Case Studies

Case Study #1: Duke Energy, DA

In 2007, Duke Energy began a 10-year smart grid program to deploy advanced metering infrastructure (AMI) and distribution automation. In 2009, Duke Energy also received Smart Grid Infrastructure Grant (SGIG) funding under the American Reconstruction and Reinvestment Act (ARRA) to support deployment activities mainly in Ohio, but also partially in the Carolinas. Duke's SGIG project had a total budget of about $555 million, with roughly $200 million of that received through the SGIG Program.

The project's distribution automation objectives were focused on improved reliability. As a result, the deployment included three main initiatives: advanced substation and line components to sectionalize circuits and enable self-healing networks that automate fault response, IVVC in Ohio to levelize voltage across the entire circuit and improve efficiency on power lines, and install a new DMS to integrate smart grid automation and control capabilities. As part of its DA project, Duke's distribution automation efforts installed multiple new devices throughout the system to improve reliability.

Duke deployed self-healing potentialities, which included and extended beyond sectionalizing capabilities on key circuits to improve reliability. Sectionalizing circuits operate hydraulic and electronic reclosers, sectionalizers, and fusing to isolate faults and prevent outages from cascading to other circuits. Self-healing circuits operate several electronic reclosers and circuits breakers to isolate a permanent fault and restore power to the unfaulted line segments. Duke's self-healing groups of field devices enabled auto-reconfiguration and enhanced reliability by rapidly restoring power after a fault was identified by the devices. As self-healing technologies were not feasible for all circuits, Duke selected circuits based on the type and number of customers on the circuit, the number of customer miles, and the circuit outage history.

In addition, Duke also assessed the need of improving utility substations through inside-the-fence deployments. These deployments included remote terminal units, circuit breakers, capacitor bank and voltage regulator controls, and smart relays. Duke also added new and upgraded components outside the fence. DA components were added to power lines on selected circuits, including upgraded network capability, capacitor bank and voltage regulator controls, new electronic reclosers, line sensors, and communication retrofits to existing reclosers. Duke's self-healing groups of field devices enabled auto-reconfiguration and enhanced reliability by rapidly restoring power after a fault was identified. Figure 3.3 shows how these components work together.

Duke's customers are seeing measurable improvements through reduced outage frequency and faster restoration after disruptions from self-healing capabilities built into 64 distribution circuits in Ohio. These self-healing capabilities

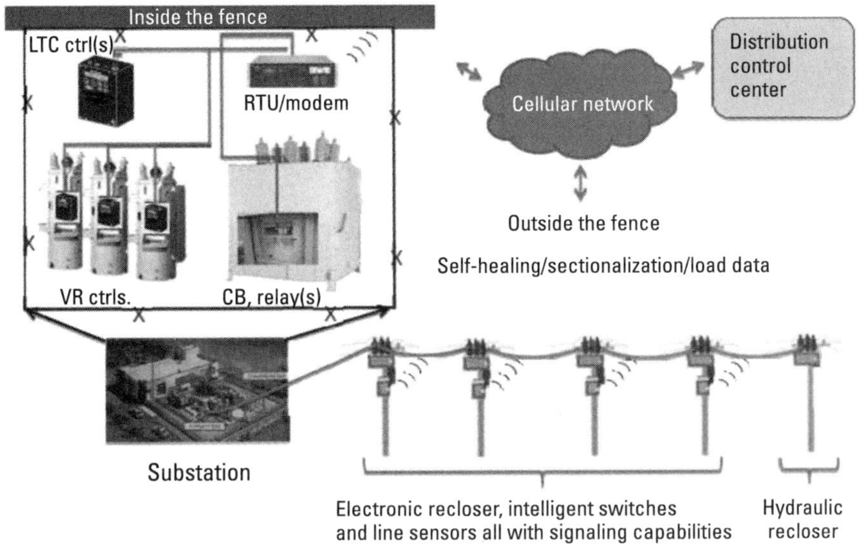

Figure 3.3 Duke Energy's DA components. (Source: Duke Energy Final Report at www.smartgrid.gov [17].)

are made possible by thirty groups of field devices that enable fault detection and automatic rapid isolation and restoration of the fault.

By August 2014, Duke's 30 self-healing device groups in Ohio had activated 84 times, reducing outage frequency and duration. Table 3.5 shows measurable improvements in system average interruption frequency index (SAIFI), which measures average interruptions per customer [17].

Table 3.5
Duke Energy's Self-Healing Teams Show Year-over-Year Reductions in SAIFI

Year	Ohio SAIFI Targets	Actual 12-Month Rolling Average Through August 2014
2009	1.50	1.30
2010	1.44	1.1
2011	1.38	1.37
2012	1.31	1.08
2013	1.24	0.98
2014	1.17	1.02
2015	1.10	-

Source: [17].

Case Study #2: Stedin, The Netherlands, DA

Stedin is one of the largest distribution network operators in the Netherlands and has a total of 21,240 secondary substations in service and serves approximately 2 million customers in the Randstad region. Randstad is one of the largest metropolises in Europe. The Rotterdam harbor district lies within its region and is one of Stedin's primary customers. In the past, Stedin has experienced power failures in its distribution grid. Not only did these outages displease customers, but they created/required expensive compensation payments for Stedin, which would not be reimbursed by the regulator.

Distribution system reliability was very important to Stedin. They needed to deliver an uninterrupted energy supply to ensure undisturbed operations to the harbor. Stedin needed to take necessary steps to minimize the impact on customers as much as possible and keep outage times as few and as brief as possible. To deliver on this mandate, Stedin needed to localize the faulty section of the distribution grid as quickly as possible so that normal operation can be resumed quickly. Stedin's main objective was to significantly reduce the SAIDI in their medium-voltage grids.

Today, there are several self-healing distribution grids in operation worldwide, but they are usually built for overhead distribution feeders, primarily by applying automatic reclosers. The Dutch medium voltage (MV) distribution grid, however, consists of underground cables, which cannot be quickly repaired in the event of faults nor can they be fixed with automatic reclosers. Therefore, Stedin developed a self-healing network solution that would work for its MV network (Figure 3.4). The self-healing algorithm it deployed is elegant, simple, and robust, and works for almost all grid structures. The applied remote terminal units (RTUs) and the developed software are very stable.

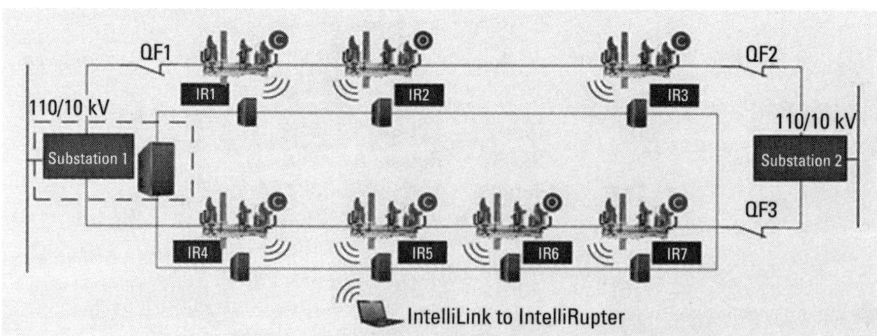

Figure 3.4 Distribution networks with local control units in each secondary substation. (Source: T&D World, http://www.tdworld.com/overhead-distribution/china-conducts-pilot-self-healing-network.)

Their solution is based on a regional controller at the substation level, which ensures automatic fault localization, isolation, and restoration. Stedin implemented the solution by upgrading its substation automation with new distribution grid automation functions. The regional controller serves as an interface to the control center, which collects data from the distribution grid and hosts the regional, centralized applications of Stedin's self-healing grid. Part of Stedin's solution consists of an intelligent local substation. Upgrade kits were either installed to modernize older ring main units (RMUs) or were replaced with new RMUs.

With the innovative self-healing grid solution installed, Stedin expects to significantly reduce its SAIDI and restore power to most of its customers in less than a minute in the event of a power outage. Stedin also expects this technology to increase customer satisfaction as well as provide considerable cost savings by minimizing the heavy contractual penalties that must be paid in case of power outages [18].

How DA Can Transform the Utility Industry

DA and the self-healing grid is the Holy Grail for the power system engineer and other utility personnel. For a long time, these technologies were either implemented as pilots or on a small set of feeders and/or circuits. However, they are now being installed across the entire field utilities on a cost-evaluated business-case perspective. This is because the technologies and their support systems are maturing and becoming cost-effective.

As the implementations are spreading to more feeders and more utilities, an important realization has begun to hit utility management. That realization is that the future cannot be a technology pure-play implementation. DA implementations must be a subset of a broad-based transformation that not only takes advantage of the technologies, but also looks at completely changing the way of doing business.

Here are some examples of how utilities are transforming by taking advantage of technologies associated with DA.

- *Self-healing grid:* In the past, when faults occurred in the grid, the first step was to send the trouble-men, or T-men, into the field to locate the fault and then identify the extent of repair needed to make the repair. With the help of DA technologies, the following changes are already being seen in the marketplace:
    - Fault locators can position the location of the fault between the right sections of the feeder. This information is made available to the op-

erator who is then able to send the trouble-men to the right location the first time around.

- FLISR devices can work either independently in the field or centrally as a part of an ADMS implementation. When these devices are enabled, the operator will wait for the schemes to operate and then, based on the outcomes, send the field crew to fix the original problem and restore the rest of the system.
- For restoration, the operator uses smart meters to confirm the restoration of customers prior to moving on to the next outage. Smart meters are also used to identify and restore nested outages.

The major impact of the self-healing grid regarding transformation is in the use of technologies by all utility personnel in reducing customer outage times and having a more accurate and timely calculation of estimated time of restoration (ETR or EToR). Both factors result in improved customer satisfaction.

- *Optimization of grid performance:* In the past, much of the system optimization was done offline by planners and asset managers, and the impact was felt months or even years later. Sometimes the impacts were never felt, because the conditions changed.

  Newer and more sophisticated DA technologies are now being implemented to optimize the performance of the distribution network. The analysis of the technologies is still done offline, but the devices are available in real time to optimize the performance of the grid under changing loading and switching configurations. The impacts of these technologies are being felt in the following areas:

  - *Improving voltage profiles:* Technologies such as VVO, CVR, IVVC, and others are able to control the voltage profile along the feeder to values between the tighter bandwidth, thereby reducing energy consumption by customers and improving their power quality. Improved voltage profiles and reduced energy consumption allow the utility to increase the energy throughput over the same feeder, thus making it capable of handling more load. This allows the utility to delay potentially massive and costly upgrades such as substation enhancements.
  - *Reducing imbalance of loads on the three phases:* Enhanced switching technologies are now allowing utilities to move loads between feeders or different phases of the same feeder. Reducing imbalance allows for a better loading profile leading to better asset utilization and, in addition, a better operational performance of the utility. This capability

also allows the utility to better respond to the variability brought on circuit loading by increased penetration of DERs.

- *Circuit reconfiguration:* Whether it is a result of unplanned outages or in the development of the clearance process for a planned outage, devices in the field are allowing utility personnel to preemptively switch customers from one feeder to other nearby feeders, in some cases, reducing outages in a significant manner for many customers.

- *Reducing costs:* Distribution automation has allowed the utility to put more sensors in the field to deliver critical data. In the past, the data was stored in various databases, mostly in an uncorrelated form. New sets of applications and systems on data analytics are being developed by vendors that take this data and deliver value-added insight to the utility, which utilizes this insight for enhanced decision-making.
  - *Reducing cost of outages:* Devices and the actions they support allow utilities to reduce the cost of outages, both planned and unplanned, by reducing the need for manual response in many cases.
  - *Moving to proactive maintenance:* The old utility saying is "the worst kind of outage always happens at 3 A.M. on a weekend requiring an emergency crew and paying them 1.5 or 2.5 times the normal hourly rate." Any improvement that allows the utility to convert an unplanned outage to a planned one reduces costs.

  An example is transformer-monitoring devices such as dissolved gas monitors allow utility personnel to monitor the health of assets and use the data to predict failures and perform maintenance or replacements ahead of that failure. Newer systems such as asset health maintenance are coming into play at utilities that both track asset health and save money.

- *Impacting climate change:* The transformational impacts of decisions from DA are also impacting climate change in new ways.
  - *Reduced truck rolls:* The example of the self-healing grid provided above is a perfect example of a side benefit that delivers reduced greenhouse gas emission. More automation in the field supported by intelligence, centralized as well as distributed, results in reduced truck rolls, which in turn results in reduced greenhouse gas emissions.
  - *Increased use of energy from renewables:* DA technology resulted in new sensors and controls such as smart inverters that allow DERs to interconnect into the distribution network. Thus, more DERS can deliver power.

Conclusions

Power system engineers and other utility personnel have been looking for technologies to help them achieve the state where circuits can detect the faulted condition and automatically act to limit the extent of the fault and restore power to those parts not impacted by the fault. They have been looking for these technologies for a long time, but the capabilities available until now were a combination of inadequate capability and high expensive. Even the communications were too expensive to implement. However, much is changing, and the DA world can take advantage of the technological breakthroughs in other industries, which also impact the electric utility industry. Key aspects of these changes include:

- *Communications:* Advances in communications technologies are now allow utilities to move away from the older generation of wired technologies into wireless. A combination of communications technologies, such as mesh, cellular, and wired, allow utilities to implement sensing and control mechanisms more easily and at a lower cost.
- *Interconnection protocols:* Proprietary protocols such as Conitel, CDC, and Landis+Gyr are now giving way to newer and more standardized protocols such as DNP3, DNP3/IP, and 61850, which allows a diverse set of industry hardware and software components to interact with each other using standard mechanisms. Many of these protocols also allow increased resilience to cyberattacks while standardizing on common internet underlying structures.
- *Automation software:* Automation implementations are also moving away from legacy and restrictive software and their implementation environments, several of which were based on line editors. An example of these restrictive software programs is Prologic. These extremely restrictive situations made it difficult to develop complex protection and automation schemes because of the limitations on their development and even more importantly on their testing. The newer environments are better from both a programming perspective and the mechanisms to code and test them.

  In addition, the newer systems also integrate with industry leading software systems such as DMS or ADMS, thereby allowing them to function from a centralized location.
- *Sensing technologies:* Sensing technologies are getting smaller, cheaper, more sophisticated, and better in their ability to network and get their power supply locally through energy harvesting. They also sense many

pieces of information at the same time. As a result, single sensor location can sense and deliver data required by multiple applications and systems thereby making it more convenient for installation as well as interacting.

- *GIS and mapping:* Increasingly more utilities are implementing GIS systems along with the associated processes necessary to keep the GIS as close to being up-to-date with the reality in the field. This ensures that the automation systems, which perform the analysis, will always have the latest and greatest connectivity model.

These changes portend well for the future of DA in the electric system distribution grid. While the self-healing grid is not in place today, advances in technology are well positioned to help the electric system attain their integration in the near-term to mid-term future [19].

References

[1] U.S. Capitol, "Remarks by the President in the State of the Union Address," February 12, 2013, https://www.whitehouse.gov/the-press-office/2013/02/12/remarks- president-state-union-address.

[2] The NETL Modern Grid Initiative, "A Vision for the Modern Grid," conducted by the National Energy Technology Laboratory for the U.S. Department of Energy Office of Electricity Delivery and Energy Reliability, March 2007.

[3] DeBlasio, D., "Toward a Self-Healing Smart Grid," *Fortnightly*, August 2013, http://www.fortnightly.com/fortnightly/2013/08/toward-self-healing-smart-grid?page=0%2C0.

[4] Office of Electric Transmission and Distribution, "Grid 2030. A National Vision for Electricity's Second 100 Years," U.S. Department of Energy, July 2003, http://energy.gov/sites/prod/files/oeprod/DocumentsandMedia/Electric_Vision_Document.pdf.

[5] McCarthy, C., "Lights On: New Technologies for a Smart Distribution Grid," *Electric Energy Online*, March 2010, http://www.electricenergyonline.com/show_article.php?mag=62&article=476.

[6] Spoonamore, S., and R. L. Krutz, "Smart Grid and Cyber Challenges, National Security Risks and Concerns of Smart Grid," U.S. White House, https://www.whitehouse.gov/files/documents/cyber/Spoonamore-Krutz%20-%20Smart%20Grid%20CyberSecurity%20Risks%20and%20Concerns.pdf.

[7] National Energy Technology Laboratory, "A Systems View of the Modern Grid, Sensing and Measurement," U.S. Department of Energy, March 2007, http://www.netl.doe.gov/File%20Library/research/energy%20efficiency/smart%20grid/whitepapers/Sensing-and-Measurement_Finsal_v2_0.pdf.

[8] Gilbert, E., L. Gelbien, and B. Rogers, "A Truly 'Self-Healing' Distribution Grid Requires Technology AND Operational Change," *Grid-Interop Forum 2009*, 2009, http://www.gridwiseac.org/pdfs/forum_papers09/gelbien.pdf.

[9] National Energy Technology Laboratory, "A Systems View of the Modern Grid, Advanced Control Methods," U.S. Department of Labor, March 2007, https://www.netl.doe.gov/File%20Library/research/energy%20efficiency/smart%20grid/whitepapers/Advanced-Control-Methods_Final_v2_0.pdf.

[10] Vadari, M., *Electric System Operations: Evolving to the Modern Grid*, Norwood, MA: Artech House, 2012.

[11] NETL, "A Compendium of Smart Grid Technologies," July 2009, https://www.smartgrid.gov/document/netl_modern_grid_strategy_powering_our_21st_century_economy_compendium_smart_grid_technolog.

[12] National Energy Technology Laboratory, "A Systems View of the Modern Grid, Advanced Components," U.S. Department of Energy, March 2007, https://www.smartgrid.gov/files/appendix_b3_advanced_components.pdf.

[13] McCarthy, C. A., "Utilities Seeking Intelligence on Electric Distribution Circuits," January 2011, http://www.sandc.com/edocs_pdfs/EDOC_067830.pdf.

[14] National Energy Technology Laboratory, "A Systems View of the Modern Grid, Resists Attack," U.S. Department of Energy, January 2007, https://www.smartgrid.gov/files/Systems_View_Modern_Grid_Appendix_A3_Resists_Attack_v20_200704.pdf.

[15] Zajkowski, S., "Live Data – The Keys to the Energy Kingdom, the Bigger Picture," *Electric Energy Online*: January/February 2014, http://www.electricenergyonline.com/show_article.php?mag=95&article=763.

[16] Uluski, R., "Distribution Automation and the Self-Healing Network," *IEEE Smart Grid*, April 2014, http://smartgrid.ieee.org/newsletters/april-2014/distribution-automation-and-the-self-healing-network.

[17] Duke Energy Final Report, "Integrated Smart Grid Provides Wide Range of Benefits in Ohio and the Carolinas," U.S. Department of Energy, Electricity Delivery and Energy Reliability, September 2014, https://www.smartgrid.gov/files/C7-Duke-Energy-Case-Study-FINAL-092914.pdf.

[18] Siemens, "Self-Healing Grid for a Reliable Power Supply," 2014, http://w3.siemens.com/smartgrid/global/SiteCollectionDocuments/References/Reference_Self-healing%20grid_Stedin_e_final.pdf.

[19] Zajkowski, S., and K. Mays, "Smart Distribution: A Self-Healing and Optimized Grid," *Electric Light & Power*, October 16, 2013, http://www.elp.com/articles/print/volume-91/issue-5/sections/smart-distribution-a-self-healing-and-optimized-grid.html

4

Energy Storage: Electric Value Chain Disruptor

Electric energy storage (EES) is a set of technologies that stores previously generated electric energy and releases that energy later. EES uses forms of energy such as chemical, kinetic, thermal, or potential that can be converted to electricity on demand. There are several energy storage technologies in different states of development and used for different purposes depending on their source and power characteristics.

Introduction

Flashlights and car batteries are examples of energy storage that have been around for a long time. Most existing electricity storage products were electrochemical in nature. They stored energy in an electrochemical reaction and released it when needed. Most electrochemical batteries had a limited time span for holding a charge, could not easily scale, and were very expensive. Some innovations, especially with the advent of lithium-ion (Li-ion) batteries, have managed to scale from cellphones and to cars. Due to advances in technology, storage is poised to become the major disruptor power for which engineers have been waiting.

Electricity is a commodity that works in the instant—generation and load need to be instantly balanced and cannot be stored. This characteristic separates it from all other networks, such as roads, communications, or even something like pork bellies. When we use the term energy storage, we are referring to the conversion of electricity into some other type of energy, such as chemical in batteries, that can be stored and converted back to electricity later.

Several pilots have been implemented over the past few years with newer technologies such as sodium sulfur (NaS) batteries and flywheel-based kinetic-energy storage. In addition, pumped-hydro-based storage is getting renewed attention by utilities worldwide to become a suitable alternative that can support the vagaries of renewable sources of energy.

Storage, in conjunction with DERs, has been identified as one of the most important areas of improvements in the industry from a need perspective as well as from the business and investor perspective. The need for electricity storage will continue to grow. Research and development by several organizations including major electric vendors and the U.S. Department of Energy (DOE) laboratories are supporting these innovations. However, until more storage-related devices come into play across different parts of the electric value chain, the utility industry will need to continue to live in a just-in-time network, with consumption and generation kept in sync in real time.

This chapter starts with a definition of energy storage and what makes it unique, followed by a section defining some of the terms used in the storage industry. Next, the chapter enumerates types of energy storage, starting with mature ones and those still in a research and development pipeline that could make a difference soon. The chapter covers applications of storage on the grid—transmission or distribution. Applications are an important aspect driven by the differing characteristics of storage. The chapter then delves into making the business case for storage from the various perspectives of the utility, investor, customer, and regulator. The chapter then concludes with an assessment of how storage will drive transformations in the grid and the utility.

Electric Energy Storage—What Is It and Why Is It Important?

Figure 4.1 shows several commodities such as food, water, gasoline, and oil and natural gas, which have an average storage capacity of more than 10% of their daily consumption. In contrast, the electricity market in Massachusetts, for example, has a storage capacity of less than 1% of daily consumption. In addition, without storage, electricity needs to be produced, delivered, and consumed nearly instantaneously for the grid to maintain balance. This requires the entire electric value chain, including generation, transmission, and distribution systems, to be sized at yearly peak consumption, despite consumer electricity demand varying significantly throughout the day and at different seasons of the year.

This need to size the entire electric value chain to peak consumption resulted in system inefficiencies, underutilization of assets, and a high cost to

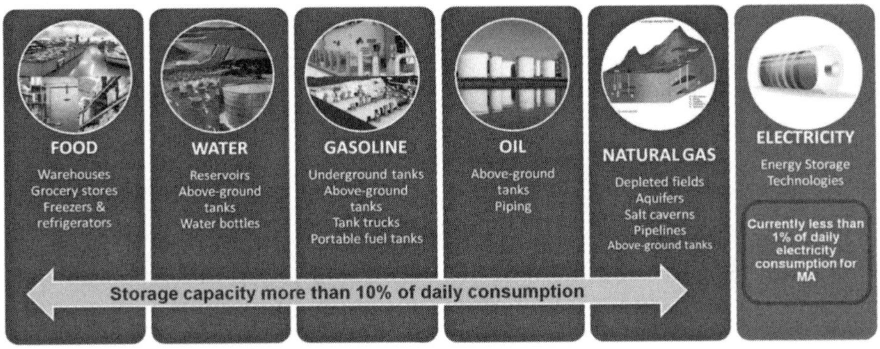

Figure 4.1 Storage and its relationship to commodity supply chains [1].

ratepayers. These high costs can be seen in the variable hourly electricity prices. During 2013 to 2015, on average, the top 1% most expensive hours accounted for 8% Massachusetts ratepayers' annual spend on electricity, to the tune of $680 million. The top 10% of hours during these years, on average, accounted for 40% of annual electricity spending, at over $3 billion. Energy storage is the only technology that can use energy generated during low-cost, off-peak periods to serve load during expensive peak periods, thereby improving the overall utilization and economics of the electric grid. Until recently, the ability to store electricity across the electric grid was limited, but recent advances in new energy storage technologies, such as grid-scale batteries, are making the wide-scale deployment of electricity storage viable [1].

Energy storage is important because of the broad range of potential value that it can provide along the electric value chain. It can improve power quality and bridge power when switching from one source of generation to another such as from wind generation to natural gas-fired generation, and at grid scale, storage can be used to decouple the timing of generation and consumption. Storage can be charged during times of low-energy cost and low utilization such as when there is plenty of wind generation and can be used to supply power during peak load periods.

As mentioned earlier, energy storage has applications along the entire electric value chain (Figure 4.2). It can be used to store energy from large-scale renewable generation for later use. Energy storage can be deployed at substations to provide power supply to equipment and computers. It can be deployed in a microgrid to supply power closer to the load. It can be installed at commercial, industrial, and residential customer premises to provide grid-independent or emergency power [2].

Storage devices are not identical. Storage technologies have characteristics such as power, energy, scale/size, and rate of discharge, which differentiate them depending on the chemistry or the mechanics of the storage system.

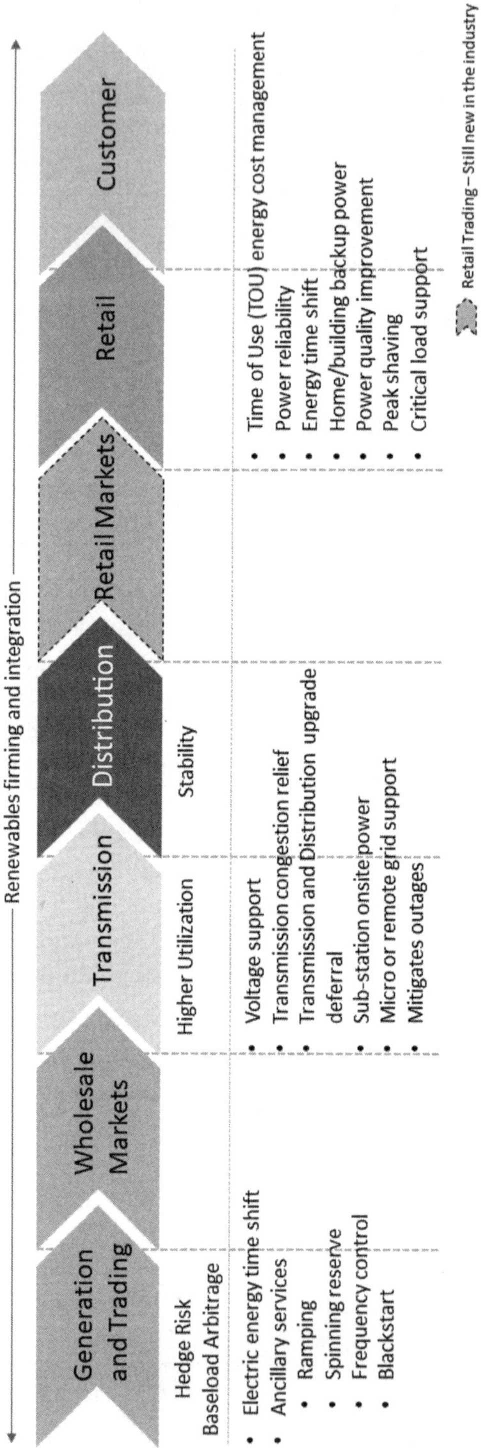

Figure 4.2 Energy storage applications along the electric value chain. (Source: [2]. © Modern Grid Solutions.)

These characteristics drive the selection of the appropriate usage of the device in the electric system, whether it is at the grid level (in such as transmission and distribution, or other applications at commercial, industrial, or microgrid locations. Figure 4.3 illustrates the continuum between high-power discharge ability and high-energy storage capability.

Figure 4.4 illustrates the next concept, assessing each technology and defining it in terms of capacity and discharge time. Capacity drives which specific technology can be used and the discharge time at the rated power determines how it can be used. For example, flywheels are small in ratings, but incredibly fast discharge rates make them uniquely positioned to provide frequency-regulation support. Pumped-hydro, on the other hand, has the advantage of the entire rating of the hydro-plant and can run at that rate for extended periods of time.

Definition of Key Terms and Concepts Associated with Energy Storage

Energy storage comes with its own specific terminology. These terms will help understand how storage works and how it can be applied in the electric system [4].

- *Capacity:* Maximum charge/discharge power and the amount of energy capable of being stored. The energy storage capacity of EES systems is often expressed in megawatt-hours (MWh).

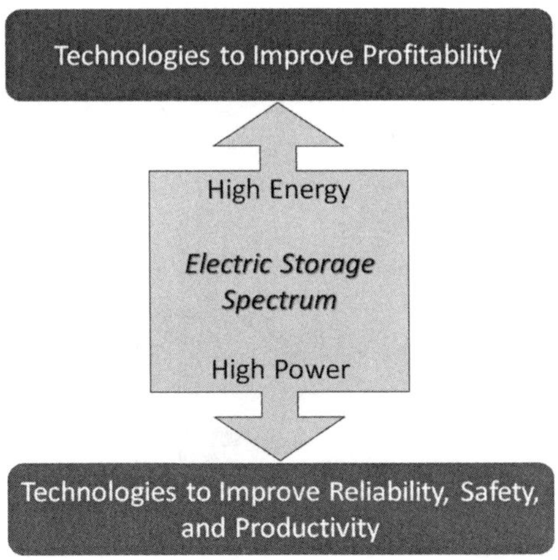

Figure 4.3 Power/energy balance characteristics of storage. (Courtesy of Modern Grid Solutions.)

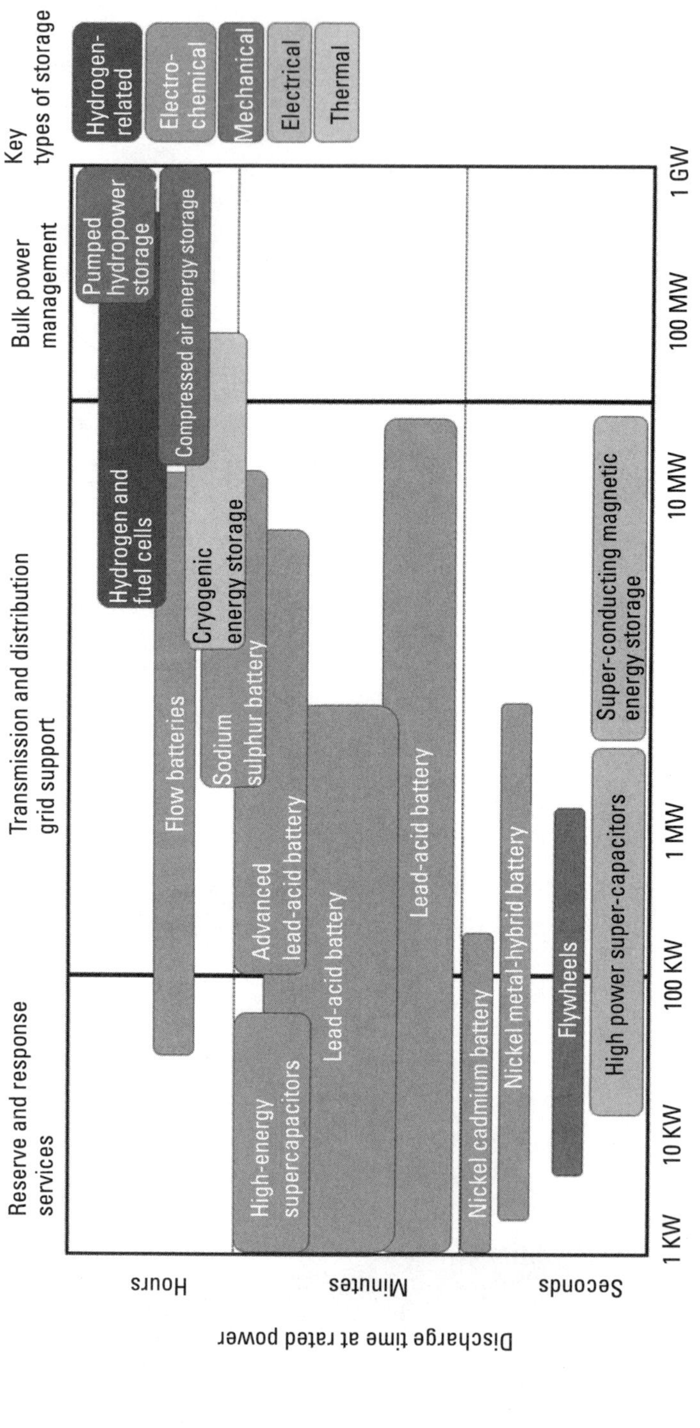

Figure 4.4 Energy storage technologies and their characteristics. (Source: [3]. Used with permission.)

- *Discharge:* Process of extracting stored energy from the storage system. In a battery, during discharge, the anode undergoes an oxidation reaction, during which two or more ions from the electrolyte combine with the anode to form a compound and release one or more electrons.
- *Discharge duration:* Total time that a storage device can be discharged at the nominal power rating. In other words, it is the amount of time that the storage system can discharge at its rated power without requiring that it be recharged. The energy released during discharge kilowatts or megawatts equals power in kilowatts or megawatts times discharge duration in hours.
- *Cycle:* Sequence of storage charging and discharging. It is also known as charge-discharge cycle. The lifetime of a storage device is measured in terms of the number of its charge-discharge cycles. Batteries generally have a lifetime cycle capacity in the range of 5,000 to 10,000 cycles, although a few advanced batteries are rated at over 100,000 cycles.
- *Charging:* Process of injecting energy to be stored in the storage system. In the case of batteries, in each cycle, the charging process stores energy in the battery in the form of potentially reactive compounds.
- *Charge rate:* Rate at which storage technology can be charged. Under ideal conditions, it is equal to the storage capacity of the device divided by 1 hour. For example, in the case of batteries, "1C" is the charge rate necessary to charge a battery in 1 hour.
- *Cycle life:* Number of charge-discharge cycles after which storage becomes inoperable or unusable for a given application. For example, in the case of a lead-acid battery, the typical service life is 6 to 15 years with a cycle life of 1,500 cycles at an 80% depth of discharge.
- *Depth of discharge (DOD):* Portion of energy discharged from a storage system relative to the amount of storage energy that can be extracted. DOD is used to describe how deeply the battery is discharged. For example, if a battery is 100% fully charged, it means the DOD of this battery is 0%. If the battery has delivered 30% of its energy, there is 70% of the energy still left; therefore, the DOD of this battery is 30%.
- *Discharge rate:* Rate at which the storage technology can be discharged. It is the discharge current specified as a C rate, divided by the theoretical current draw under which the storage device would deliver its nominal rated capacity in 1 hour. In the case of a battery, a 1C discharge rate would deliver the battery's rated capacity in 1 hour.
- *Electrode:* Electrical conductor through which an electric current enters or leaves a conducting medium. An electrode can be an electrolytic so-

lution, solid, molten mass, gas, or vacuum. An electrode in an electrochemical cell is referred to as either an anode or a cathode. The anode is the electrode where electrons leave the cell and oxidation occurs, indicated by a minus symbol, and the cathode is the electrode where electrons enter the cell and reduction occurs indicated by a plus symbol. Each electrode may become an anode or a cathode depending on the direction of current through the cell.

- *Electrolyte:* Chemical compound that, when fused or dissolved in certain solvents, usually water, will conduct an electric current. The dissolved electrolyte separates into cations and anions, which disperse uniformly through the solvent. Electrically, such a solution is neutral. If an electric potential is applied to such a solution, the cations of the solution are drawn to the electrode that has an abundance of electrons, while the anions are drawn to the electrode that has a deficit of electrons. The movement of anions and cations in opposite directions within the solution produces a current. An electrolyte is used in electrochemical batteries.
- *Energy density:* Amount of energy that a storage system can store per unit volume occupied by the system. In energy storage applications, the energy density relates the energy stored to the volume of the storage facility, such as a fuel tank. The higher the energy density of the fuel, the more energy may be stored or transported for the same amount of volume.
- *Power density:* Level of power a storage system can provide per unit volume occupied by the system. Also referred to as volume power density or volume specific power, it is the amount of power—time rate of energy transfer—per unit volume. In energy storage applications such as batteries and fuel cells, power density refers to a volume.
- *Ramp rate:* Rate at which power output can be changed. It is typically expressed in units of megawatts per minute (MW/minute). Storage technologies with fast ramp rates are more valuable and are employed to provide frequency regulation and response.
- *Response rate:* Ramp rate that a storage device can achieve under normal operating conditions expressed in MW/min. Essential parameter to check the feasibility of the storage device to respond to changes in load.
- *Response time:* Total time required for a storage system output to transition from no discharge to full discharge. In other words, it is the time for a storage device to respond from no load to a step change in load.
- *State of charge (SoC):* An indicator of how much charge there is in a storage device, expressed in percentage points (0% = empty and 100% = full).

Electric Energy Storage Types and Applications

There are various types of energy storage [2, 3].

Pumped Hydro

Pumped-hydroelectric energy storage is a large, mature, and commercial utility-scale technology currently used at many locations around the world. Pumped hydro is a hydro power plant with a reservoir at the top and at the bottom. It employs off-peak electricity to pump water from the low reservoir at the bottom to the high reservoir at the top. When electricity is needed, water is released from the high reservoir through a hydroelectric turbine into the low reservoir to generate electricity.

Projects may be sized up to 4,000 MW and operate at 76% to 85% efficiency depending on design. Pumped-hydro plants have long lives, on the order of 50 years, and fast response times, making them suitable across the entire energy value chain and for a broad range of applications such as voltage and frequency regulation, spinning reserve, nonspinning reserve, energy arbitrage, and system capacity.

There is a significant drive across the world to convert all hydro plants that have the capability into pumped-hydro plants.

The primary downside of pumped-hydro is that specifics of geography and topography determine where they can be installed. Even some existing hydro plants are incompatible to change to pumped hydro. The geography must allow the creation of both a low reservoir and a high reservoir, so there are many places where they cannot be used. Other significant challenges to building new pumped-hydro plants include licensing, environmental regulations, and uncertainty in long-term electric markets.

Compressed Air Energy Storage

Compressed air energy storage (CAES) uses off-peak electricity to compress air and store it in a reservoir, either underground in a cavern or aboveground in pipes or vessels. When electricity is needed, the compressed air is heated, expanded, and directed through a conventional turbine generator to produce electricity. Improved second-generation CAES systems are being designed that have potential for lower installed costs, higher efficiency, and faster construction time than first-generation systems. In one type of advanced second-generation CAES plant, a natural gas-fired combustion turbine (CT) generates heat during the expansion process and two-thirds of the electricity generated is produced during the compressed air cycle.

Underground CAES storage systems are highly cost-effective, with storage capacities of up to 400 MW and discharge times of 8 to 26 hours. Owing to its huge energy and power capacity, CAES finds applications in energy arbitrage, frequency regulation, and ancillary services.

There are two existing CAES plants today. One is in Huntorf, Germany, and the other in McIntosh, Alabama. These two plants are based on conventional technology and not the improved version as described above.

The primary downside of traditional CAES is that it uses turbomachinery to compress air to approximately 70 bars before storage, which, in the absence of intercooling the air, would heat up to around 900K. This makes it impossible and highly expensive to process and store the gas without intercooling. Researchers are working on advanced CAES technology such as isothermal CAES which tries to overcome some of these limitations of traditional CAES [5].

Sodium Sulfur Storage

Sodium sulfur (NaS) batteries use molten sulfur as the positive electrode and molten sodium as the negative electrode. These active materials are separated by a solid, ceramic electrolyte that conducts sodium ions. During discharge, positive sodium ions flow through the electrolyte to combine with the sulfur, forming sodium-polysulfide. Electrons flow through the external circuit of the battery to create a potential difference. Charging of the battery releases positive sodium ions from the sodium-polysulfide, sending them back through the electrolyte to recombine as elemental sodium. To keep the electrodes in the molten state, the battery is kept at a temperature of roughly 570°F. NaS modules also utilize an electric heater to keep the reactants molten and thermal insulation to reduce heat losses.

The estimated life of a NaS battery is ~15 years and support about 4,500 cycles at 90% depth of discharge. NaS is a commercial energy storage technology with applications in the electric utility distribution grid support, wind power integration, and high-value service applications on islands.

NaS battery technology has been installed, used, and tested at over 190 sites in Japan. More than 270 MW of stored energy suitable for 6 hours of daily peak shaving have been installed. The largest NaS installation is a 34 MW/245 MWh unit for wind stabilization in Northern Japan. The demand for NaS batteries as an effective means of stabilizing renewable energy output and providing ancillary services continues to expand.

U.S. utilities have deployed 9 MW of NaS batteries for peak shaving, backup power, firming wind capacity, and other applications. Projections indicate that development of an additional nine MW is in progress [6].

The primary downside of NaS revolves around the need for high temperatures for operation in the form of molten Na. There are also safety concerns because of the need to protect the Na from moisture. Another common problem with the technology is that in some combinations it tends to form undesirable Na2S2 solids with higher depth-of-discharge operation [7].

There is significant research being conducted to produce NaS batteries that are stable at room temperature.

Flow Batteries

Flow batteries are a relatively newer class of electrochemical devices. They can store large amounts of electrical energy, from tens of kWh to tens of MWh, and deliver it slowly over several hours or within milliseconds or minutes as high-power pulses. In a flow battery, each electrode is immersed in a different electrolyte. The two electrolytes are separated by an ion-exchange membrane. There are different types of flow batteries, each use different chemicals to store energy. Examples include: vanadium reduction-oxidation (redox), zinc (Zn)-bromine, iron/chromium, and Zn/Air. The primary operational advantage of flow batteries, which use reversible (redox) reactions, is power or energy output can be optimized in real time.

The versatility of flow batteries makes them suitable for a wide range of applications such as peak shaving for transmission and distribution upgrade deferral, small load leveling, and backup power.

The largest-capacity flow battery system in North America is a 1 MW/4 MWh vanadium redox flow battery installed by Avista in Pullman, Washington. The system supports Washington State University's smart campus operations by providing load shifting, frequency regulation, and voltage regulation [8].

In addition, a vanadium demonstration project for the U.S. Army is underway at Fort Devens, in Worcester County, Massachusetts. In Canada, Ontario's Independent Electricity System Operator has plans for a 5-MW system that will use vanadium redox technology [8].

Some of the challenges plaguing flow batteries include toxicity of materials, stability, need for cooling systems, and safety concerns.

Flywheel

Flywheels store kinetic energy within a rapidly spinning rotor or disc made of advanced high-strength materials, which is electrically charged and discharged through a generator. Flywheels charge by drawing electricity from the grid to increase rotational speed and discharge by generating electricity as the wheel's rotation slows. High-velocity flywheels are operated in vacuum vessels to reduce friction.

Flywheels have a very fast response time of 4 ms or less are sized between 100 kW and 1,650 kW, and may be used for short durations of up to 1 hour. They have very high efficiencies of as much as 93%, with lifetimes estimated at 20 years. Owing to their high-power density, flywheels are typically used in power quality and uninterruptible power supply applications. Research is underway to develop more advanced flywheel systems that can store large quantities of energy.

A major storage project is the flywheel frequency-regulation plant setup by Beacon Power in Stephentown, New York. It is the world's first grid-scale flywheel energy storage plant. The storage system has 200 flywheels operating in parallel to provide 20 MW with a less than 4-second response time. The system provides frequency-regulation service to New York Independent System Operator (NYISO) [9, 10].

The primary concern with flywheels is removal of heat that results from rotation, although vacuum containment and magnetic bearings significantly reduce friction in flywheels, thereby reducing the amount of heat that must be removed. The trade-off is the vacuum containment and magnetic bearings can make it difficult to remove the remaining heat.

Li-ion

Li-ion is a type of rechargeable battery in which a lithium ion moves between the anode and cathode. The ion moves from the anode to the cathode during discharge and from the cathode to the anode when charging.

Rechargeable Li-ion batteries are a commercial and mature technology used in consumer electronic products from cellphones to laptops. Li-ion is also positioned well to be the leading technology platform for plug-in hybrid electric vehicles (PHEV) and all electric vehicles (EVs) that will use larger-format cells and packs with capacities of 15 to 20 kWh for PHEVs and up to 50 kWh for all EVs. Given their attractive cycle life and compactness, in addition to high DC-to-AC efficiency that exceeds 85% to 90%, Li-ion batteries are also being seriously considered for several utility grid-support applications such as distributed energy storage system (DESS), community energy storage, transportable systems for grid support, commercial end-user energy management, home backup energy management systems, frequency regulation, and wind and photovoltaic smoothing.

Today Li-ion batteries are the technology of choice in the electric car industry. Nontraditional players such as Volkswagen and Tesla are creating an even greater demand through the former's bid to invest $15 billion in a new battery factory and the latter's plans to build half a million EVs per year by 2018. These forces will lead to an unprecedented boom in Li-ion in the near term.

Supercapacitor

Capacitors allow for the storage of energy in an electrostatic field, which is due to the separation of electric charges. Electrochemical capacitors obtain the highest level of performance. These are also referred to as supercapacitors, ultracapacitors, or electric double-layer capacitors. A direct voltage is applied between the electrodes, which are separated by an electrolyte. A polarized layer is formed to provide charge separation over very small distances between each of the electrodes and the electrolyte ions. This results in energy stored in the form of an electrostatic field.

Supercapacitors have a long life cycle, which allows for many charge and discharge processes, and a high efficiency of as much as 95%. Supercapacitors provide power at very short response times and over a few seconds of duration. Their high-power density makes them suitable for applications related to power quality, emergency bridging power, and smoothing power fluctuations.

Japan employs large supercapacitors. About four installations of 4-MW systems are completed in commercial buildings to reduce grid consumption at peak-demand times and ease of loading.

The Long Island Rail Road (LIRR) trial in New York is testing a 2-MW supercapacitor bank against flywheels that deliver 2.5-MW of power. LIRR is one of the busiest railroads in North America and is testing these systems to prevent voltage sag during train acceleration and to reduce peak power usage [11].

Some key limitations of super capacitors include:

- Low-energy density—holds a fraction of a regular battery;
- Linear discharge voltage, which prevents it from using the full energy spectrum;
- Low cell voltage, which requires series connection of several units with voltage balancing;
- High cost per watt.

Superconducting Magnetic Energy Storage

Superconducting magnetic energy storage (SMES) devices store energy in a magnetic field created by a direct current circulating through a superconducting coil. Cooling from either liquid helium or nitrogen is required to attain superconducting characteristics. Increasing the inductance and increasing the direct current enhance the energy storage capacity of SMES.

The lifetime of SMES does not depend on the number of charge and discharge operations. SMES can provide high power within seconds and have efficiencies on the order of 90%. SMES systems are particularly interesting for high-power applications. Owing to the need for refrigeration and the high cost of superconducting wire, SMES is currently used for short-duration energy storage applications such as power quality and emergency bridging power.

SMES systems are developed and currently are being tested and while the systems are commercially available, their reliability has yet to be established.

An important project to improve the application of SMES at grid level is being sponsored by U.S. DOE, under the Advanced Research Projects Agency-Energy (ARPA-F). They aim to develop 1 to 2 MWh commercial scale SMES system that is cost-competitive with lead-acid batteries. The team includes Brookhaven National Laboratory, University of Houston, ABB, and Super-Power [12]. A 30-mJ, 8.4-kWh SMES unit with a 10-MW converter has been installed and commissioned at the Bonneville Power Administration substation in Tacoma, Washington. This system is capable of absorbing and releasing up to 10 MJ of energy at a frequency of 0.35 Hz [12].

Some key technical challenges with SMES are low-energy density, the need for a cryogenic system to maintain lower temperatures, and health concerns associated with a strong magnetic field.

Key Storage Technologies on the Horizon

There are several up-and-coming storage technologies.

Planar-Stacked Na-Beta (Planar Na-β) Batteries

Na-β batteries (Na-β batteries or NBBs) use a solid β-alumina electrolyte membrane that selectively allows Na ion transport between a positive electrode, such as (metal halide and a negative Na electrode). Current NBBs are constructed using ~3-mm-thick tubular electrolyte elements requiring operating temperatures of about 350°C to attain sufficiently rapid Na+ transport and adequate power characteristics in the battery. The high operating temperature complicates thermal/heat management and creates material-stability issues that limit the overall cycle life, resulting in increased costs.

A new approach uses a planar, stacked, modular battery design that employs thinner electrolyte materials. Planar designs also allow greater stacking efficiency, resulting in a 30% increase in energy density. Manufacturing processes are simplified and process yields are improved with planar fabrication, leading to the reduction of fabrication costs by a factor of 10 over current tubular batteries. A prototype battery is expected to be developed over the next 3 years. With commercial deployment, the next generation Na-β battery could enable integration of more intermittent renewable energy sources such as wind and solar to become base-load generators [13]. The main concerns for this technology include the need for more robust seal development and difficulty in automating the manufacturing processes involving high-temperature casing materials [13].

Overall, this technology is in the process of transitioning from laboratory to grid-scale deployment.

Reversible Metal Hydride Thermal Storage for High-Temperature Power Generation Systems

Reversible metal hydride thermal storage is an advanced chemical heat storage technology using metal hydride as a heat storage material. Metal hydride material can reversibly store heat, as hydrogen cycles in and out of the material. In this storage system, metal hydrides remain stable at high temperatures, 600°C–800°C. A high-temperature tank releases heat as hydrogen is absorbed and a low-temperature tank stores the heat until needed.

In solar thermal storage systems, heat can be stored in these materials during the day and released at night—when the sun is not out—to drive a turbine and produce electricity. In nuclear storage systems, heat can be stored in these materials at night and released to produce electricity during daytime peak-demand hours. Thus, the technology provides an alternate way of storing the heat from solar or nuclear that can be dispatched later.

Most utility-scale solar power plants only run at approximately 25% of their capacity because they cannot generate power at night. Thermal energy storage makes it possible to increase this capacity up to 60% to 75%.

Similarly, nuclear power plants produce a constant output of power. Thermal energy storage could help increase this output during times of critical peak demand.

This technology is currently in the pilot stage. Once fully deployed, it has the potential to create a low-cost thermal energy storage system that can store 10 times more energy than current state-of-the-art thermal energy storage systems [14].

HybriSol Hybrid Nanostructures for High-Energy Density Solar Thermal Fuels

HybriSol is an advanced thermal energy storage device that captures energy from the sun; this energy can be stored and released later when it is needed most. Within the device, the absorption of sunlight causes the solar thermal fuel's photoactive molecules to change shape, which allows energy to be stored within their chemical bonds. A trigger is applied to release the stored energy as heat, and it can then be converted into electricity or used directly as heat. The molecules then revert to their original shape and can be recharged using sunlight to begin the process anew.

This device is currently in the proof-of-concept phase. When deployed, they can also be used without grid infrastructure for applications such as deicing, heating, cooking, and water purification [15].

Disrupting Energy: Energy Storage and Its Applications to the Grid

Electric power systems, for the most part have followed a standard value chain: generation, transmission, distribution, and consumption. This has been the status quo for over 100 years and consequently, the entire design of the power system has been focused on transmitting power from the generator to the consumer, until storage came along.

Storage is the first power system component that can be both a generator and a consumer; more importantly, this behavior can be controlled by the user. When it consumes, the power system charges the storage mechanism and stores electricity. When it generates, it discharges and releases the electricity into the grid. The variety and versatility of storage technologies make them suitable for numerous applications to the grid creating a disruptive scenario changing the landscape of choices available to the electric utility.

Provide Firm Peak Capacity to the Grid

Electric power follows a cycle that has peaks, generally during early to mid-afternoon, and valleys, in the middle of the night. A different cycle exists for weekdays and weekends and even by season. For this reason, electric utilities maintain peaker plants, called peakers, that are designated to provide power during times of peak demand and exist solely for their ability to provide power for short periods of time. Peakers are not designed to run continuously. This is mainly due to their greenhouse gas emissions' characteristics and costs. They are still basically small power plants (mostly combustion turbines) that require constant maintenance and personnel to manage and run them.

Energy storage systems, such as battery storage, have the potential to take over many of the functions currently performed by peaker plants. Certain storage systems already outperform conventional plants in several areas, but need improvement from a pricing perspective to become more competitive. For example, combustion turbines, which have traditionally been used for peak capacity, typically take a few minutes to dispatch, while storage can switch from charging to discharging faster. Storage is also significantly more flexible than simple cycle combustion turbine peakers, with an operating range that is typically two to four times the operating range of a combustion turbine [16, 17].

Provides Backup Power to Homes, Businesses, or Utilities During Outages

When homes and businesses lose power, they generally either stay dark or have small generators installed at the home/business and run either on natural gas or gasoline, called gensets, to deliver backup power. Gensets pollute, due to their greenhouse gas emission characteristics. They can also be loud. The limitation of gensets is that they burn fuel and generate power. This by itself is important, but in the new age of distributed generation (DG), to only generate and not store, is limiting from an investment perspective.

Battery storage systems can be used to charge using electricity generated from solar panels or when utility rates are low and power homes and businesses during peak periods such as mornings and evenings. They also fortify homes and businesses against power outages by providing a backup electricity supply. Utilities can deploy storage as backup at substations and control centers. Ultimately, energy storage can be instrumental for emergency preparedness because of its ability to provide backup power as well as grid stabilization services. Over time, energy storage may be a significantly better alternative than the current gas-power backup generators at homes and businesses.

Cuts Peak Demand Charges to Customers

For commercial and industrial customers, the monthly electricity bill has two parts: energy and demand charges. The demand charge defines the generation capacity set aside for delivering power to that customer and the energy charge defines the amount of energy consumed by that customer. Demand charges are based on the highest 15-minute average usage recorded within a given month or year as appropriate. If a facility uses a lot of power over short periods, then demand charges will comprise a larger part of the electric bill, which for commercial and industrial customers is typically be between 30% and 70%.

Storage allows these customers to use behind-the-meter storage to reduce peak demand and thereby reduce demand fees by charging the storage mechanism during off-peak periods and discharging it during peak periods thereby reducing the peak power consumed from the grid. This storage can be procured either through purchase or deployment of storage units onsite. Technologies such as vanadium redox batteries and other battery systems are being used for this purpose [18].

Helps Offset Negative Impacts of DG

A significant percentage of DG is from renewable sources, whose output is unreliable and can impact grid stability in negative ways such as variability of generation, unpredictability of when it will happen, and the potential to cause two-way power flow in a distribution system designed to have power flow in only one direction. As penetration of DG and related assets increase, their potential to cause overloading of distribution assets, disruption of protective relaying, power quality issues, and bidirectional flow of power leading to worker safety concerns also increase.

Coupling DG with storage can offset many of the issues identified above. For example, rapid variability of output from renewable DG such as rooftop solar and wind can be offset by adding storage. That storage can charge when excess power is generated and discharge when the power output of the DG is reduced. Storage systems can adjust their output quickly to reduce or eliminate the variability caused by DG and provide a more consistent power output to the distribution grid.

Provides Frequency Regulation and Reserves

The utility grid and balancing authority have a responsibility to maintain a balance between generation and consumption in a specific geographic area, also known as control area [19]. For this to happen, the utility contracts either directly or through market mechanisms for reserves of different kinds—generally based on location and the speed of their ability to respond. The fastest of these is called frequency regulation, which is also the most expensive because of the need to respond on a second-by-second basis.

Historically, bulk generation has provided most of the frequency-regulation service. For this to occur, generators need to maintain some spare capacity, so they can deliver energy almost instantaneously when called upon.

Storage with high-power characteristics can provide this service more efficiently from several perspectives:

- Storage can provide frequency regulation up with increased discharge and can provide regulation down via reduced discharge.
- It is much easier to site storage where it is needed.
- It does not consume generator capacity, which could be used for other purposes.

When charged, storage can provide reserves merely by being ready to discharge. Typically, reserves are needed infrequently, so storage used for reserve capacity is rarely discharged. This gives storage an advantage over generation-based spinning reserves, which need to be already spinning and ready to pick up load at a moment's notice [20].

Serves as Critical Resource in Microgrids

Microgrids are a new and upcoming energy construct. Microgrids bring generation and consumption together into one part of a grid, which can exist in a self-sustained manner either for short or extended periods of time. The concept of microgrids was designed in part to increase the penetration of renewable energy resources. Microgrids are normally attached to a larger grid to allow access to more resources at scale, but go micro when the macrogrid has issues. Microgrids are a collection of loads and resources that can isolate as needed. They are less likely to tolerate inefficient usage of infrastructure assets than the macrogrid.

As a result, energy storage is a critical ingredient for smart microgrids. Energy storage facilitates use of renewable energy resources to power microgrids, by acting as a buffer to either absorb excess generation, or discharge energy to meet minimum load requirements. Energy storage can provide both frequency regulation and smooth fluctuations in renewable energy output, thereby preventing voltage fluctuations. It can also supply power during outages by charging when renewable generation is at its peak and discharging when needed. When a microgrid disconnects from the grid and switches to a backup power source, there is typically a 5 to 15-second lag as diesel generators spin up. Storage can serve to bridge this gap by allowing critical loads to be maintained during the period the microgrid transitions to its alternate energy source [21].

Mitigates Congestion

In an electric grid, where electricity must flow over transmission and distribution lines, there is potential for congestion in locations where consumption is high, but there is not enough line capacity to bring supply to the consumer. Physics will force electricity to flow to that point of congestion, but the lines and associated components will get overloaded, and depending on the protection equipment available, can lead to equipment disconnecting. Given that the capacity of the lines receiving power from generation to the load is constrained causing the congestion, often the only solution(s) in the past was to site new generation within the congestion zone, a long-term solution at best, try to redispatch the generation to change the pattern of flow of electricity, or to disconnect the load.

Storage offers a new set of solutions for grid congestion. Storage is easier to site in congested locations compared to siting a new generator. It can also store when there is no congestion and discharge power into the congested zone, thereby alleviating the need to transfer more electricity over the lines through the congested zone.

Making the Business Case for Storage

The complexity of energy storage, and the various options for their implementations along the electric value chain, makes defining the business case much more complicated. This is driven by the variety of technologies and their applicability and because of the diversity of the stakeholders driving implementation decisions.

Utility Viewpoint

The electric utility makes money when it installs equipment on the grid. That income is determined through the rate case, which needs to be approved by the state Public Utilities Commission. For the utility, the business case for storage comes from the following:

- Installing storage-related devices in its distribution network to alleviate peak-time congestion, thereby eliminating the need for substation enhancements or adding an extra line to bring in more power;
- Installing storage-related devices to provide regulation support in the transmission network because the utility could either not purchase regulation reserve from the market or enough reserves were not available in the existing generation within the footprint and/or regulation reserve was needed in a specific location due to power flow constraints and a new generator could not be sited or was deemed too expensive within that location;
- Installing storage-related devices in the distribution network to offset a larger-than-normal influx of DERs in a specific neighborhood with the intention of smoothing out the potential generation variability and to offset any possible power quality issues;
- Electric codes must be adapted for more resources down in the system to allow for proper isolation;
- The utility also has a role in its customers' implementation behind the meter. This is mainly to ensure that customers' implementations are not harming the grid in any way, whether operationally or from a safety point of view.

The utility case becomes difficult because of the disrupting aspect of the storage device being both a generator and a consumer. In some states, the regulated utility cannot engage in generation and are also not allowed to install storage devices. There is also the complication of the utility exploiting its role as an incumbent and having an unfair advantage in siting storage. In increasingly more places, however, utilities can make the business case for storage and are installing devices to support their mandate of delivering reliable power to their customers at the lowest rate possible.

Investor Viewpoint

The business case for the investor can be a little more complex in terms of their role as investor in technology or investor in the implementation. In terms of the need to focus on the pure business value that can be delivered for their investment, however, the business case is simple.

- *Investor in technology:* Investing in the future of the technology and making a bet that one technology is either better than alternatives or has a better chance of solving a more important problem. This is important because storage technologies are relatively nascent and it is still unknown whether the technology will pan out or if it will succeed based on market needs. The decision process here is based more on the investor's intuition about the potential success for a specific technology and the focus of their portfolio mix.
- *Investor in implementation:* Generally made with mature technologies and the type of technology implemented is defined by the type of service required by the consumer. In this case, the consumer can be a utility or the wholesale market operator. In these cases, the decision on the off-take of the storage device's services is most often defined ahead of time in a contract, thereby lessening the risk profile for the investor.

Customer Viewpoint

The customer's viewpoint has its own complications and is generally based on the type of customer.

- *Commercial/industrial:* The existence of the demand charge for these two classes of customers moves their focus from energy consumed to demand or the need to reserve capacity from the utility to serve one's needs. Implementing storage at the commercial location allows the customer to reduce their peak demand and hence the demand charge, which allows them to reduce their utility charge for an entire year.
- *Residential:* The residential customer does not have the demand charge and hence the motivation is different. If the customer installs some form of local renewable resource, storage could be added to either smooth out the generation variability or store the generation for when electricity is needed and the renewable source cannot generate. This also supports the individual customer's need to go net-zero. Net-zero is defined as when net electricity consumed from the grid over a specified period, generally 1 year is zero or completely off-grid.
- There are also a specific subset of residential customers who install DERs supported by storage to justify their need to go green. This class of customer does not do it from a business case perspective.

Regulator Viewpoint

The regulator comes into play during one of two instances.

- When working with the utility in approving the storage installation as requested by the utility. This interaction focuses primarily on whether the installation requested by the utility is in the best interests of the customer.
- When working with the utility to institute rule-making regarding customer or other nonutility installations of storage on the grid. Here, the regulator's role is not to determine if or why the customer is installing the storage device, but to define the regulations that need to be in place to drive the requirements of the installation and, more importantly, to make sure that the regulations are not too severe or onerous for the customer.

Case Study

Case Study #1: Southern California Edison—Tehachapi Wind Energy Storage Project

The Tehachapi Wind Energy Storage Project (see Figure 4.5), funded by the Southern California Edison (SCE) and federal stimulus funding awarded by the DOE as part of the American Recovery and Reinvestment Act of 2009, demonstrates the application of Li-ion battery and smart inverter technologies to integrate vast amounts of variable energy resources. The $55 million project is sited at the Tehachapi Wind Resource Area, in Tehachapi, California, to store energy from its existing 5,000 wind turbines and any future additions. It is one of the largest wind resource areas in the world with an additional 4,500 MW of wind resources that came online in 2015.

The goals of the project include:

- Validating the performance and effectiveness of lithium-ion technology;
- Demonstrating the integration of intermittent resources;
- Developing a smarter, more efficient electrical grid;
- Advancing the market readiness of utility-scale storage.

The storage system employs Li-ion battery of 8 MW with a 4-hour duration and a smart inverter system. The battery energy storage system is used for frequency regulation, spinning, nonspinning, replacement reserves, ramp

Figure 4.5 Tehachapi wind resource area. (Tehachapi wind farm credit: Don Barrett/flickr, content licensed under Creative Commons Attribution 4.0 International license.)

management, and energy price arbitrage. The performance of the project is measured with 13 specific operational uses, which include providing voltage support and grid stabilization, decreasing transmission losses, diminishing congestion, increasing system reliability, deferring transmission investment, optimizing the size of new renewable-related transmission, providing system capacity and resource adequacy, integrating renewable energy (smoothing), shifting wind generation output, frequency regulation, spin/nonspin replacement reserves, ramp management, and energy price arbitrage. Most of the operations either shift other generation resources to meet peak load and other electricity system needs with stored electricity or resolve grid stability and capacity concerns that result from the interconnection of variable energy resources. In this project, SCE demonstrates the ability of lithium-ion battery storage to provide nearly instantaneous maximum capacity for the supply-side ramp rate control to minimize the need for fossil fuel-powered backup generation.

The benefits of the project included improved power quality, increased system reliability, integration of more clean, renewable energy, fostering of energy independence, and the creation and retaining of jobs [23].

Case Study #2: Beacon Power—Flywheel Energy Storage Projects

The flywheel frequency-regulation plant setup by Beacon Power is the world's first grid-scale flywheel energy storage plant, located in Stephentown, New York as seen in Figure 4.6. The storage system has 200 flywheels operating in paral-

Figure 4.6 Flywheel frequency-regulation plant in Stephentown, New York [9]. (Courtesy Energy.gov.)

lel to provide 20 MW with a less than 4-second response time. The system provides frequency-regulation service to NYISO. It can provide 20 MW of up-regulation and 20 MW of down-regulation in immediate response to NYISO's automatic generation control signal. The system can operate at a 100% depth of discharge with no performance degradation over a 20-year lifetime, for more than 100,000 full charge-discharge cycles. The storage contributes to only 10% of NYISO regulation market capacity currently but to over 30% of area control error correction, with as much as 95% accuracy.

Beacon Power has a second 20-MW frequency regulation plant located in Hazle, Pennsylvania. This plant currently provides 20 MW of frequency regulation service to Pennsylvania Jersey Maryland (PJM). The plant includes 200 flywheels and reached full commercial operation in July 2014. The capex for this project dropped by approximately $10 million compared to the Stephentown project.

Both projects are paid for by providing frequency regulation service to the relevant independent system operator (ISO). These projects are enabled by FERC order 755, a ruling from the Federal Energy Regulatory Commission to substantially increase the value that Beacon's flywheel plants can earn for their services. The ruling calls for the country's interstate grid operators to institute market systems that pay more for fast-responding sources like flywheels and batteries than for slow, fossil-fueled power. Rates are based on a pay-for-performance tariff [9, 10].

How Storage Will Transform the Utility Industry

Power system engineers have been waiting for their entire professional lives for the disruptive technology of energy storage.

While energy storage comes in many different forms, with the electrochemical mechanism the most common in the marketplace, the key characteristic making it the ultimate disruptor is its ability to act as both a consumer and a generator.

- It is a consumer when it takes on electricity to load up on the charge.
- It is a generator when it discharges electricity back into the grid.

There are no other devices with such a capability in the power system arena. Every other device in the grid is a consumer, a generator, or a conductor. The ability to disrupt comes from consuming excess energy in the grid and delivering energy back to the grid when and where there is a need.

Why is this important? Consumption of electricity follows a profile during the day with a couple of peaks and valleys. This type of consumption pattern forces utilities around the world to commit their generating units based on a cost/location profile. As a result, there are always some generators, also known as peaking plants, that deliver electricity for only a few hours a day, a week, or a month leading to a very inefficient use of their capacity. As a result, the overall cost of power supply goes up to cover the fixed and operating costs of these peaking plants, some of the more inefficient plants in the system. This activity also impacts the infrastructure capacity in terms of the amount of power available to conduct for the gird including for lines, transformers, and circuit breakers. If the lines become limited in their ability to conduct electricity, then load (consumption) on the other side of the device will need to be curtailed.

Over the past few years, demand response was considered the ultimate solution for this issue. For demand response to deliver results, however, consumers must curtail their consumption, either voluntarily or involuntarily under remote external control. This has not yet proven to be very successful, primarily because customers have not yet changed their consumption behavior, especially during utility peak periods or congestion times.

Energy storage can deliver value by:

- Consuming energy—charge up—when prices are low and the deliver energy—discharge—when prices are high or there is congestion in the grid.
- Placement in locations close to consumption to avoid congestion and also supporting needs at transmission and/or the distribution level.

From a location and a permitting perspective, it is generally considered far easier to install storage at any location on the grid, based on need as compared with generators, which can only be installed at specific locations. It can also relieve congestion for local load.

Storage, when added to generation from distributed renewable sources, allows generation to become predictable and controllable, relieving the power system operator from dealing with the variability associated with these sources. The value, as a result, comes from the ability of the storage device to smooth out the load profile, thereby allowing better use of the full capacity from all forms of generators and lines. It also makes more use of renewable energy supply by allowing devices to deliver generation when they are capable.

Conclusions

As DERs become more affordable, energy storage could potentially become the ultimate disrupter to the electric utility. Storage is unique in that it can generate and consume electricity on command. In effect, it is a new kind of asset that utilities are trying to define and understand how to capitalize on from a value-delivery perspective.

This flexibility allows the user to store electricity when it is in excess or cheap and release the electricity back when it is in greater demand or higher cost. Storage takes a load profile with peaks and valleys and flattens it, thereby reducing the cost of energy and allowing for greater penetration of renewable energy sources.

Storage will allow DERs to go further in, assisting customers with greater independence from the grid.

Key aspects of these changes include:

- *Technology not yet mature:* Storage technologies that are currently in widespread use in the marketplace have been around for a while and all have limitations, even within their specific use profile. There are major innovations currently on the drawing board that will be available in the marketplace in the next 5 to 6 years. This chapter also presents several technologies that are currently in play as well as those that are in the research and development pipeline.
- *Technology survivability:* There are several mature storage technologies that will most likely continue to exist in a similar form. Over the past few years, however, technologies have been identified that are either in pilot form or in implementation. New technologies are in research and development and may be ready for commercial implementation soon. It is fair to assume that some of the technologies will fail and fall by the wayside, while others will emerge and succeed.

- *More than a utility play:* Just like DERs, storage is not solely of interest to utilities. Storage installations will take place across all segments of the energy value chain. This also means the stakeholders who implement storage will run the gamut from utilities and investors to customers, and aggregators. Stakeholders can be expected to utilize storage in ways we have not yet imagined.
- *Centralized and/or distributed implementations:* The breadth and depth of storage technologies will allow for their implementation at the wholesale level as well as the retail level. Examples such as kinetic-energy-based flywheel storage can work at the wholesale market level delivering regulation services, whereas examples such as Li-ion-based storage provides storage at the residential level.

As the industry designs newer technologies and approaches for storage, identification of different modes of storage have the potential to change the dynamics of how that storage can be used. Examples of these new modes are using gas to drive new storage options like fuel cells and/or gensets and using heat in the residential water heater to offset the consumption of electricity or gas depending on the time of day. These kinds of innovations will move the utility forward in lowering greenhouse gases and improve quality of life.

References

[1] Energy Storage Initiative, *State of Charge: Massachusetts Energy Storage Initiative Study*, September 16, 2016, http://www.mass.gov/eea/docs/doer/state-of-charge-report.pdf.

[2] EPRI, "Electricity Energy Storage Technology Options," December 2010, large.stanford.edu/courses/2012/ph240/doshay1/docs/EPRI.pdf.

[3] Barbour, E., "Energy Storage Sense: Energy Storage Technologies Explained," http://www.energystoragesense.com/energy-storage-technologies/.

[4] Energy Storage Association, "Glossary," 2017, http://www.energystorage.org/energy-storage/glossary.

[5] Energy Storage Association, "Compressed Air Energy Storage," 2017, http://www.energystorage.org/compressed-air-energy-storage-caes.

[6] Energy Storage Association, "Sodium Sulfur (NAS) Batteries," 2017, http://www.energystorage.org/energy-storage/technologies/sodium-sulfur-nas-batteries.

[7] Cowie, I., "All About Batteries, Part 9: Sodium Sulfur (NaS)," *EE Times*, July 14, 2014, http://www.eetimes.com/author.asp?section_id=36&doc_id=1323091.

[8] Maloney, P. "How Technical Advances Are Driving Flow Batteries Closer to Mass Commercialization," *Utility Dive*, January 12, 2016, http://www.utilitydive.com/news/how-technical-advances-are-driving-flow-batteries-closer-to-mass-commercial/411923/.

[9] U.S. Department of Energy, "American Recovery and Reinvestment Act (ARRA): Grid-Scale Flywheel Energy Storage Plant," October 2012, https://energy.gov/sites/prod/files/Beacon.pdf.

[10] Wesoff, E., "Flywheel Energy Storage Lives on at Beacon Power," *Greentech Media*, May 31, 2013, http://www.greentechmedia.com/articles/read/Flywheel-Energy-Storage-Lives-On-at-Beacon-Power.

[11] "BU-209: How Does a Supercapacitor Work?" *Battery University*, April 21, 2017, http://www.batteryuniversity.com/learn/article/whats_the_role_of_the_supercapacitor.

[12] Holla, R. V., "Energy Storage Methods – Superconducting Magnetic Energy Storage – A Review," *Journal of Undergraduate Research*, Vol. 5, No. 1, 2015, http://www.jur.phy.uic.edu/issue8/JUR-1401009.pdf.

[13] U.S. Department of Energy, "Sodium-Beta Batteries," October 2012, http://www.energy.gov/sites/prod/files/NaS_0.pdf.

[14] ARPA, "Metal Hydride Thermal Storage," September 29, 2011, http://www.arpa-e.energy.gov/?q=slick-sheet-project/metal-hydride-thermal-storage.

[15] ARPA, "Solar Thermal Energy Storage Device," September 29, 2011, http://www.arpa-e.energy.gov/?q=slick-sheet-project/solar-thermal-energy-storage-device.

[16] Hales, R. L., "Where Battery Storage Will Take Over from Backup Power Plants," *Energy Post*, November 19, 2014, http://www.energypost.eu/battery-storage-will-take-backup-power-plants/.

[17] Lyons, C., "Guide to Procurement of Flexible Peaking Capacity: Energy Storage or Combustion Turbines?" *Energy Strategies Group*, October 7, 2014, http://www.energystrategiesgroup.com/wp-content/uploads/2014/10/Guide-to-Procurement-of-New-Peaking-Capacity-Energy-Storage-or-Combustion-Turbines_Chet-Lyons_Energy-Strategies-Group.pdf.

[18] Neubauer, J., and M. Simpson, "Deployment of Behind-the-Meter Energy Storage for Demand Charge Reduction," *NREL*, January 2015, http://www.nrel.gov/docs/fy15osti/63162.pdf.

[19] Energy Storage Association, "Energy Storage Technologies," 2016, http://www.energystorage.org/energy-storage/energy-storage-technologies.

[20] Energy Storage Association, "Grid Operations Benefits," 2016, http://www.energystorage.org/energy-storage/energy-storage-benefits/benefit-categories/grid-operations-benefits.

[21] S&C Electric Company, "The Role of Energy Storage in Smart Micro-Grids," July 29, 2013, http://www.sandc.com/edocs_pdfs/edoc_076176.pdf.

[22] Edison International, "Tehachapi Storage Project," 2016, http://www.newsroom.edison.com/gallery/album?id=5420bb5cee750e68c201f1d6&t=photo&p=100&s=order.

[23] U.S. Department of Energy, "Southern California Edison Company Tehachapi Wind Energy Storage Project," May 2014, http://www.energy.gov/sites/prod/files/2015/05/f22/SoCal-Edison-Tehachapi-May2014.pdf.

5

Distributed Energy Resources: Challenge of Integrating Supply and Demand Diversity

Distributed generation is the set of small-scale power generation technologies located close to the load being served, capable of lowering costs, improving reliability, reducing emissions, and expanding energy options.

Introduction

A lot has changed for distributed energy resources (DERs). No longer are they reminiscent of Don Quixote tilting at windmills. Many new technologies exist alongside the modernized wind generator.

The Reform the Energy Vision (REV) initiative in New York defines DER as a set of technologies that include photovoltaic (PV) cells, battery storage, fuel cell, wind, thermal, hydro, biogas, cogeneration, compressed air, flywheel, combustion generators, demand response (DR), and energy efficiency [1–3].

These new technologies are causing a dramatic shift in how electricity is generated, transmitted, and consumed. Instead of depending on large remotely located generators to produce and transmit power, it is now generated at or close to delivery voltages. This process reduces dependence on remote generators and long transmission lines to bring power from generating centers to the load centers.

The basics of many of the technologies, such as solar and wind, are not new. However, they are changing in ways that we could not even imagine a few years ago, and their costs are coming down dramatically. Their performance is improving in a similar manner. In addition, newer technologies, such as storage and DR, are joining in to make this a compelling set of technologies, which cannot be ignored.

Equally importantly, these technologies have the potential to shift how utilities procure their generation, moving them from large centralized generation sources to those of a distributed nature at lower voltage levels. While the penetration of DERs is still low, this shift is already causing disruption in some places in ways that we have not seen before. The disruption is both technological and financial and has the potential to impact traditional utility business models in ways that we cannot imagine.

However, it is important to note that the term DER does not define either one set of technologies or one set of behaviors. The technologies run the gamut from supply sources, which use traditional fuel sources, and those that are based on renewable sources, storage of energy, and even mechanisms to reduce energy consumption.

DER is typically applied to smaller-scale power generation technologies located close to the load being served. Locating them close to the load can reduce delivery losses and additional investment in the distribution system. While the renewable options are intermittent, pairing these resources with storage helps to improve their reliability profile while simultaneously reducing emissions.

DERs tend to be small and modular compared with central power plants. Due to their distributed and modular nature, they are flexible and usually quieter, emit less, and often require no water compared to large power plants. Not all DERs are based on renewables, although most are. Several are also based on natural gas and other nonrenewable sources.

The complexity of a DER installation will vary based on customer needs. It may be grid tied consisting of generation, storage, and an energy management system. These can be either owned by a utility or third party or even as simple as a roof-top solar installation, which sells back to the grid what it does not consume locally. DERs may or may not rely on renewable sources that are dispatchable.

The chapter starts with defining DERs and their characteristics. Some rely on renewable sources and some are not. Within the challenges and benefits of DERs, I have included both business and technical challenges from the perspective of major stakeholders. Then the chapter moves on to the components necessary for the smooth integration of DERs and the economics associated with them. This includes assessing different mechanisms currently available and a postulation of their evolution. There are also key considerations, which must be kept in mind as utilities implement and integrate DERs around the world. These key considerations are discussed in the dos and don'ts section. Furthermore, the need to equalize supply and demand is explained in the inset, "Always Match Supply and Demand".

Always Match Supply and Demand

Electricity has a unique characteristic. Apart from energy storage, to be discussed later, supply and demand must always be matched.

This need began when the grid moved from a DC electric power system to an AC electric power system. This is when frequency was introduced. To understand frequency, we need to understand the concept of a control area or balancing authority.

A control area, or balancing authority, is a fundamental self-contained unit of transmission grid operation. The control area is essentially an electrical system with generation and load bounded by interconnection or tie-lines, which are metered.

A control area is analogous to a lake. In this lake, water coming into the lake is equivalent to the power being generated, and water leaving the lake is the power being consumed. The level of the water in the lake needs to be constant and represents the frequency of the system, which must also be kept constant.

The control area operator maintains and controls the generation levels to manage its load and schedules interchange of power with other neighboring control areas. The control area also contributes to the frequency regulation of the interconnection. For each of the interconnections to operate safely, reliably, and dependably in providing service to its customers, the interconnection must be continuously monitored and controlled. This monitoring and control function is distributed among the control areas that comprise the interconnection.

Except for an islanded system, control areas rarely exist independent of each other. The transmission lines that connect control areas are called tie-lines. The amount of power that flows between each control area is held to strict schedules. All participants need to ensure that they do not overproduce or underproduce. Extending the lake analogy, power flowing between control areas becomes another part of the same equation and still requires the level of the lake to be constant.

The next tranche of supply of energy is expected to come from DERs and this supply will continue to grow to levels of significance. As a result, learning to integrate DERs into the grid both from both a technical and market force perspective is critical to the electric utility of the future.

So let's get some concepts right, starting with defining the different types of DERs and their characteristics.

Types of DERs and Their Applicability

There are various types of distributed generation. Their implementation and application are often dependent on locational aspects and somewhat on a broader business case. The more well-known of these technologies are described next.

Microturbine

A microturbine is a type of combustion turbine that produces both heat and electricity on a relatively small scale. They are derived from turbocharger technologies found in large trucks or the turbines in aircraft auxiliary power units. Most microturbines are composed of a compressor, combustor, turbine, alternator, recuperator, and generator (illustrated in Figure 5.1). They are generally small in terms of power output, and typically used in residential and commercial applications. They can be operated as base, peaking, or backup units and can be run on natural gas, propane, diesel, or multifuel [4].

Some of the advantages of microturbines include: few moving parts, compact systems, good efficiencies in co-regeneration, low emissions, the ability to utilize a variety of fuels—including waste fuels, low investment costs, and low maintenance costs.

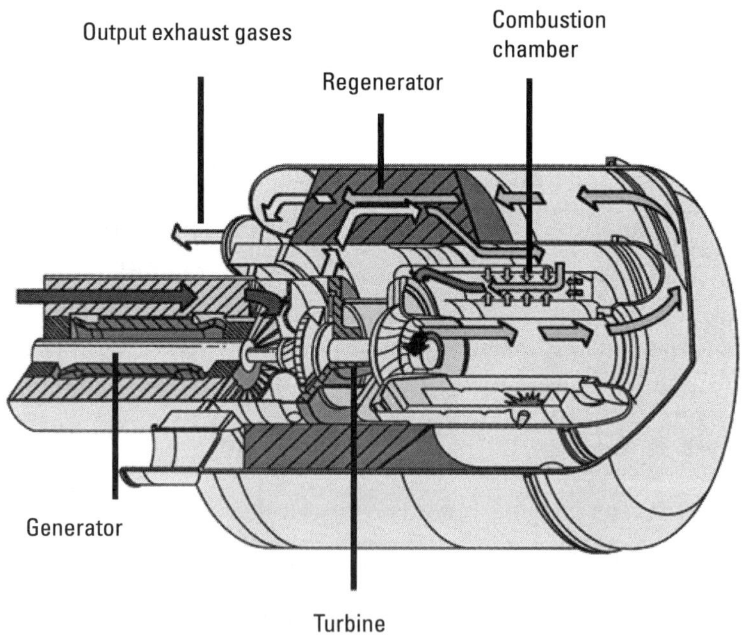

Figure 5.1 Microturbine. (https://commons.wikimedia.org/wiki/Category:Micro_gas_turbines, Creative Commons Attribution-ShareAlike 3.0 Unported License ["CC BY-SA"].)

Some of their disadvantages include: high operating rpms (90,000 to 120,000), low fuel-to-electricity efficiencies, and reduced power output and efficiency with higher ambient temperatures.

Combustion Turbine

The combustion turbine (shown in Figure 5.2) operates like a jet engine: they draw air in at the front of the unit, compress it, mix it with fuel, and ignite it. The hot combustion gases expand through turbine blades connected to a generator to produce electricity. Typically, they burn natural gas, but can also use low sulfur fuel oil if needed. Their power output can range from 3 to 15 MW. Combustion turbines are designed to start quickly to meet the demand for electricity during peak operating periods.

Some of the advantages of combustion turbines include: very high power to weight ratio; less vibration compared to reciprocating engine; start easily and quickly; the ability to work in changing load conditions; higher efficiency compared to IC engines; and the ability to develop uniform torque.

Some disadvantages include: need for an external unit to start the turbine; reduced output as most of the power drives the compressor and reduced overall efficiency of turbine due to loss of heat in exhaust gases.

Internal Combustion Engine

A reciprocating, or internal combustion (IC) engine (shown in Figure 5.3), converts energy contained in a fuel into mechanical power, like an automobile engine. A generator is attached to the IC engine to convert the rotational motion into power. Their capacities range from 5 KW for residential to roughly 7 MW for commercial/industrial applications. The most common application is to provide back-up power to residential and commercial/industrial locations.

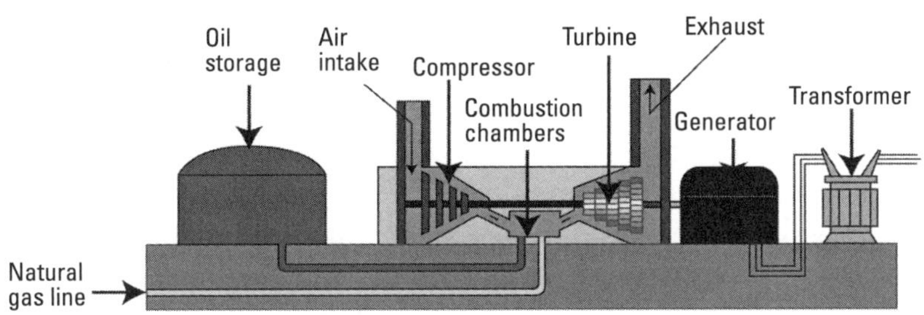

Figure 5.2 Combustion turbine. (Source: Tennessee Valley Authority (tva.com, Commons Archive) [Public domain], via Wikimedia Commons, https://commons.wikimedia.org/wiki/File%3ACombustion_turbine_diagram.svg.)

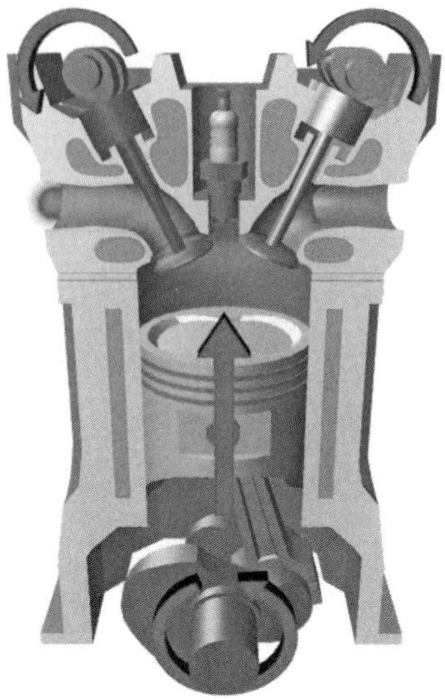

Figure 5.3 Internal combustion engine. (http://www.gnu.org/copyleft/fdl.html or CC-BY-SA-3.0 [http://creativecommons.org/licenses/by-sa/3.0/], https://commons.wikimedia.org/wiki/File%3AFour_stroke_cycle_exhaust.jpg.)

Some of the advantages of an IC engine over an external combustion engine include: smaller and lighter so it occupies less space; it can be started instantly; it is quite safe to use; and higher efficiencies of about 40%.

Some of their disadvantages include: an inability to use solid fuels—cheaper compared to liquid or gaseous fuel; they have reciprocating parts so balancing them is a problem; and they are susceptible to mechanical vibrations.

Stirling Engine

A Stirling engine (shown in Figure 5.4) is a heat engine operated by the cyclic compression and expansion of air or other gas, at different temperature levels. This creates a net conversion of heat energy to mechanical work. They can run directly on any available heat source (solar, geothermal, biological), and can be very efficient when combined with heat recovery. They are generally found in small sizes, 1 to 25 KW and are currently being produced in small quantities for specialized applications in space and marine industries.

Some of the advantages of Stirling engines include: silent operation; high efficiency due to co-generation, reliability and easy maintenance; the ability to use a multitude of hot sources (gases, wood, sawdust, waste, solar or geothermic energy); and the ability to achieve complete combustion and therefore reduce air pollution.

Figure 5.4 Stirling Engine Zephyris (GFDL [http://www.gnu.org/copyleft/fdl.html] or CC-BY-SA-3.0 http://creativecommons.org/licenses/by-sa/3.0/ https://commons.wikimedia.org/wiki/File%3AStirling_Engine.jpg.)

Some of the disadvantages include: not yet competitive in terms of price; the variety of models prevents standardization; heat transfers with a gas are delicate and often require bulky apparatuses; and lack of flexibility, which creates an inability to adapt to the fast and effective variations of power.

Fuel Cell

A fuel cell is an electrochemical device that uses oxygen and hydrogen to produce electricity (illustrated in Figure 5.5). It operates like a battery, but does not run down. Individual fuel cells are stacked in a module to produce the desired energy output. With heat and water as the primary by-products, fuel cells are virtually pollution free and have more than two times the efficiency of traditional combustion technologies. As long as fuel is supplied to the fuel cell, energy in the form of heat and electricity will be produced. The input fuel to the fuel cell varies from hydrogen to methane and so on. Their output capability varies depending on the technology applied and their use is being investigated in military applications and long-distance trucking.

Some of the advantages of fuel cells include: higher efficiency than diesel or gas engines; silent operation; nonpolluting—the only by-product at point of use is water; hydrogen can be produced anywhere there is water and a source of power; operating times are much longer than with batteries; and easy maintenance due to few moving parts in the system.

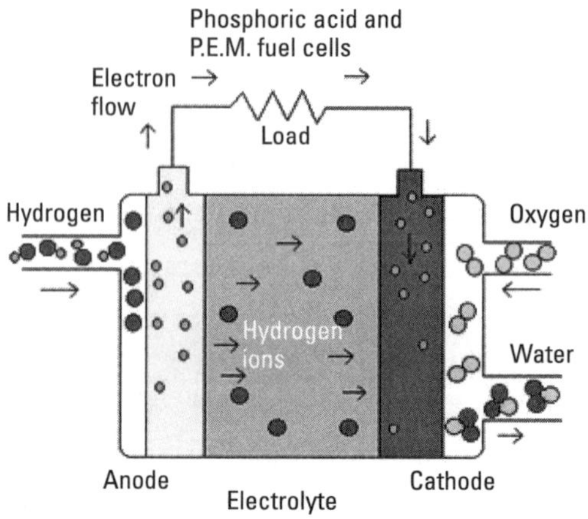

Figure 5.5 Fuel cell. (http://reich-chemistry.wikispaces.com/o%27brien.o%27kane.wiki.spring.2011 Creative Commons Attribution Share-Alike 3.0 License.)

Some of the disadvantages include: they are expensive; hydrogen is difficult to store and highly flammable; and hydrogen is hard to transport.

Solar Power Technologies

Among DERs, solar is one of the most popular technologies and the most discussed. There are two main types of distributed solar power technologies—photovoltaic (PV) as illustrated in Figure 5.6, and concentrated solar power (CSP).

- PV cells, convert sunlight directly into electricity. PV cells are solid-state devices assembled into flat plate systems and mounted on roof tops or other sunny areas. They are used in both residential and commercial applications.
- CSP performs this conversion indirectly. CSP uses mirrors or lenses with tracking systems to focus a large area of sunlight onto a small surface. Concentrated light converts to heat, producing electrical power, which drives a heat engine (usually a steam turbine) connected to an electrical power generator. CSP can be used to provide industrial process heating or cooling, such as in solar air-conditioning. The author anticipates CSP systems will grow fast and may even scale faster than PV [2].

Some of the advantages of solar power technologies include: clean, green energy; can be deployed anywhere there is abundance of sunlight; have few or no mechanically moving parts-moving parts in cases of sun-tracking mechanical bases; reduced cost of solar panels, and silent operation.

Figure 5.6 Solar farm. (https://www.australiansolarquotes.com.au/2012/02/13/300-million-grant-for-australian-solar-farm/ Attribution 3.0 Unported [Creative Commons BY 3.0].)

Some of the disadvantages include: intermittent energy; requires additional equipment (inverters) to convert direct electricity (DC) to alternating electricity (AC) to connect to the electric grid; land-mounted solar installations require relatively large areas for deployment; and relatively low efficiencies—between 14% and 25%.

Wind Turbines

Wind turbines convert mechanical energy (wind) to electrical energy (shown in Figure 5.7). The fan blades and turbines are placed on a tall tower to harness high velocity wind without turbulence from obstacles such as trees, buildings or local terrain. The turbine is connected to generator and power electronics responsible for either forming an independent grid, or grid synchronization. The output power from a single wind turbine can range in size from a few kilowatts for residential applications to more than 5 MW for commercial applications.

The advantages include: energy source (wind) is freely available, environmentally friendly, and no green-house gas emissions. In addition, the technology is mature and there are multiple manufacturers leading to lower cost and more choice. The disadvantages include: site permitting requirements; intermittent energy production; and the need for a large footprint per kilowatt hour generated.

Hybrid System

Developers and manufacturers of DERs are looking for ways to combine technologies to improve performance and efficiency of these diverse sources of generation. Several examples of hybrid systems include: a wind and solar hybrid

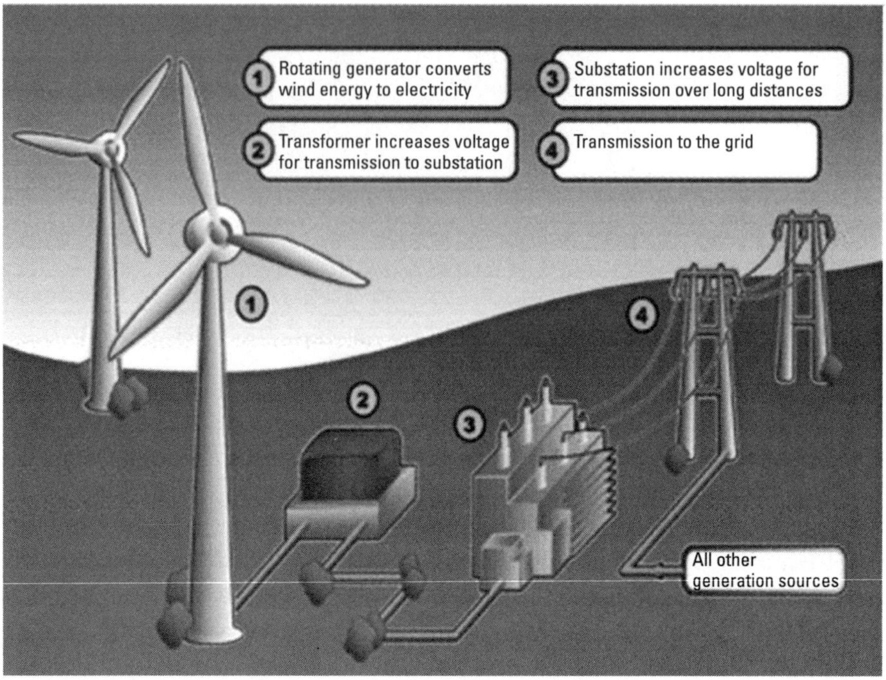

Figure 5.7 Wind turbine. (http://opecmuller7-8.wikispaces.com/Renewable+Energy+ Sources Creative Commons Attribution Share-Alike 3.0 License.)

power system; solid oxide fuel cells combined with a gas turbine or microturbine; Stirling engine combined with a solar dish; wind turbines with battery storage and diesel back-up generators (shown in Figure 5.8) ; engines (and other prime movers) combined with energy storage device such as flywheels.

Some of the advantages of these hybrid systems include: the ability to provide increased system efficiency as well as a greater balance in energy supply; the ability to use various fuel sources such as solar, wind, chemical, and hydrogen, depending on the combined technologies; and the ability to create systems of various sizes according to requirements.

The largest drawback for these hybrid systems lies in the fact these technologies are still in their infancy, so combining them has not yet passed the pilot phase.

Electric Energy Storage

Electric energy storage or energy storage is a set of technologies capable of storing previously generated electric energy and releasing it later. Energy can be stored as chemical, kinetic, thermal, or potential energy in a form later con-

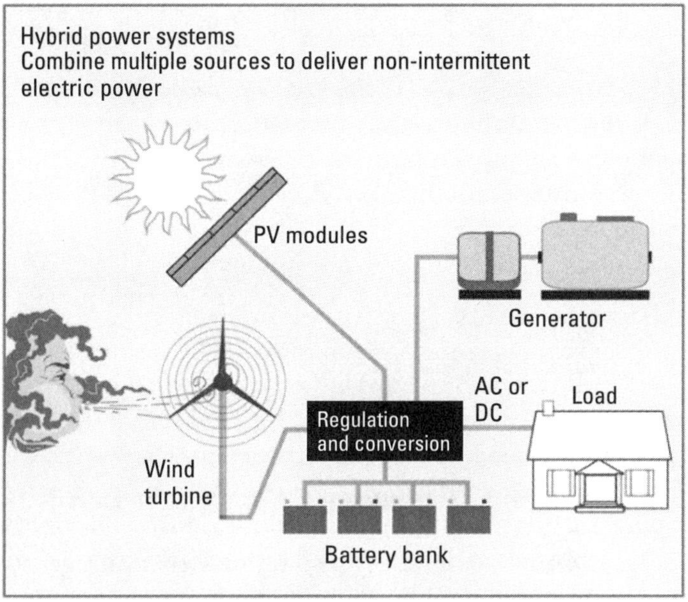

Figure 5.8 Hybrid system. (By DOE [Public domain], via Wikimedia Commons https://commons.wikimedia.org/wiki/File%3AHybrid_Power_System.gif.)

verted to electricity. There are various technologies in different states of development and used for various purposes depending on their energy and power characteristics. Some of the best-known technologies used to store energy include lithium-ion, flywheel, compressed air energy storage, and sodium sulfur (NaS) battery in addition to others.

More information on this technology is covered in Chapter 4.

Demand Response

DR is the ability to change electric usage by end-use customers from their normal consumption patterns in response to variations in the price of electricity over time. DR is used for a variety of reasons including peak shaving, reducing peak-time wholesale prices, and ancillary services. DR can also be in response to incentive payments designed to induce lower electricity use at times of high wholesale market prices or when high usage jeopardizes system reliability. Examples of price-based programs include time of use, critical peak pricing, and peak-time rebate. Examples of incentive-based programs include direct load control and curtailable service.

Additional chapters in this book cover DR in greater detail.

Microgrid

A microgrid is a group of interconnected loads and DERs within clearly defined electrical boundaries that acts as a single controllable entity with respect to the grid. Most microgrids routinely incorporate different forms of DERs. Customer supply needs are provided for in a microgrid through planned outputs designed in relation to each other. Power generated from DERs is converted through bidirectional smart inverters, installed to serve load or the grid. A microgrid can connect and disconnect from the grid, enabling it to operate in both grid-connected or island mode. Microgrids are generally designed to meet the needs of its consumers. Microgrids can be deployed at university campuses, commercial and industrial locations, military bases, communities, and islands.

More information on this microgrid technology is covered in Chapter 6.

Technical and Business Challenges of Supply from DER

DERs have the potential to create new and unforeseen challenges to the existing distribution utility franchise. Penetration of DERs will impact the roles of various stakeholders. To better understand how the challenges from these new technologies will impact the existing franchise model, let us review them from the viewpoint of each of the major stakeholders.

Utility Viewpoint

Utilities are asset management companies and service providers that install and manage large quantities of infrastructure as well as provide a service to their customers. The utility is guaranteed a rate of return on that investment by the regulator, which monitors the quality of service provided to the customer. The regulator then sets the rates and penalties for the utility, which effectively guarantee a rate of return, provided that the utility properly manages the assets that it installed and operates.

> *The efficiency of the panels and seeing how much we really did generate was probably the most surprising. It's pretty exciting to get an electric bill every month and see a zero.*
>
> —Rebecca Amato, homeowner, Oakland, California
> https://us.sunpower.com/home-solar/success-stories/amato-family-oakland-california/

Some of the challenges that DER presents to the utility are:

- *Reduced revenues:* As the penetration of DERs increase, more people will generate power locally and take less energy from the grid/incumbent utility. This change will force the utility to recover their investment for the T&D assets installed from across a smaller amount of kWhs (kilowatt hours) sold, resulting in increased possibility for stranded assets.

- *Customer becomes a supplier (prosumer):* The basic utility equation, over time, has been that which the utility generates, the customer consumes. For the first time, this equation will no longer be valid under all scenarios, because of solar PV and other DER-based production. The customer has the potential to generate more than their needs and sell back to the grid under the right regulatory mandate. Under these circumstances, the customer will expect to get paid for the energy delivered to the grid altering the utility-customer contract.

  Thus, the new generation of customers will fit into new categories, including the following:

  - *Individual customers:* Both residential and commercial/industrial may deploy DERs behind the meter and either deliver excess power back to the grid for a payment or store the generated energy locally for use later. This is one of the fastest growing segments.
  - *Microgrids:* Take advantage of DERs and the availability of new technologies to manage and control university, industrial, and commercial campuses by installing local sources of generation and manage much of their own supply and demand internally. These same mechanisms supported by new control technologies allow them to manage their energy usage and control their interaction with the grid.
  - *Aggregators:* Aggregators are a new class of entities that learned from the success of the sharing economy of companies such as Uber and Airbnb and bring those concepts to the electric grid. Fundamental to this business model is the ability of the aggregator to take the production and/or consumption of multiple customers in either contiguous or separate location and use this diversity of generation or load to deliver multiple commodities—energy and/or ancillary services—to the wholesale and retail energy marketplace.

    This is still in a nascent stage, so examples of plans and ideas are vast and varied.

  By delivering power back into the grid, the prosumer creates reliability issues by converting a power system designed for one-way flow of power into one in which power flows both ways. This creates reliability issues, but also a host of other issues, including the safety of line crews and problems with protection systems.

- *Maintaining grid reliability with less control:* Whether it is into the transmission network or the distribution network or the increased customer control of new generation sources, the utility is no longer the sole determiner of either the placement of generation sources on the grid and/or the timing for the supply of power to the grid.

The utility will likely remain in charge of maintaining reliability on the transmission and distribution system. In support of this mandate, utilities will need to establish interconnection and safety requirements allowing the system to be operated reliably while not erecting unreasonable barriers to entry. The new generation resources in the distribution system will potentially impact the need for distribution system upgrades. These needs will be included in the interconnection agreements, transparent to all stakeholders. The regulator will need to sign off on these requirements.

- *Volatility of supply and negative effects on grid reliability:* A significant percentage of DER comes from renewable sources. Most renewable sources of supply tend to have a component of variability. Wind must be blowing for a wind farm to generate and the sun must be shining for the solar cell to generate. Thus, their output can quickly change from zero to full capacity and to levels in between, impacting grid stability in a variety of negative ways. The utility and/or the balancing authority need to plan for these volatilities through existing mechanisms, such as more spinning reserve.
- *Variety and high cost of new technologies:* The availability, type, cost, and diversity of behavior of these technologies are numerous, and selecting from them can be daunting. In addition, the technologies are new and many of them are only in pilot phase. Not all will survive and move into production. Those that do survive must be relative from a cost and performance perspective. Utilities prefer selecting proven technologies to solve critical problems.

Utilities need new control and dispatch strategies, which allow the grid operator to integrate these new technologies and still maintain a safe and reliable grid.

Investor Viewpoint

The onset of DERs producing economic value in the distribution system can also be found for entities other than the distribution grid owner or operator. Already in states such as Connecticut and New York, we are seeing innovative business models designed by investors as they define new approaches to install DERs on customer premises. From an investment point of view, markets that provide clear rules and methods for real time and forward price signals will be the ones to attract the most investment. These will create the best DER business models.

DERs are generating interest in two groups of investors.

- *Investors in DER technologies:* Everyone is aware of this group. They bring new technologies to the market, whether it is the latest and most efficient solar PV panel, storage mechanism, fuel cell, or something else. Their primary role is in selling their technology for installation to utility, residential, or commercial customers through a variety of sales channels.
- *Investors who provide new services to utility customers:* This is a newer group, which looks at a variety of business models to deliver new services to their customers. Some of the more innovative business models are listed here:
  - Providing solar PV installations to residential customers in exchange for the rights to harvest and sell that electricity to the incumbent utility for a price shared with the residential customer. The residential customer owns the PV installation for a set number of years.
  - Aggregating the DER output from multiple customers to interact either at a retail market (if it exists) or even at the wholesale market if sufficient critical mass has been achieved. The advantage of this model is that based on the diversity of their portfolio, the aggregator can interact in multiple markets—energy and ancillary services.

The investor landscape is still in its infancy and may take on new roles and value streams as the DER technologies become more mature and the utility and retail markets become more active.

Customer Viewpoint

All discussions about DERs begin and end with the customer. DERs, at their core, have led the revolution to change the utility-customer compact by allowing the customer to move from a pure consumer of energy to one that also can deliver power to the utility. The interaction is between the utility (business) and the grid (technical). The key challenges faced by the customer include:

- Getting paid when sending power back into the grid. When the customer sends power into the grid, they like to get paid for that energy at the time of transmission.

- Fair value paid for the net energy produced and sent back to the grid requires the development of good economic price signals. The value of the energy produced must be based on the value when and where it is sent to the grid.
- Technologies are required to measure how much energy was sent back to the grid and when. In addition, tariffs or market mechanisms are needed to define the amount of payment based on the energy or other services provided by the customer.
- There is a considerable debate on the installation costs of net metering, which is currently being socialized across the entire customer base. It is also based on the payment of energy at full retail rates, without adjustment for time of day or location.
- Despite their diverse characteristics, most DERs either generate in DC or in varying frequency and voltage AC. Integrating them into the grid can be challenging and may require DC-to-AC interfaces for their interconnection.

Regulatory requirements such as DER interconnection standards and policies are evolving and need to be in place to allow markets to develop in an orderly fashion.

Regulator Viewpoint

DERs create a difficult situation for the regulator. On one side, they must keep the customer's interest in mind. At the same time, they need to make sure that the utility can recover the cost of its investments made on the grid. They cannot be favoring specific customers of any specific income segment. A common situation with DER installation is that people with higher income are more likely to be able to afford DER installation, but the cost of net metering and associated equipment are socialized across the entire customer base.

The regulator has many options available to them to resolve the challenges:

- *Unbundling:* With this option, the utility separates out the cost of the energy delivered from the cost of the infrastructure being provided. This way, the customer only pays for the true cost of the energy being delivered to their premises but pays for the privilege of being connected to the grid.

  The customer has the freedom to completely disconnect from the grid and not pay either the energy charge, when their own generation is offline, or the connection charge.

- *Provide new services:* Allow utilities to partner with their customers that have DERs installed in their premises. These services could take on a range of options from managing and maintaining the DERs to taking on the extra energy fed into the grid. Each of these options may have the potential to bring new revenues into the utility, while allowing the grid to potentially benefit from DERs such as solar or PV, DR, and/or storage.
- *Installing DERs on customer premises:* The utility could be allowed to use ratepayer funds to install DERs on customer premises and use the excess energy to assist in delaying or avoiding grid upgrades, thereby saving money for the ratepayer and supporting and encouraging an environmentally beneficial generation mix.

For any of these to work, the utility and the regulator need to work closely with each other to design policies and regulations supporting the economics of installing DERs, while allowing a business structure to incentivize the utility to encourage DERs to be installed, all while still delivering on their reliability mandate.

Benefits of DERs

Despite the obvious obstacles that DERs bring with them, they also come with several tangible benefits that depend on the specifics of type and implementation (Table 5.1).

Benefits from grid-connected DERs are varied, but include reduction in peak power requirements, provision of ancillary services (e.g., operating reserves, regulation reserve, black-start, and others), and emergency power supply. DERs also provide savings in several categories, such as energy costs from reduced T&D losses, reduced congestion costs, deferred generation or T&D capacity, land use effects, and reduced vulnerability from terrorist attacks [5, 6].

The benefits from grid-disconnected DERs include grid-disconnected DERs such as within a microgrid control system and the ability to operate autonomously, strengthen grid resilience, and helping to mitigate grid disturbances, for example, remote microgrids deployed on islands due to the impracticality of connecting to the main grid. These remote microgrids facilitate the mass introduction of renewable energy sources, such as solar and wind power generation. Advanced power electronics employed in microgrids address issues with intermittent renewable DERs, voltage buildup, surplus power, and lack of frequency adjustment capacity. These remote microgrids enable the electrification

Table 5.1
Benefits and Services of Distributed Generation

		Benefit Categories							
		Energy Cost Saving	Savings in T&D Losses and Congestion Costs	Deferred Generation Capacity	Deferred T&D Capacity	System Reliability Benefits	Power Quality Benefits	Land Use Effects	Reduced Vulnerability to Terrorism
DER Services	Reduction in Peak Power Requirement	√	√	√	√	√	√	√	√
	Provision of Ancillary Service, Operating Reserves, Regulation, Black-Start, Reactive Power	√	√	√	√	√	√	√	√
	Emergency Power Supply	√	√	—	—	√	√	—	—

of remote villages in several countries in Africa powering cell phone towers and other technologies.

Building Blocks to Reliably Integrate DERs into the Grid

The power generated by DERs must be controlled to serve the diverse needs of the distribution grid without requiring excess reserves provided by other sources of generation. Depending on the type of power generated and how it is transmitted into the grid, appropriate dispatch and control mechanisms use and support different mechanisms.

Interface

DER technologies generate either AC or DC. AC is often generated with varying frequencies. To connect to a grid that requires power at a specific voltage and a standard frequency, DERs need advanced power electronics to convert their output power to meet those specifications.

- DER systems such as wind, microturbine, or IC engine, generate AC output, often with variable frequencies. An AC-DC converter is needed in this case.
- DER systems such as PV and fuel cells generate DC output. A DC-DC converter is typically needed to change the DC voltage level.

A DC-AC inverter is then needed to convert the resultant DC source to grid-compatible AC power [7, 8].

Sensing and Controls (Distribution Automation)

The broad variability of energy produced from DERs also results in the need for advanced sensing and control of their output. As the output of the DERs change, the distribution system operator needs to have a real-time understanding of the amount of energy generated at any one time so any excess or deficit can be managed. This management is done by exercising control on the existing generation or bringing new generation into the grid [9, 10].

Smart Inverters

The need for dynamic control increases as utility grids see increasingly more DERs. The intermittency of these DERs is difficult for a utility to deal with. This is where the inverter comes in. For utilities and grid operators, the capabilities of the inverter are vital to reliably isolate and integrate solar and other renewables onto the grid. Current inverter performances fall short and disconnect at the first sign of grid disturbance. However, advancements in inverter technologies have given rise to a new type—the smart inverter.

At a fundamental level, the inverter converts DC input to AC output, which, for a grid-tied system, enables the supply of real power to the grid. Other standard functions include power transfer optimization, voltage conversion, grid synchronization, disconnection, anti-islanding protection, and storage interfacing. These capabilities ensure DERs efficiently supply power to the grid while also providing key safety features. Functions beyond this scope qualify an inverter as being smart. These include reactive power control, voltage, frequency ride-through, advanced two-way control capabilities, storage integration, and data streaming. These new inverters enable levels of PV penetration that would otherwise not be possible. These advanced features will enable the next phase of growth in the global distributed renewables industry [11].

Microgrid Management Software

Traditional approaches for integrating DERs onto the utility (macro) grid focus on the impacts of a few relatively small number of micro-sources on grid performance. Microgrids integrate a variety of DERs and present to the distribution grid as a single self-controlled entity. This is enabled by advanced power electronics, such as smart inverters, that control the interface between the DERs and their local AC system. A microgrid control system monitors and controls the various smart inverters, and power that flows to and from the grid [12]. The system manages any imbalance between load and generation within the microgrid, all while taking the specific network characteristics into account and not depending on the need for generator-based inertia to balance the generation-load imbalance. It monitors and controls issues related to loss of aggregation, micro-source generation, protection, and safety.

Economics of DER

The economics of DER are complex and evolving. Much of the complexity is centered around the inability to determine the value of the products and services they provide. DERs have a temporal and a locational aspect to them, making their value variable in two dimensions. For implementations such as DR, it is also difficult to validate how much energy was not consumed from the grid and when. However, it is difficult to put a price on the variability of the delivery of power into the grid. Their valuations also change due to the variable nature of their generation, if from a renewable source, and other services such as ancillary services, needed to support their generation and interconnection into the grid.

- *Incentives for installation:* In several states in the United States and in other countries around the world, there are incentives that provide rebates for installations for some DERs, especially solar PV and wind. These are generally one-time payments, either rebates or tax incentives, which significantly reduce the cost of installation.
- *Tariff basis:* In several states in the United States and in other countries around the world, there is also a fixed rate of payment ($/kWh) when DERs deliver power into the grid. Use of net metering is quite common in these situations to determine the amount of energy delivered back into the grid. This payment amount is determined by the regulatory process and is generally set up to encourage greater penetration of renewable-based generation.
- *Wholesale markets:* Rulings at many regional transmission organizations (RTOs) in the United States and around the world have made DERs eligible to play in wholesale markets offering various services based on their characteristics. For example, because of the U.S. Federal Energy Regulatory Commission (FERC) order 755, energy storage devices such as flywheels can now provide frequency regulation to markets [8]. The characteristics of a flywheel storage that enable it to release high power over a very short duration, make it suitable to provide a critical reliability service such as regulation. Similarly, FERC orders 745 and 719 allow Independent System Operator New England (ISO-NE) to allow DR to provide energy, capacity, and nonspinning reserve [7, 13]. In addition, FERC order 2006 allows California to have a regulatory mechanism in place to procure 1,000 MW of distributed renewable generation to be offered in their wholesale market.
- *Retail markets:* Moving from wholesale to retail markets, the state of New York has launched the Reform the Energy Vision (REV) effort investigating the incorporation of market mechanisms to encourage the incorporation of DERs into the distribution grid [1, 2]. The New York effort plans to implement the following key mechanisms: creation of a distribution system platform provider that will manage the grid reliability and market for specific jurisdictions, a futures market clearing price to be used as investment signals to investors, and a real-time market clearing price to pay for actual delivery of DER-based energy into the grid. The exact mechanisms for calculating the market clearing prices have not yet been defined, but it is safe to say similar structures will be implemented in other jurisdictions and each one of them will develop their own pricing mechanisms, following a path like that of wholesale markets around the world.

To summarize, the economic impacts of DERs have not yet been fully realized or defined. Much of today's implementations are focused on payments mandated within the rate base or by state law. It is believed that this will change at some point in time to a market-based calculation of the rate on which the DER generation is compensated. While New York is taking the first step with their REV implementation, other states are expected to follow.

Dos and Don'ts of Integrating Distributed Generation

The radially designed nature of the distribution grid—at least in North America—was designed for a power flow in one direction only: from the substation to the consumer. However, the fact that DERs are installed at the customer site leads to a bidirectional generation grid for which they are not designed. These bidirectional flows impact both the settings of the grid protection systems and the safety protocols in place to protect linemen working on the feeders.

The most important dos and don'ts for installing DERs at the customer site are as follows:

- Do work in grid-connected and grid-disconnected mode. This means that when the grid is down due to an outage, the system should be able to disconnect itself from the grid. This would prevent a situation of reverse power flow into the grid, potentially causing harm to the people working on the grid.
- Do install net metering so the utility can track how much power was returned to the grid and when.
- Do inform the utility what kind and capacity of DER that a customer installs on their premise. The utility can then account for the DER in its planning process, for system protection and other aspects of monitoring and operating the grid.
- Do implement the right level of sensing and control, based on the size, capacity, and type of DER. This would allow the utility to either take advantage of the production or control it, while maintaining the reliability of the network.
- Don't implement a DER installation at the residence without informing the local utility.
- Don't implement a DER installation at the residence without the right set of safeguards so that utility line crew is safe when they are working on the feeder.

Utilities work in conjunction with their regulators to design interconnection policies and procedures based on the type of connection and the type of DER being installed.

Case Studies

Case Study #1: New York State—Reform the Energy Vision and DER Interconnection

New York State leads the nation in developing new policies to encourage and reward consumers for the use of new technologies for controlling energy use.

Reform the Energy Vision (REV) is an initiative that aims to fundamentally transform the way electricity distribution used in New York State. REV's objective is to build a bridge to a cleaner and more efficient and affordable energy system. It plans to achieve this by:

- Creating the power grid of the future and enabling customers to better manage and reduce their energy costs;
- Focusing on system efficiency, total bills, carbon emissions, technology innovations, resiliency, and competitive markets around customers;
- Addressing issues such as rising electric bills, reliability, resiliency, emission reductions, jobs, and the low-income electric divide.

The initiative broadly aims to lower energy costs, create opportunities for economic growth, and protect the environment.

Specific to distributed generation, the initiative includes wider deployment of DERs, such as microgrids, roof-top solar, and other on-site power supplies, and storage. REV's Phase 1 mandates the state's Investor Owned Utility (IOU) to use standardized interconnection requirements as a framework to streamline DER interconnection. New York's IOUs such as Consolidated Edison, National Grid, PSEG Long Island, and others have adopted the New York Standardized Interconnection Requirements (SIR) and streamlined their current interconnection approval processes to reduce administrative burden, increase transparency, and adequately prepare for greater amounts of DER deployment. Most of these utilities have either implemented or are in the process of implementing an online portal for handling DER applications, tools, and software for linking together customer project management, work management, billing, and metering systems. Thus far, the clear majority of interconnection applications received by these utilities have been for small residential systems, generating below 50 MW. These accounted for almost 99% of all received applications (see Figure 5.9).

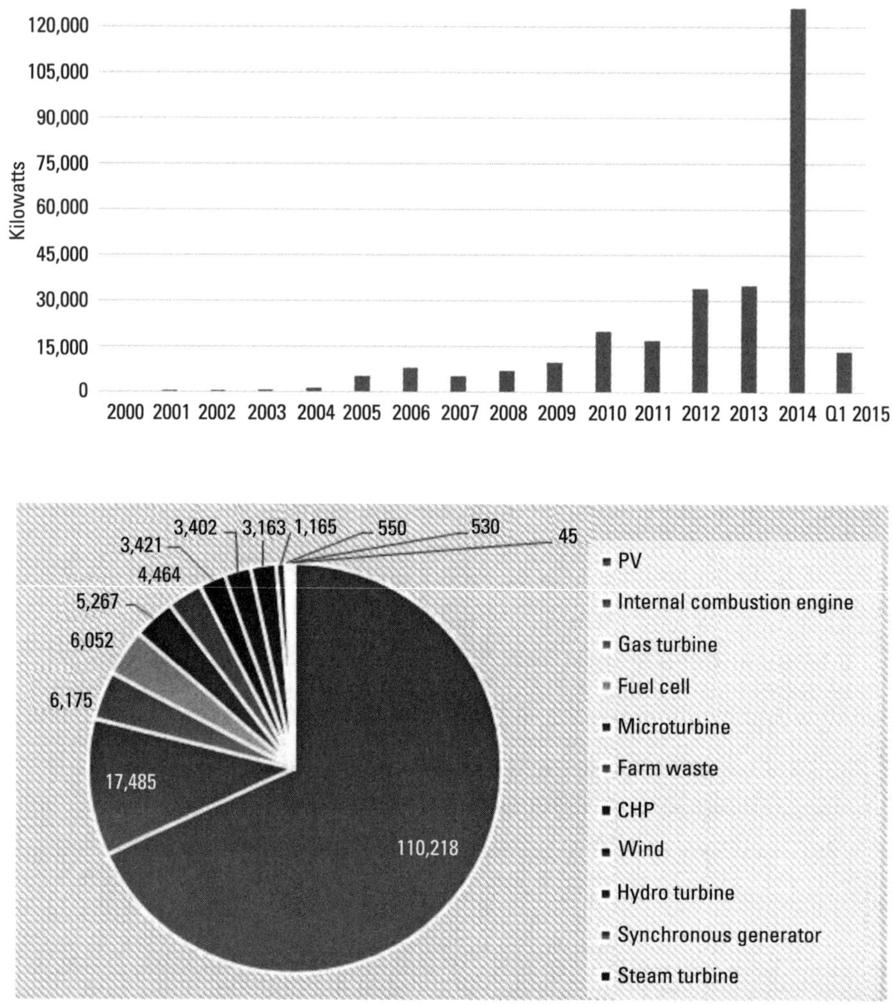

Figure 5.9 Distributed generation deployed by NY IOUs. (Source: New York Department of Public Service, Interconnection of Distributed Generation in New York: A Utility Readiness Assessment Final Report, http://www3.dps.ny.gov/W/PSCWeb.nsf/96f0fec0b45a3c6485257688006a701a/dcf68efca391ad6085257687006f396b/$FILE/83930296.pdf/EPRI%20Rpt%20-%20Interconnection%20of%20DG%20in%20NY%20State-complete%20-%20Sept%202015.pdf.)

These utilities expect to see an increase in applications for larger sized systems in response to Net Metering Transition Plan and NY-Sun MW Block program [11]. The NY-Sun Incentive Program was designed to deliver certainty and transparency to solar industry and their customers regarding incentive levels and eliminate them when market conditions support them. The MW block approach allocates megawatt targets to specific regions of New York, breaking them down into blocks Once all blocks within a region/sector are fully subscribed, the incentive is no longer available to that region/sector.

Case Study #2: City of Fort Collins, FortZED Renewable and Distributed Systems Integration Project

Zero Energy District (FortZED) is a renewable and distributed systems integration (RDSI) demo project funded in part by the U.S. Department of Energy. FortZED creates as much thermal and electrical energy locally as it uses. The mission of FortZED is to transform the downtown area and the main campus of Colorado State University into a net Zero Energy District through conservation, efficiency, renewable sources, and smart technologies. The primary objective is to substantially increase the use of renewables and DERs for supplying power during peak load periods. The project was conducted in two phases. In phase 1, the target was to achieve a 20%–30% peak load reduction on two feeders through an integrated system of 5 MW of mixed distributed resources.

As part of the project,

- The City of Fort Collins deployed a 500-kW conventional generator, 92-kW thermal storage, 5-kW PV array, 62-kW HVAC and DSM, and 2 × 10 kW Ford Escapes (PHEVs).
- Colorado State University deployed 80-kW thermal storage, 80-kW fan variable speed drives, 21.6-kW water fountain pumps, 3.6-kW hot water heater controls, 6-kW daylight control, and 950-kW conventional gensets.
- InteGrid Lab deployed 2 × 80 kW natural gas gensets, 300-kW natural gas genset, 2 × 80-kW microturbines, 100-kW wind turbine simulator, and 10-kW fuel cells.
- Larimer County deployed 10-kW generation and 1.5 kW of load shed.
- The New Belgium Brewing Company deployed 200-kW PV arrays with inverters, 292-kW methane-based combined heat (CHP), 650-kW methane-based CHP, 135-kW thermal storage, and 160-kW load shedding potentials.

Phase 2 was a concentrated effort to demonstrate that a microgrid could work with remote sites across the Fort Collins distribution system. It paved the way for a greater understanding of how to successfully integrate solar energy sources how to manage DR on a microgrid [14] (Figure 5.10).

RDSI, also known as the FortZED Jumpstart project, went live in 2011, and has since provided numerous insights. This project showcased innovative techniques for managing energy in Fort Collins and beyond.

Findings from this project helped the city of Fort Collins to pave the way for utilities of the twenty-first century and significantly impact the future of renewable energy in Fort Collins, the state of Colorado, and even the nation. This

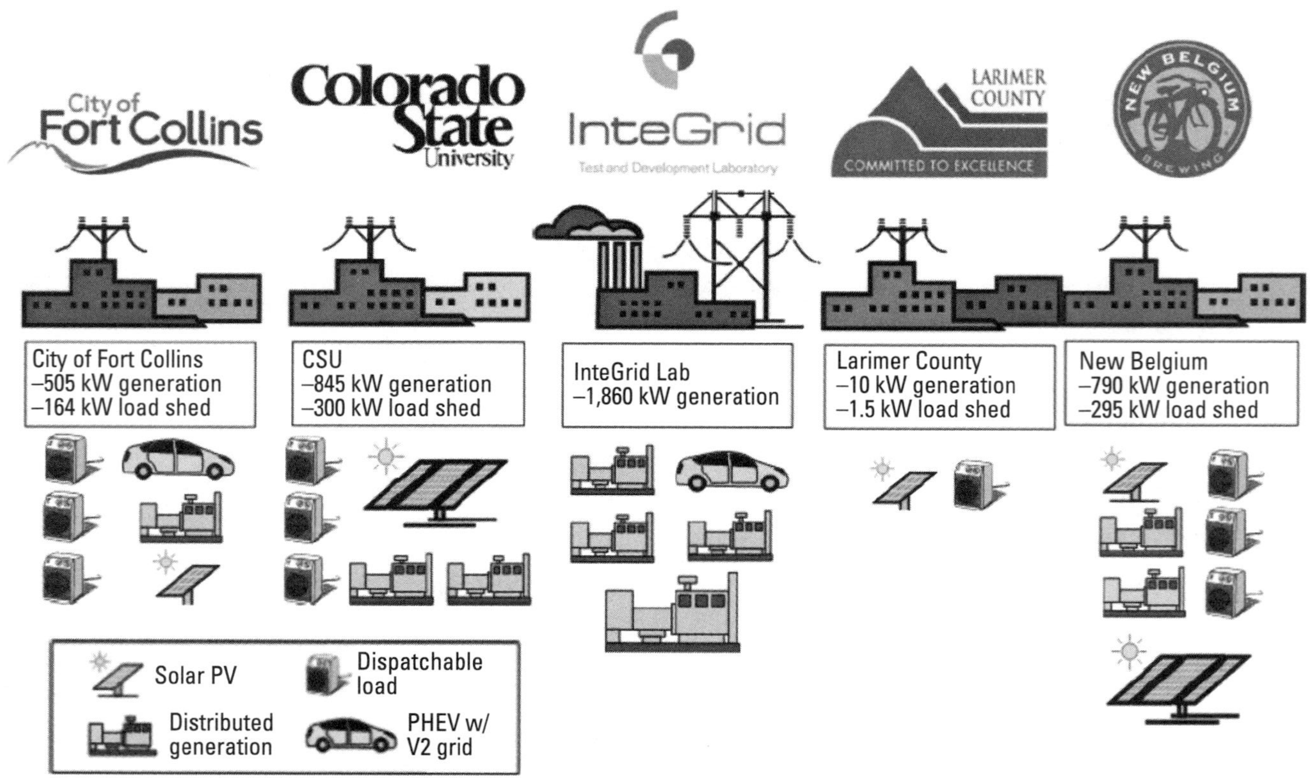

Figure 5.10 FortZED RDSI project demo sites and DERs. (Source: FortZED project. Used with permission.)

project, which is one of only nine smart grid demonstrations in the country to receive funding from the U.S. Department of Energy, has demonstrated that smart grid strategies can save utilities money. It has shown how peak electric demand can be reduced without affecting local air quality.

Transformational Impacts of DERs to the Utility of the Future

DERs will do to distribution what FERC 888/889 and PURPA did to transmission. Key aspects of these changes include:

- *The consumer is now also the supplier:* The utility's primary relationship with the consumer needs to change completely. The two entities need a more collaborative relationship. They must depend on each other. The financial relationship between the customer and the utility will also change. Today, the utility sends a bill and the customer pays. Since the customer now sells energy to the utility, billing will be more complex and sometimes the utility may end up paying the customer.
- *How the utility handles reliability will change:* Presently, the utility, whether it is a part of a wholesale market or not, has a predictable source of electricity from generators that are dispatchable and whose outputs are predictable. With the increase in penetration of DERs, especially from renewable sources, this will change.
- *The advent of retail markets:* Lessons learned from the work done at the New York's REV effort have taught us that developing the right price signals for real-time settlements and for incentivizing construction is important to ensure DER growth.

  For this to work effectively, it will also require further disaggregation of the distribution side of the utility, such as how transmission was impacted during FERC orders 888 and 889. A new entity, a distribution system operator (DSO) (or distribution platform provider [DSP], as defined in the NY REV), independent from the asset owner, may need to be created. This new entity will operate the grid and run the retail markets.
- *How the utility procures energy will change:* As indicated above, energy from renewable sources will enter the grid when generating, not based on dispatch orders from the utility or the wholesale market. The utility will need to become better at forecasting the outputs of DERs based on location, but still be ready to deliver the high level of reliability expected of them.

- *The utility has more options to procure supply:* With DERs becoming cheaper and more diverse and existing closer to the load, the utility has more options to procure supply and solve local problems locally instead of looking to all problems centrally.

Utilities will need to transform themselves to handle these new sources of energy and their owner or operators. DERs will impact almost every aspect of utility operations from system operations to asset management and power system planning to the management of the back office.

Conclusions

Change is not new to the utility industry. Starting with Public Utility Regulatory Policies Act (PURPA) of 1978, the Environmental Protection Agency (EPA) of 1992, and FERC orders 888/889, which drove wholesale deregulation, and moving forward in time to retail in states such as Texas, the utility industry has seen a lot of changes over the last 40 years. PURPA brought to play the independent power producer. FERC orders 888/889 and Texas deregulation brought about separation between generation and transmission. The last front in this journey of change is distribution.

The final takeaway is that distributed (and renewable) sources of power have the potential to make our grid more secure, reliable, and resilient by moving us away from the dominant dependence on fewer centralized sources of power and lead us toward more sources of power. This new paradigm could provide us with a better ability to withstand cyberattacks, weather-related disasters, and physical man-made attacks. The future is bright. The utility industry is resilient and our personnel very creative and intelligent. The solutions defined may not be perfect on day one, but will continue to evolve with increased penetration of new and unpredictable sources of energy supply.

DERs portend a series of changes to the grid that Tesla and Edison never imagined when they envisioned the electrical energy industry. DERs are about to change everything we know about the utility forever [15, 16].

References

[1] "Reforming the Energy Vision (REV) Working Group II: Platform Technology," New York State Department of Public Service, July 8, 2014, https://www3.dps.ny.gov/W/PSCWeb.nsf/96f0fec0b45a3c6485257688006a701a/853a068321b1d9cb85257d100067b939/$FILE/WG%202_Platform%20Technology_Final%20Report%20&%20Appendices.pdf.

[2] New York State Department of Public Service, "Staff Proposal Distributed System Implementation Plan Guidance," October 15, 2015, http://www3.dps.ny.gov/W/PSCWeb.nsf/All/C12C0A18F55877E785257E6F005D533E?OpenDocument.

[3] New York State Department of Public Service, "Interconnection of Distributed Generation in New York State: A Utility Readiness Assessment," September 2015, http://www3.dps.ny.gov/W/PSCWeb.nsf/96f0fec0b45a3c6485257688006a701a/dcf68efca391ad6085257687006f396 b/$FILE/83930296.pdf/EPRI%20Rpt%20-%20Interconnection%20of%20DG%20in%20NY%20State-complete%20-%20Sept%202015.pdf.

[4] Renew Wisconsin, "An introduction to distributed generation interconnection," March 2004, http://www.renewwisconsin.org/wind/Toolbox-Applications%20and%20forms/Interconnection/Introduction%20to%20Distributed%20Generation.pdf.

[5] FERC, "The Potential Benefits of Distributed Generation and Rate-Related Issues That May Impede Their Expansion," February 2007, https://www.ferc.gov/legal/fed-sta/exp-study.pdf.

[6] Rawson, M., "Distributed Generation Costs and Benefits Issue Paper," *California Energy Commission*, July 2004, http://www.energy.ca.gov/papers/2004-08-30_RAWSON.PDF.

[7] FERC, "Demand Response Compensation in Organized Whole-Sale Energy Markets," March 15, 2011, https://www.ferc.gov/EventCalendar/Files/20110315105757-RM10-17-000.pdf.

[8] Vadari, D. M., "Smart Inverters: Revolution or Evolution?" *Grid Insights by Intel*, January 26, 2016, http://blogs.intel.com/energy/smart-inverters-revolution-or-evolution.

[9] Thomas Beach, R., 'The benefits and costs of solar distributed generation for Arizona public service," *Solar Energy Industries Association*, May 8, 2013, http://www.seia.org/research-resources/benefits-costs-solar-distributed-generation-Arizona-public-service.

[10] Kramer, W. S. C., "Advanced Power Electronic Interfaces for Distributed Energy Systems," National Renewable Energy Laboratory, March 2008, http://www.nrel.gov/docs/fy08osti/42672.pdf.

[11] Vadari, D. M., "Smart Inverters: Revolution or Evolution?" *Grid Insights by Intel*, January 26, 2016, http://blogs.intel.com/energy/smart-inverters- revolution-or-evolution/.

[12] "Microgrid Management Software to Control Distributed Power Generation," *Transmission & Distribution World*, February 4, 2015, http://tdworld.com/ energizing/microgrid-management-software-control-distributed-power-generation.

[13] FERC, "Frequency Regulation Compensation in the Organized Wholesale Power Markets," October 20, 2011, https://www.ferc.gov/whats-new/comm-meet/2011/102011/E-28.pdf.

[14] "Case Studies and Lessons Learned," *FortZED*, March 11, 2015, http://fortzed.com/case-studies-and-lessons-learned/.

[15] Vadari, M., "Setting the Context: Defining Competitive Threats Embedded in DER Technologies," *Transmission & Distribution World*, March 29, 2016, http://tdworld.com/distribution/monopolies-competition?page=7.

[16] Vadari, M., "The Key to Unlocking Tomorrow's Energy," *Transmission & Distribution World*, September 29, 2015, http://tdworld.com/transmission/key-unlocking-tomorrow-s-energy?page=6.

6

Microgrids: Fragmentation of the Grid

A microgrid is a group of interconnected loads and distributed energy resources within clearly defined electrical boundaries acting as a single, controllable entity with respect to the grid. A microgrid can connect and disconnect from the grid-enabling operation in both grid-connected or island-mode.

Introduction

Microgrids are not new. They have been around for a very long time. In fact, one could argue the first set of electricity grids were all microgrids, at least until they were all connected together to form today's interconnected grid.

As time progressed from the first days of the interconnected grid, for a long time, microgrids were relegated to a niche status due to their limited use in remote industrial locations or remote regions of the world where it was either uneconomical to provide them with access to centralized generation through the transmission network or adequate infrastructure just did not exist. Over time, they have been moving toward college campuses, industrial campuses, and so on.

This move from niche to more mainstream status has been significantly influenced by an increased need for individuals and commercial entities to take more control of their energy destiny by the option of being connected to the grid or not.

Microgrid popularity has also significantly influenced several major factors:

- The advent of distributed energy resources (DERs) such as solar photovoltaic cells, wind turbines, and energy storage has allowed local energy generation and a potential option for independence from centralized generation sources which otherwise require connection to the main grid at all times.
- Significant improvements in automation such as remote sensing and control capabilities have become proficient in performing more functions at lower costs.
- Control algorithms have improved substantially enabling a full range of generation-load balance management within the microgrid.
- Access to cheap and ubiquitous communications required to support automation and implementation of the results of the control algorithms has increased.

These influencing factors are leading not just to increased presence of microgrids, but also to the development of microgrids at a smaller scale (sometimes known as nanogrids) with potential for each residence to become its own micro (nano) grid.

This chapter starts by defining microgrids and looking at their varying configurations. It then identifies key drivers leading to the development and thriving of microgrids. The chapter lists and explains the key features of microgrids and their enabling technologies. The chapter then moves to discussing the next generation in microgrids and the possibility of redefining the design and operation of today's grid. The chapter also takes a deep dive into a case study focused on outages, showing how today's grid will respond as compared with a hypothetical future next generation grid that instead of supporting the existence of microgrids actually embraces them by incorporating them into the core itself. It also provides a roadmap to achieving this new state. The chapter concludes by identifying the microgrid challenges and how microgrids are contributing to the transformation of the utility business.

So let us start with a microgrid definition and also how is it different from our normal, bulk power macrogrid.

Defining Microgrids

Microgrids are a key component of the smart grid. The Galvin Electricity Initiative [1] defined microgrids as modern, small-scale versions of the centralized electricity system. They achieve specific local goals established by the community being served, such as reliability, carbon emission reduction, diversification of energy sources, and cost reduction. Like the bulk power grid, smart microgrids generate, distribute, and regulate the flow of electricity to consumers, but do so locally. Microgrids are slowly making their way into the modern grid, mainly showing up in industrial campuses, university campuses, and military bases. By their very definition, they are designed to either operate independently bringing together a variety of sources of generation to service their load or to be fully connected into the grid.

A comparison of a microgrid versus today's macrogrid is shown in Table 6.1.

These microgrids can take one or more configurations based on business objectives:

1. *Classic separable microgrid:* This is the classic microgrid that is a part of the main distribution grid but can be separated when needed either in an emergency or otherwise. This configuration has a need for higher generation margin driving the path to reliability.
2. *Defensive microgrid:* While this configuration is basically designed for energy independence over short-term or mid-term time frame, there is a greater need to focus on power reliability and quality. Key features include supplementary generation, storage, and power conditioning.
3. *Demand response microgrid:* This configuration is more about a business microgrid in which aggregators could deliver power to single customers or groups of customers. Given the greater diversity and potential load dispersion, we envisage this configuration to enable DR and could be used for large residential complexes (dense urban demand response).

Table 6.1
Comparison of Features of Microgrid versus the Macrogrid

Microgrid	Macrogrid
Localized network part of centralized grid	Centralized grid
Ability to disconnect from macrogrid and function autonomously, as an island	Large interconnected network
Incorporates local DERs	Centralized generation and long-distance transmission lines
Local choice of source of generation	No local choice of generation; mix of centralized energy sources
Low-voltage operation and control	Mostly high-voltage operation
Supports combined heat and power systems	Recovery of waste heat not supported as part of network
Service closer to load and delivers heterogeneous power quality and reliability	Power quality is homogeneous

4. *Controlled separation island:* This configuration is an increasingly important one in that it allows the distribution grid to split into a controlled set of independently survivable islands. It also creates an interesting set of issues associated with who manages the island and whether or not it could be extended across utility boundaries.
5. *Recovery island:* This is basically the flip side of the controlled separation island focused on stitching the system back together by creating islands of supply and demand, thereby moving toward a faster path to recovery.
6. *Island-like feeder (or dependent muni/coop):* The island-like feeder exists in that perfect spot between the customer and transmission and represents real economic potential for utilities through smaller-scale feeder optimization delivering smart grid benefits.
7. *Real island:* This is the other end of the microgrid continuum in which the microgrid is forced to be formed due to either physical (island) or geopolitical boundaries and has similar requirements of the classic separable microgrid.

The microgrid of today, as it is being designed and as it appears to be visualized by the Galvin Electricity Initiative, is more focused as a niche application of the distribution grid for specific outcomes. This is somewhat limiting.

It can be expected that the microgrid will evolve to a state where the entire distribution grid will be an interconnected network of microgrids, each with its own distinct characteristics, and some capable of surviving independent of the grid in a sustained manner over time.

This version of the microgrid, if implemented properly (see Figure 6.1), will allow the utility to design their entire distribution system in such a way that when problems happen, the system will separate along predetermined boundaries, creating a system of microgrids, some of which may be connected to each other and some that may not. Some of these microgrids may have a balanced load/supply and some may not. Those that do will be stable, still providing power to their customers, and those that do not will not supply power to their customers. However, since these are predetermined, the utility can swing into action quickly and restore power faster [3].

Key Drivers for a Microgrid

Improve Power Reliability and Quality

The biggest physical threat to the grid is extreme weather events. The share of power outages caused by extreme weather is on the rise. Storm resilience is the

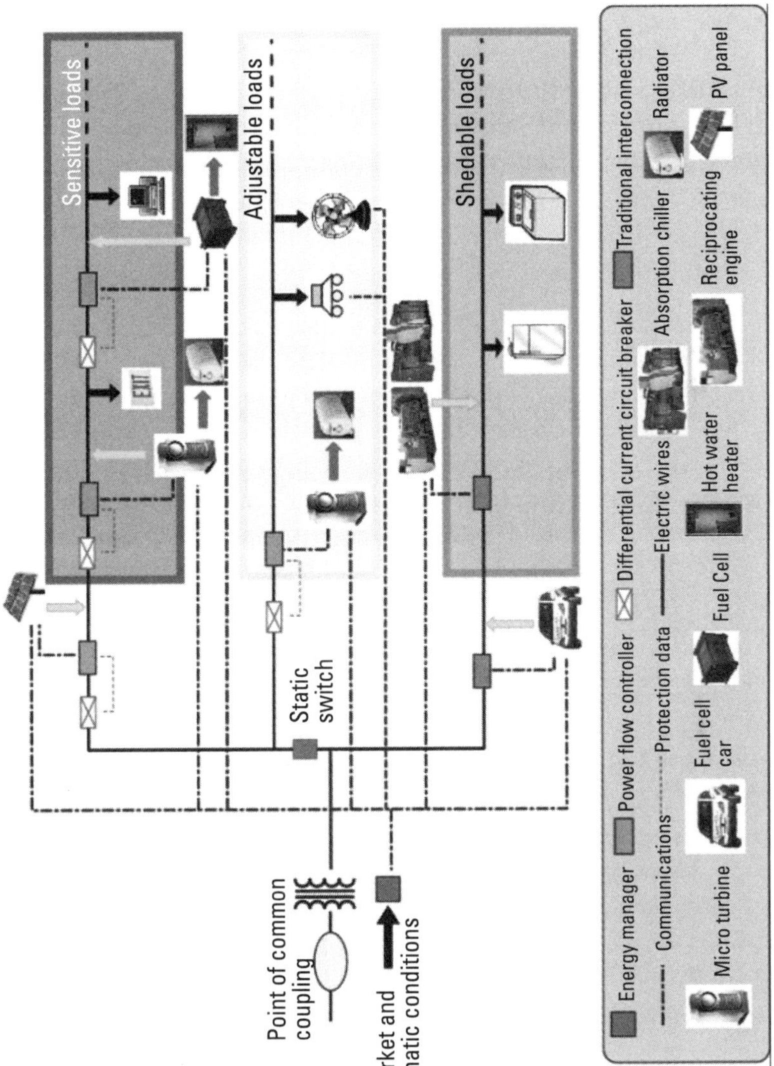

Figure 6.1 Microgrid architecture. (Source: www.lbl.gov [2]. Used with permission.)

biggest driver for microgrids. The ability of microgrids to aggregate and control local power sources and loads in an optimal fashion independent of the grid makes them suitable to ride through weather events. Once the event has passed, microgrids can switch back to grid power seamlessly without disrupting service to customers. The advanced power electronics used in monitoring and controlling microgrids enable transitioning between grid and local resources, keeping the quality of power optimal.

Reduce Need for Infrastructure Upgrades

Production of energy closer to the location of consumption reduces the need for construction of centralized generation, transmission, and, to some extent, distribution lines. It many cases, utilities are looking for ways to optimize assets and get the most value out of existing infrastructure. This freed-up capital has led to incremental investment in smarter grid technologies employing advanced power electronics and operational systems such as microgrids.

Reduce Energy Costs

The demand for lower-cost, locally available energy supplies is especially true for remote sites, such as islands, military bases, and isolated communities. These remote areas typically rely on expensive, distant centralized generation, with power flowing over long congested transmission lines. Having microgrids that avail local generation sources slashes energy costs for these customers and reduces demand charges, which for a commercial customer could be substantially more than the energy charge.

Integrate Renewables

Solar PV and wind turbines have become some of the least expensive options for energy generation, accounting for a significant percentage of new generation in recent years. Microgrids provide an opportunity for communities and businesses to be self-sufficient by utilizing their roof-top PV and small wind turbines in their backyards. Advances in smart inverter technology have further improved the business case for microgrid deployment which can actively balance intermittent renewable supply and demand in real time.

Reduce Carbon Emissions

The U.S. Clean Power Plan requires states to slash carbon emissions, but leaves it to states and utilities to figure out mechanisms for achieving this. Utilities see renewable microgrids as a way to demonstrate their progress. There are also several state and city initiatives with ambitious targets. These include San Diego and Hawaii with 100% renewable energy targets.

Military bases have an especially high incentive to move to microgrids. The drivers include high cost of gasoline (almost $400 per gallon at the front lines) as well as associated high carbon emissions and a large global oil footprint.

Increase Customer Participation

Local generation comprises roof-top solar, small wind turbines, microturbines, fuel cells, gas or diesel cogeneration, and small batteries and are often owned by customers. These prosumers (combination of consumer and a producer) look for ways to hook up to the grid to sell overgeneration from their resources. Microgrids through their ability to dynamically manage generation and load provide a mechanism to safely pool and dispatch these various resources. Customers can also find opportunity to control their energy costs in DR.

Security

As utilities move to cloud-based operational systems and network their smart grid technologies, cybersecurity becomes increasingly critical. Protection of both physical and virtual grid assets becomes paramount. Part of a realistic security plan is to deploy systems that can work off the grid in the event of a cyberattack on grid systems. Microgrids' ability to function in islanded mode and keep power up for critical infrastructure such as hospitals and military bases when the larger grid is down is of great value.

Key Features of a Microgrid

Self-Sustaining Electricity Infrastructure

Typically, one or more conventional generation assets comprise the core of the microgrid. These include diesel or gas generators, and other distributed power systems from renewable or nonrenewable sources, such as solar photovoltaic or fuel cell systems. All of these sources are local and in close proximity to the load they serve, and within the electric boundaries of the microgrid. Containing all required resources within its discrete boundaries makes the microgrid self-sufficient when it disconnects from the grid.

Intelligent Distribution System

One of the distinguishing features of a microgrid is the ability to disconnect from the utility grid to provide autonomous power in response to demand needs or external events, such as power outages or other emergencies. Having a single point of connection to the larger utility grid, microgrids' advanced power electronics in its point of common coupling (the point on electrical system where multiple customers or loads can be connected) can sense loss of power on the larger grid and seamlessly switch to local resources.

Redundant Distribution

Microgrids can operate more efficiently than the larger grid by stacking energy services. Microgrids can deploy solar and storage to provide power to the grid, storage in case of emergency, and grid services such as voltage regulation based on its power production. They can switch to alternate sources of power rapidly and seamlessly without interrupting or degrading power quality.

Self-Healing Distribution

The microgrid senses loads and fault conditions and can reroute power to as many critical areas as possible given any situation. Sensors in the field can rapidly detect, respond, and communicate with the microgrid controller. In that sense, it is self-healing. A number of microgrids operating in islanded mode in the event of a larger grid outage can have the overall effect of healing the grid while working, albeit functioning independently. Microgrids can also take the pressure off the grid while power is restored, by slowly transitioning load back to the grid rather than in a manner that causes demand to peak.

DR Capability

One of the defining features of a microgrid is demand management. DR in its various flavors (e.g., direct load control, time of use, real-time pricing) can engage customers to increase or decrease energy consumption based on grid conditions. This enables demand to be a lever in the supply-demand balance equation, making the microgrid more resilient to the larger grid conditions.

Sustainable Energy Systems

Green buildings and complexes can leverage lower carbon generation sources such as solar, wind, and biogas to reduce their carbon footprint and be more self-sufficient. Inclusion of combined heat and power systems further increases the energy efficiency of these systems by recycling the thermal energy for heating and/or cooling purposes, thereby making the microgrid more economic to operate.

Technology-Ready Infrastructure

Designing microgrids with the latest power electronics devices such as smart inverters, intelligent electronic devices, and electric vehicle chargers enables them to incorporate a greater variety of resources both on the supply and demand sides. Smart inverters provide for greater integration of renewables without degrading power quality. Electric vehicle charging stations and vehicle-to-grid technology enable greater penetration of electric vehicles, which can also provide energy to the grid during peak demand.

Microgrid-Enabling Technologies

Figure 6.2 illustrates the key technologies that enable a microgrid.

Distributed Generation

Distributed generation (DG) is the set of small-scale power generation technologies located close to the load being served, improving reliability, reducing emissions and expanding energy options. DG characteristics include:

- Sources are small and modular compared to central power plants;
- Located closer to consumer loads;
- Flexible due to their distributed and modular nature;
- Fed mostly by natural gas or renewable energy;
- Quieter and less polluting than large power plants;
- Can be simple or complex.

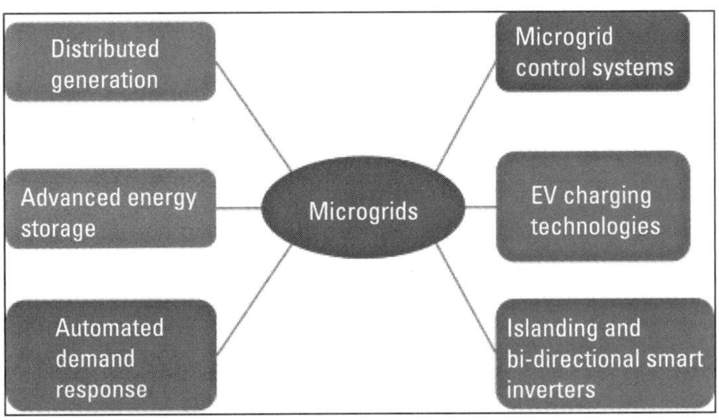

Figure 6.2 Microgrid-enabling technologies. (© Modern Grid Solutions. Used with permission.)

DG could be a stand-alone backup generator that is owned by the consumer. Complex DG may be grid-tied unit consisting of generation, storage, and an energy management system that is either owned by a utility or a third party. It is important to note that not all types of DG are renewable in nature [4].

DGs provide microgrid operators with the opportunity to install generation closer to load, within the same network. These generators allow the microgrid operator and/or the constituents greater say and control in the source of their generation. The location, the type of microgrid, and the operator determine the specifics and characteristics of the DG sources that are implemented.

Electric Energy Storage

Electric energy storage is a set of technologies capable of storing previously generated electric energy and releasing that energy at a later time. It uses forms of energy such as chemical, kinetic, thermal, or potential energy to store energy that will later be converted to electricity. There are various technologies in different states of development and used for various purposes depending on their energy and power characteristics.

Energy storage has various applications. It can be used to improve power quality. It can be used as bridging power when switching from one source of generation to another, such as from wind generation to natural gas-fired generation. Storage can also be used to decouple the timing of generation and consumption. Storage can be charged during times of low-energy cost and low utilization as in when there is plenty of wind generation and used to supply power during peak load periods.

Energy storage allows microgrid operators to take advantage of power output and generation variability from renewable sources by storing it when in excess and releasing it when needed. Depending on the nature of the load profile, the design may include one or more types of storage installed within the confines of the microgrid.

DR

DR as defined by the U.S. Department of Energy is: "Changes in electric usage by end-use customers from their normal consumption patterns in response to changes in the price of electricity over time, or incentive payments designed to induce lower electricity use at times of high wholesale market prices or when system reliability is jeopardized."

Energy balance within a microgrid is critical especially when it is disconnected from the grid. When a microgrid is disconnected from the grid, in essence, it becomes a balancing authority on its own requiring the instantaneous balancing of load and generation. Most microgrids include a certain amount of critical load and other noncritical load. As a result, it is almost always important to ensure that power supply to critical loads be maintained under all conditions. So microgrids with an installation of distributed renewables require a certain amount of controllable load that may need to be dropped to ensure stability of the load-generation balance.

DR is a key enabler of microgrids because it enables the operator to either apply preselected load reduction or other control strategies automatically, without the need for manual intervention.

Electric Vehicle Charging Technologies

Electric vehicle (EV) charging technologies are another key enabler of microgrids. An electric vehicle is a vehicle where the propulsion system contains one or more electric motors that contribute, partly or entirely, toward providing the motive force to drive the vehicle. Electrification of transportation is the use of hybrid electric and all-electric vehicles instead of all-petroleum vehicles. In addition to the propulsion mechanism, a key component of the EV is the battery. The battery is charged from the grid and the discharged energy is used by the propulsion mechanism. EVs, whether powered by batteries, fuel cells, or gasoline hybrids, also have within them the energy source and power electronics capable of producing electricity that powers homes and offices.

When connections are added to allow electricity to flow from car to grid, it is called vehicle to grid power, or V2G. Cars pack a lot of power. A typical EV can put out over 10 kW, enough to power several houses. The key to realizing economic value from V2G are grid-integrated controls to dispatch according to power system needs. With V2G, electricity flows back to the grid from EVs thereby acting as an energy source. V2G is seen as a technology with great potential to assist microgrids.

The importance of EVs within a microgrid is somewhat similar to that of a typical storage device. When energy is available in excess, then it can either be sent back to the main grid or one or more of the EVs in the microgrid can take advantage of the excess energy and store it in the batteries. Similarly, when the microgrid is deficient in energy, V2G technology can be used to siphon some of the power off and be used to power some of the more mission-critical needs.

Microgrid Control Systems

The power generated by DERs such as DG, energy storage, DR, and EVs can be useful power only when they can be controlled to serve the diverse needs of the microgrid.

DER technologies generate either AC or DC. Thus, they need advanced power electronics to convert the power into useful power that can be directly interconnected with the utility grid for use in consumer applications.

DG systems such as wind, microturbine, or internal combustion engine, generate AC output, often with variable frequencies. An AC-DC converter is needed in this case. DG systems such as, PV and fuel cells, generate DC output. A DC-DC converter is typically needed here to change the DC voltage level. A DC-AC inverter is then needed to convert a DC source to grid-compatible AC power. Bidirectional smart inverters are needed to convert power generated from DERs to serve microgrid load or connect to the larger grid. A microgrid control system is needed to monitor and control the various smart inverters and flows to or from the grid [5].

In addition to converting all the loads and generation to a common base of either DC or AC and same voltage and frequency, microgrid controllers also perform an extremely important role of driving a broad variety of mission-critical functions. These mission critical functions could be one of many: (1) stabilizing the microgrid from a power system perspective (maintaining a steady voltage and/or frequency across the networks), (2) manage the interconnection with the main grid (if appropriate), (3) monitor the instantaneous power generation load balance within the microgrid and use it to make control decisions on loads to curtail, generation to pick up, and others, and (4) if appropriate using a financial basis such as price signals to make these decisions. Some of the newer controllers are also designing the concept of artificial inertia being superimposed on loads and generators to make them behave as regular loads and generators in a normal electric grid.

Islanding and Bidirectional Smart Inverters

As utility grids see increasingly more DERs, there is a greater need for dynamic control. The intermittency of these DERs is very difficult for a utility to deal with. This is where the inverter comes in. For utilities and grid operators, the capabilities of the inverter are vital to reliably integrating solar and other renewables onto the grid. Current inverter performances fall short and disconnect at first sign of grid disturbance. However, advancements in inverter technologies have given rise to a new type: the smart inverter, shown in Figure 6.3.

Smart inverters' advanced features enable reactive power control, voltage, frequency ride-through, advanced two-way control capabilities, storage integration, and data streaming. Bidirectional smart inverters are needed to convert power generated from DERs to serve microgrid load or connect to larger grid.

As identified above, microgrids also need inverters to ensure that all the components are working harmoniously with respect to voltage, frequency, DC, and/or AC. In addition to the functions identified here, smart inverters also perform additional roles along the interface to the rest of the grid by providing better control capabilities and also in providing VAR support and so on.

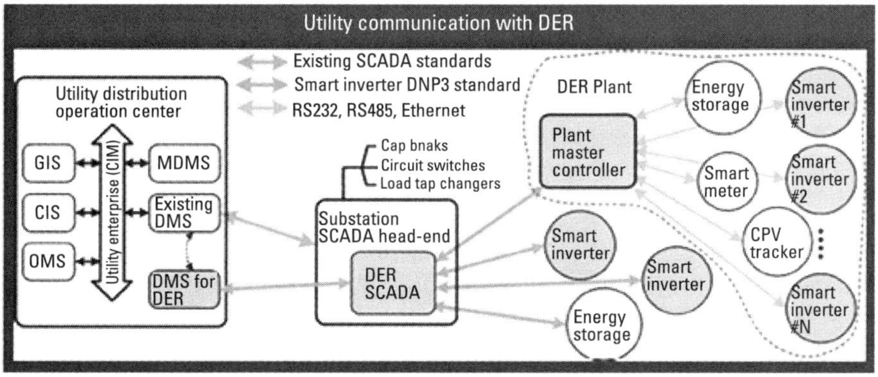

Figure 6.3 DERs and smart inverters. (Source: www.energy.gov [6].)

The Next Generation Microgrid

The next generation microgrid can be identified through several key characteristics:

- *Accommodate different sources of energy:* The microgrid needs to have the control capability to handle different types of energy supply all the way from traditional fossil-based to distributed and/or from renewable sources. It also needs to be able to handle energy sources further up the spark spread[1] like direct gas influx to a load or gas-fuel-cell to electricity.
- *Can be self-sufficient in terms of energy balance, at least for short periods of time, and possibly on a continual basis:* The control mechanism within the microgrid should be able to handle balancing the supply-demand equation within itself or as a part of a larger grid on a continual basis.
- *Has advanced self-healing capabilities:* The self-healing capabilities associated with the microgrid include features like the ability to decouple itself from the main grid automatically under certain system conditions, the ability to reconfigure and reroute power through different feeders upon incidence of a faulted condition, the ability to drop one or more loads depending upon their criticality at that time, and so on.
- *Uses demand and supply in a similar fashion and can automatically optimize between them:* Somewhat related to the concepts of self-healing, the microgrid also has the ability to optimize between demand and supply

1. The spark spread is the theoretical gross margin of a gas-fired power plant from selling a unit of electricity, having bought the fuel required to produce this unit of electricity. (*Source:* https://en.wikipedia.org/wiki/Spark_spread.)

so both are balanced in real time. This would be in stark contrast to today's grid where demand generally drives the need for supply.

- *Must have the ability to respond to market prices and market opportunities:* The lessons learned from efforts in New York and California have led the industry to believe that retail level markets are coming. Any microgrid design needs to have the ability to interact with these markets.

For these characteristics to materialize, the design and implementation of the distribution grid may need to change in a dramatic manner. For the change to be institutionalized, the following key requirements need to be in place:

- Adequate sensing and controls need to be installed into the grid, both at the micro level and the macro level. The sensors need to be placed in such a way that one can get full visibility into all aspects of the grid. The controls need to be such that between automatic and remote controls, grid operators can perform sectionalizing the gird, rerouting of power from source to load, open and close breakers either automatically or remotely, voltage/VAR optimization, demand side management, and so on.
- Adequate optimal control algorithms and systems need to be available to support the operation of the system, either as multiple independent microgrids or one or more larger interconnected distribution systems. This is critical to operating in a stable manner in several predetermined or dynamic conditions.
- Adequate focus on the model will start becoming critical as we move to multiple microgrids. The model should be in a location easily accessible by most software systems so they are all working off a single version of the truth. The models should also be complete enough so each of the systems has the requisite amount of detail they would need.
- A combination of centralized and decentralized control systems that will allow each microgrid to be managed and controlled as an independent grid and also be able to coordinate when they are all working together as one large interconnected grid.
- Appropriate software-based planning and design applications to assist in the design and implementation of these microgrids to either function independently or together.
- Appropriate software-based operational applications to assist in ongoing real-time operations of these microgrids to function either independently or together.

- Adequate visualization of the state of the grid. Managing a grid of this kind can be an extremely complex problem because of the tremendously large number of moving parts, all of which could react in real time. To support this, the distribution system operator needs to have full access to the data and algorithms, as well as the visualization to understand the state of the system at any time, the state of the various microgrids as well as the options available to move from one state to the other.

Table 6.2 presents a comparison of the pros and cons of the next generation microgrid.

Analyzing Storm Restoration Under a Next Generation Microgrid Scenario

One of the bigger problems in the distribution grid is outage management and emergency restoration. An outage can be anything from a sustained lack of electric power to temporary loss of power and even to flickering of power (which is more of a lack of power quality) [3]. When any of these happen, it disrupts life. When an outage happens, the utility is not able to deliver on its mandate.

Outages happen due to a variety of reasons, some of which are listed below:

- A line gets overloaded and either burns down due to overheating or the overload leads to a relay or circuit breaker tripping which in turns leads to one or more circuits being disconnected from supply.
- A line gets overloaded and the overheating leads to the line sagging so much that it may touch a tree or something else, thereby short-circuiting it to the ground leading to relays tripping and disconnecting parts of the system from the supply.

Table 6.2
Comparison of the Pros and Cons of the Next Generation Microgrid

Pros	Cons
Infinite amounts of flexibility by functioning either separately as an independent microgrid or together as a singular distribution grid.	Today's microgrid is not flexible. It is designed to be more of a custom implementation and is in place for special circumstances.
The reliability outcome should be higher, which means unless an outage directly impacts the customer, the actual loss of power should be temporary.	The costs will be higher given the sophisticated level of controls and sensors that are required to keep it running properly.
Can accommodate different combinations of generation (renewable and otherwise) in a very flexible manner.	Somewhat limited capability to support multiple combinations of generation.
Combination of localized and/or centralized control of the grid.	Most often, the microgrid is controlled locally and the rest of the distribution grid is controlled centrally.

- A storm blows over an area and trees or tree limbs fall on electric lines, thereby severing them, resulting in disconnecting from supply.
- Others, such as equipment failures, car accidents knocking down poles, animals getting into equipment and causing short circuits.

This is only an indicative list. Such incidents can cause outages ranging from one home to those extending to well beyond a single utility's jurisdiction.

One key takeaway in regard to outages in today's grid is that they happen where the disaster strikes, but, more importantly, they extend to locations beyond until the first protective device identifies the fault and then opens the circuit. As a result, each disaster results in a faulted area profile that is different from the previous one. This last characteristic makes it difficult for utilities to restore the system back from a major outage. As a result, for each outage, the utility needs to (many times) start from scratch in trying to figure out where the problem is, fix it, and then bring power back for its customers. There are thus an infinite number of scenarios or problems that can arise. This needs to change.

Let us consider a typical storm scenario:

As most hurricanes (category 3 and above) come up the East Coast of the United States, they are accompanied by heavy rains and heavy winds, sometimes reaching to greater than 100 mph. When this kind of a hurricane hits land, the severe winds tend to bring down trees, which can impact subtransmission and distribution lines creating a storm/emergency scenario. This situation happens quite often in the Eastern and Midwestern states and tends to impact several hundreds of thousands or millions of people every year.

Let us analyze the storm/emergency scenario and the typical utility response that accompanies it.

- When trees hit the lines, the lines tend to either break or sag under the weight of the branch, bringing it very close to potential human contact.
- When lines break, they tend to impact all downstream loads, thereby creating localized or widespread outages.

  This situation happens because much of the distribution network tends to be radial in nature and when something breaks or opens up, all downstream loads have no access to power.

  This also happens because much of the power is provided from large centrally based generation and distributed generation is still rare. (Even with DG, it is often not designed to operate in an island model.)

- When they do not break, a live line may be sagging under the weight of a tree branch, causing a dangerous situation that will need to be reported to the utilities as well as fire/safety officials.
- Depending upon the level of sensing or controls available, the distribution control center operator will be aware of some of the outages through their SCADA system. However, unless there is an automated metering infrastructure (AMI) system fully integrated into the outage management systems (OMS), the operator will generally have little understanding of the extent of the outages unless customers start calling to inform of their loss of power.
- As they learn about the full extent of the outage, the utility starts prioritizing the repairs and restoration. Every utility has its own prioritization sequence and it is generally based on criticality, number of customers lost, main line first before lateral lines, and so on.
- As equipment needs to de-energized and/or reenergized, the control center operator will need to develop switching orders to assist the field crew in isolating the equipment for reconnecting to the grid. This is critical because if too much load is picked up, the lines may trip again due to an overloaded situation. Repairs are made as needed and sometimes equipment may need to be replaced due to major damage. The reconnecting process is quite a delicate operation because of the lack of sensing and understanding of the exact amount of load that will be picked up when the lines are reconnected.
- The system is slowly stitched back together and the system comes back up and running. The new state of the network may not be a long-term situation. The system is (over time) brought back to the normal state where it can function on an ongoing basis.

While the sequence of steps may be different from place to place, it is important to notice the overall general characteristics driving how a typical utility responds to a storm and how it recovers from it.

Now let us analyze how this scenario should unfold in a next generation microgrid.

- When the system gets impacted, the breakdown of the system should happen along the microgrid boundaries.
- Each microgrid will be responsible for maintaining its own supply/demand balance leading to the potential need for drawing down on load until each microgrid is capable of maintaining its own load.

- When this happens, the number of people who will lose power for extended periods of time will be smaller because each stable microgrid will have sources of supply within itself and not be totally dependent on the centralized supply, as is the case today.
- Fixing the system will be done in parallel on a microgrid-by-microgrid basis and, hence, can be done much faster. Also given that the microgrid operator will have local sources of generation, it is anticipated that restoration should be faster and some may have never lost power at all.
- As each microgrid is up and running, the system operator will slowly get the microgrids connected to each other thereby focusing on bringing the entire system up and running.

Table 6.3 compares the outage response between today's grid and the next generation microgrid.

The next generation microgrid brings several core capabilities not just for storm restoration but also for the normal day-to-day management of the power system and in its ability to provide better power more reliably.

Evolving to the Next Generation Microgrid: A Road Map

The move to the next generation microgrid is an evolution from today's situation to a sophisticated set of sensors managed by advanced control systems capable of working in a centralized/decentralized manner delivering the most

Table 6.3
Comparison of Outage Response Between Today's Grid and the Next Generation Microgrid

Today's Grid	Next Generation Microgrid
The system pretty much breaks down based on where the storm hits the system. The outage pattern will look different every time. Less availability of DG, in fact two-way power can be a problem. Restoring the grid is a different every time and can make it difficult for a utility to develop specific process steps to recover. When external crews are brought in to help in storm restoration, it becomes difficult because of the broad inconsistencies in how utilities are designed and what needs to be done to complete the restoration.	The system will break down along microgrid boundaries, thereby expecting each individual microgrid to take on the responsibility to repair itself and bring its customers up and running in the most optimum way and be ready to get reconnected to the rest of the grid. Heavy reliance on DG is critical for the reliable operation of an independent microgrid. Restoring the grid can become more predictable since it is a matter for each microgrid to come up on its own and the system operator can then stitch the system back. It is easier to bring in external crews in a next generation microgrid because of a greater level of consistency in how these independent networks are interconnected.

reliable power possible at the lowest cost and at the same time focusing on taking maximum advantage of localized/distributed renewable sources of supply.

The phases, as illustrated in Figure 6.4, will evolve as follows:

1. *Formation of natural microgrids:* The first phase will be today's distribution clumps, which have natural boundaries. College/university campuses, business parks, company campuses (e.g., Google campus in Mountain View, California) remote mining/industrial towns, and military bases are all good examples. Through the auspices of efforts like the Galvin Initiative, college campuses have already begun to look at mechanisms to become less dependent on the local electric utility. Over time, we believe that this will also include (on a global stage) the plethora of apartment/condo complexes. These entities are already setting themselves up to work independently if and when the local utility is not able to deliver power reliably.
2. *Evolution of localized mechanisms:* In the second phase, these natural microgrids will start investing in newer and more sophisticated sensors, controls, and smart grid technologies like DG, storage, and DR. It is a combination of these mechanisms supported by a full-featured control capability that will allow the microgrid to sustain its energy needs in the short term and in a mission-critical mode, even for extended periods of time.

   The military, for many of its bases, is thinking of something like this in both onshore and remote bases.

   During this phase, much progress will be made in DG (possibly more of PV-based) and storage. DR will become the mechanism to drive sustainability if and when the microgrid gets disconnected from the main utility, or even to manage their costs through a targeted management of the demand charge.

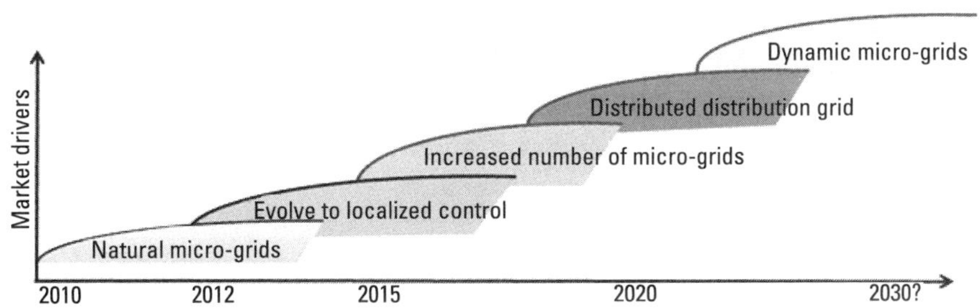

Figure 6.4 Road map to the next generation microgrid. (© Modern Grid Solutions. Used with permission.)

3. *Formation of more microgrids:* As the success of phase 2 becomes more popular and known, newer microgrids will begin to come into play. In this phase, we will start seeing smaller cities and housing subdevelopments moving in this direction. This could also be considered as the second phase of municipalization (at least in the United States) in which increasingly more of the smaller cities will not completely secede from the local utility but look for more independence and better control of their costs.

   Depending upon their constituency and location, they will also spend more money and effort on getting local sources of energy (PV, wind, and so on), storage (community energy storage or larger storage devices at scale), or DR.

4. *The movement to a distributed distribution grid:* As increasingly more locations move to a microgrid, it is easy to imagine a distribution grid consisting mostly of a group of microgrids that function primarily as a loose-knit configuration of a single interconnected grid including both transmission and distribution. However, upon need (or in a storm, emergency, or restoration scenario), it has the ability to split into a set of predefined microgrids fully capable of functioning independent of each other at least for some period of time, possibly in a reduced capability mode.

   Such a configuration will need a complex control system mechanism with a centralized distribution management system (DMS) or DMS-like system, which is capable of either functioning in a single DG mode or being able to function in a distributed grid/control mechanism either managed from a central location or from a cloud-like scenario with each disconnected microgrid controlling its own reliability operation through a common set of tools.

   There is a certain level of sophistication that needs to be in place before this type of a system can be operational, from sensors to controls to visualization tools to an overall control center mechanism capable of functioning either in a centralized manner or in a distributed manner.

5. *The movement to dynamic microgrids:* Until now, most microgrids that were discussed in this chapter were all of the static kind, meaning that their configuration was predefined.

   The next level of sophistication and on the evolutionary path is the need to create these microgrids in a dynamic manner. Moving to dynamic microgrids would result in creating microgrids on the fly and could even be based on finding the best portion of the network with a supply-demand balance and capable of staying together for a longer period of time.

To bring it all together, as we look at this evolutionary path to achieving the next generation microgrid, it is easy to see this path and the challenges that will be faced and also the benefits that will come along, such as greater reliability, greater energy independence, more control at a local level, a more beneficial impact on the environment, lower long-term cost stream, and a smaller carbon footprint.

Challenges of a Microgrid

Technical Challenges in Islanded Operation

Voltage and frequency control is one of the challenges with microgrid operation in islanded mode. The purpose of voltage and frequency control is to ensure both voltage and frequency remain within predefined limits around the set point values by adjusting active and reactive power generated or consumed. In the operation of the microgrid, a challenging task is to operate multiple sources of DG on the island, where it is not always possible with active and reactive power control. Fast and accurate voltage and frequency control are fundamental requirements for successful operation of an islanded microgrid. This is enabled by controllable loads and DG sources such as PV, fuel cells and microturbines that can participate in voltage and frequency control via their integration through smart inverters [7].

Another challenge facing microgrids is protection. Once a microgrid disconnects from the larger grid, it is important to ensure the loads, lines and DG on the island are adequately protected. The bidirectional flow of power due to DG sources is the main concern. It requires the use of directional overcurrent relays in protection schemes. Further, these protection schemes need to be able to support both grid connected and autonomous operation [8], preferably seamlessly.

Need for Advanced Power Electronic Interfaces

The microgrid comprises of a number of active resources such as DG, energy storage, DR, and EVs. These resources change the distribution grid from passive to active. Each of these resources tends to exhibit different operational characteristics and independently generating and consuming energy based on their separate profiles. As a result, greater coordination is required via the use of advanced power electronic interfaces. This is achieved through the use of advanced power electronic devices such as smart inverters, as illustrated in Figure 6.5, controlled by smart controllers.

At a fundamental level, the inverter converts DC input to AC output, which, for a grid-tied system, enables the supply of real power to the grid. Other standard functions include power transfer optimization, voltage conversion, grid synchronization, disconnection, anti-islanding protection, and storage interfacing. These capabilities ensure that power is efficiently supplied to the grid while also providing key safety features. Functions beyond this scope are what qualify an inverter as smart. These include reactive power control, voltage, frequency ride-through, advanced two-way control capabilities, storage integration, and data streaming. Bidirectional smart inverters are needed to convert power generated from DERs to serve microgrid load or connect to a larger grid.

To be useful in interconnecting microgrid resources, these power electronic inverters need to be small, consume less power, and even be easier to manage, control, and integrate with other similar devices.

Availability of Renewable Energy Resources

Many of today's microgrids are located in remote areas and they are generally difficult to connect to the distribution grid. Hence, some parts of the generation come from generators that locally burn fossil fuels such as diesel or natural gas. Other parts of the generation also come from natural local resources such as wind, sun, and water. Getting power from renewable resources are also a major driver for microgrids not located in remote locations and are very often due to the need of the operators and/or owners to be greener and to have a lower carbon footprint.

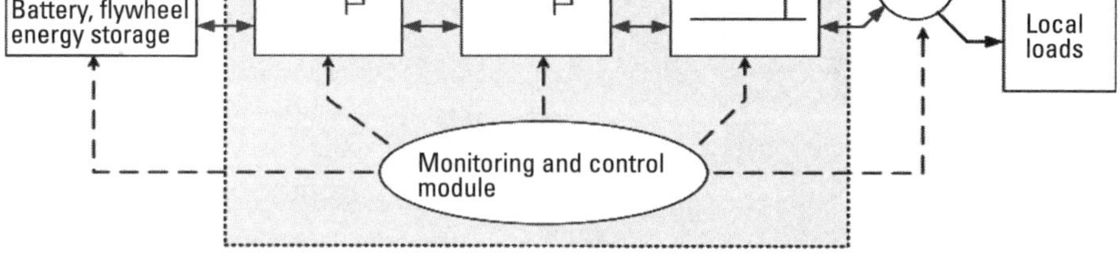

Figure 6.5 Microgrid control system. (Source: https://www.nrel.gov/docs/fy08osti/42672.pdf. Used with permission.)

Microgrids offer a mechanism that enable integration of these distributed renewable sources. They can employ technologies that can integrate renewable energy into generating systems that relied wholly on fossil fuels. Hybrid systems can provide power to these isolated areas, making them less dependent on fluctuating fossil fuel prices.

The increased levels of dependence on intermittent renewables poses its own technical challenges with voltage stability, and requires methods that can bridge these gaps all the time while allowing the stakeholders to enjoy a normal life.

Scalability

Over time, demand for microgrids is increasing and their numbers are growing at a rapid pace both in North America and the world. In addition, their scale is also increasing. This increase in scale requires tapping into additional renewables sources or local generation. It could also include nonconventional sources of energy such as EVs and deployment of distribution energy storage.

From a microgrid perspective, this requires the development of better inverters and controllers capable of managing larger capacities of resources both from a consumption and production perspective. As the load and generation size increases, the controllers also need to become more sophisticated to be able to handle these grids with little to no inertia.

Advanced Automation for Monitoring and Control

Microgrids have a variety of generation sources; some conventional and some nonconventional (or renewable-based). For the microgrids to work effectively, in addition to the inverters and controllers that have been discussed earlier, they also require advanced monitoring and control mechanisms.

The goal for intelligent monitoring and control of the microgrid is to have real and reactive power control while regulating line voltage and frequency. In a small system with little to no inertia, this is not easy. A master controller is deployed that uses SCADA to monitor and regulate frequency and voltage in microgrids. A combination of hierarchical control performed by master controller, intelligent switching, and energy storage implementation helps to maximize the use of renewable resources and improve voltage stability and grid reliability [9]. Further, the single point of common coupling should enable it to connect or disconnect from the grid without adversely affecting power quality and reliability on the microgrid as well as the larger grid.

Dependence on Energy Storage

A major driver for the newer microgrids is to move generation from conventional to renewable sources to meet demand. Generation from renewable sources can vary sometimes dramatically over short periods of time.

Energy storage is increasingly seen as the mechanism to smooth out the generation profile from renewable sources. In addition, PV and other renewable DG technologies also require a voltage source to synchronize. Traditionally, this has been accomplished with a backup generator. Energy storage provides a similar voltage source but without the emissions of a diesel generator. Other value additions of energy storage include peak shaving, smoothing power flow, and VAR support. Islanding and black-start assistance require further support from storage, especially in the case of renewable DG microgrid systems [10].

Case Studies

Case Study #1: DTE Energy, United States Advanced Implementation of Energy Storage Technologies

DTE Energy is a Michigan-based diversified energy company that provides electric service to 2.1 million residential, business, and industrial customers in southeast Michigan.

DTE Energy has worked with selected subrecipients, consultants, contractors, and vendors to demonstrate the use and benefits of distributed energy storage, often referred to as community energy storage (CES) (see Figure 6.6), in a utility territory and to test the ability to integrate secondary-use EV batteries in the CES demonstration.

This is the first large-scale utility community energy storage project with an aggregated capacity of 1 MW. Its 21 energy storage systems were managed by a distributed energy resource management system (DERMS). This DERMS was created to allow aggregation of any asset within the DTE energy service territory using a utility industry protocol (DNP3).

This project installed 18 S&C Electric (S&C) supplied 25-kW/50-kWh CES units, a 500-kW Li-ion battery storage device integrated with a 500-kW solar system and two repurposed (secondary use) energy storage systems using Fiat Chrysler Automobile (FCA) 500e EV batteries.

CES entails utility deployment of modular, distributed energy storage systems (DESS) located on the utility distribution system close to residential and business end customers. The concept of CES is that a DESS can behave as a large battery system with additional reliability benefits to the end-use customer by creating a small microgrid for the customer downstream from the CES. This local microgrid continues to serve customer load, in case of loss of power on the larger grid.

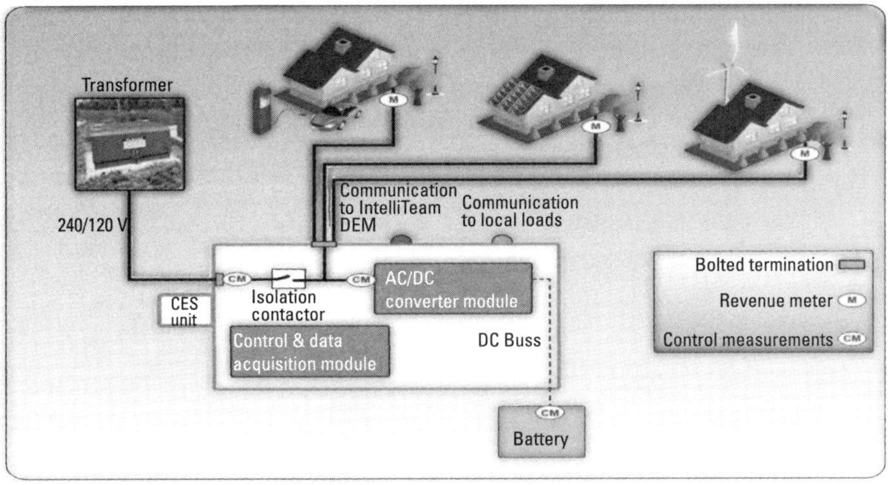

Figure 6.6 Community energy storage (CES) and microgrid. (Source: https://www.smartgrid.gov/files/OE0000229_DTE_FinalRep_2016_03_16.pdf.)

The project objectives are to integrate the CES units into the electric utility system and determine the performance of the CES and the control system and the development and integration of CES devices from secondary-use battery [11].

Case Study #2: Nice Grid, France—Smart Solar District

Nice Grid is a pilot project of a smart solar district in Carros, France. The ambition of Nice Grid is to study all aspects of the smart grid concept, the smooth integration of DER into the local LV grid: renewable generation with rooftop solar, electricity storage, and load curtailment with smart home equipment.

The pilot involves 1,500 customers, 200 solar rooftops, and 100 batteries equivalent to 2 MWh of storage capacity. The duration of the project is 4 years with a budget of 30 million euros.

Its main objectives are as follows:

- Optimize the operation of an MV/LV electrical network with a major integration of solar power generation and electricity storage capacities;
- Test the islanding of a microgrid based on solar power generation and electricity storage;
- Guarantee continuity of supply and reduce potential constraints on the high-voltage transmission lines of the area;
- Encourage the customer to be proactive in managing his or her production, consumption, and storage of electricity.

The expected outcomes of the project are (see Figure 6.7):

- Manage solar power generation on the LV grid;
- Manage efficiently the local load-shedding capacities;
- Provide reliable electricity production and consumption forecasts;
- Study the prosumers acceptance of DR programs;
- Succeed in storage and solar-based islanding.

How Microgrids Are Affecting Utility Transformation

Microgrids, as the definition goes, are a group of interconnected loads and generation sources within a clearly defined electrical boundary that has the potential to act as a single controllable entity with respect to the grid. From a business perspective, this means that microgrids have created an opportunity for a customer or a group of customers to get their own sources of energy supply and either reduce or eliminate their dependence on the grid and the electric utility.

Microgrids represent, for the first time, an opportunity for a utility's customers to break away from the utility. As a result, the microgrid's impacts to the utility's business construct may have far greater implications than the technical challenges that have been presented. This means that as microgrids become more popular, more customers may have an opportunity to leave the utility and define their own energy future. If customers leave the utility, it creates a situation for the utility's assets to become stranded as it delivers less power to fewer customers on the same infrastructure cost.

However, this also represents an opportunity for the utility to take advantage of the vast array of technologies available to it in addition to access to cheaper and centralized resources and an infrastructure that is designed to deliver reliable power to its customers. This means that the utility also needs to transform itself to respond to these changes. Examples of transformation options available to the utility include:

- Making special accommodations for customers who require greater reliability and are willing to pay more for it. This could include mechanisms such as managing the microgrid for the customer, creating a situation in which the specific needs of the customer such as mix of renewable-based generation and/or local generation can be met.
- Redesigning the distribution grid from the bottom up as a combination of microgrids as explained under the definition of the next generation microgrid, thereby delivering to both the specific needs of their customer and, at the same time, delivering a more reliable and resilient grid.

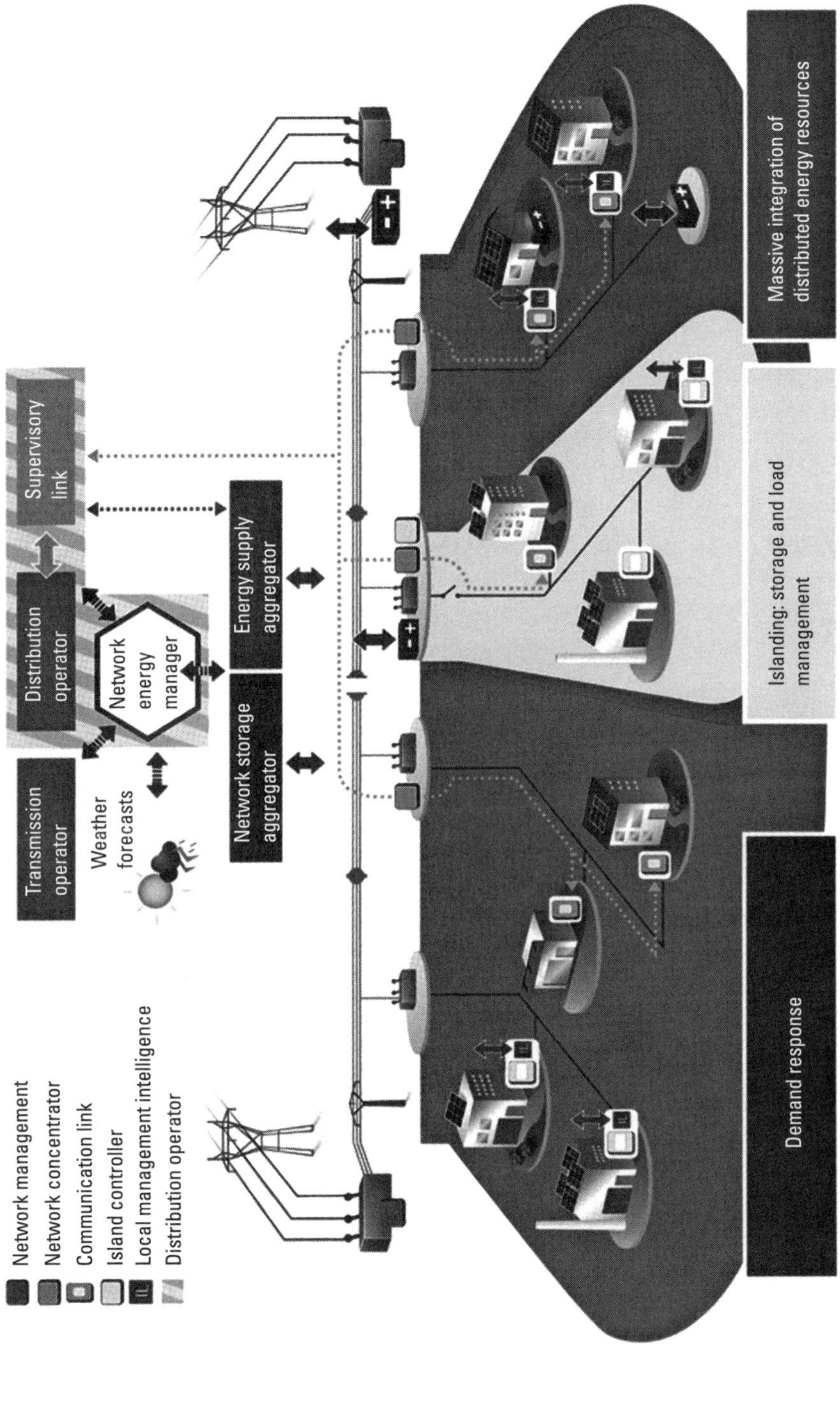

Figure 6.7 Nice Grid demo. (Source: http://www.grid4eu.eu/project-demonstrators/demonstrators/demo-6.aspx.)

While the technological challenges are significant from a microgrid perspective, the business challenges to the utility can be even more important. A group of customers within a defined electrical boundary form a microgrid, and if they are still a customer of the utility, every aspect within a utility will need to change, all the way from asset management and customer service to grid operations and field services.

Conclusions

Microgrids until now have been relegated to a niche status because of their normal use being in remote industrial locations and moving toward use in college campuses and so on. While this move is happening slowly and steadily, microgrids are positioned to move from this niche status into more of a mainstream status due to their unique ability to look beyond microgrids by themselves as they exist right now. However, it is believed conditions are evolving to a new state in which microgrids (or a new manifestation of what exists today) will move from today's niche status to become the underlying backbone of the electric grid of the future. This move will completely transform today's distribution grid in almost all aspects.

Microgrids present a real opportunity. The major reason for its existence is so that all or much of the control is local, thereby potentially increasing the chance for more reliable power delivery to the end user. This may or may not be more economic; everything known so far has confirmed that centralized sources of power are much cheaper. However, if one takes the conversion process of coal into energy and the losses from transmission and distribution, the overall efficiency factor from source to consumption is about 33%. This means that the barrier to entry is not that high for other sources of energy if their price or performance is brought up a little more. However, right now, much of the justification for microgrids is based on reliability, which means that for more reliable power, people are willing to pay more. This is even more prevalent in second-world and third-world countries where the power supply is not that reliable—virtually every apartment complex, hotel, and commercial complex has their own localized control mechanism to keep critical loads operational upon loss of utility-side supply, which at times can happen several times a day.

As the technologies such as storage and forms of distributed renewables become cheaper, it is easier to see more and more sources of supply at the last mile of the distribution system. This means that there is a possibility, at least in the case of a storm or emergency scenario, for parts of the grid to be split away from the main grid and stay capable of operating in a self-sufficient manner for extended periods of time at least until the main grid can be brought back into a viable state. We call this the dynamic microgrid. In this situation, the grid, under normal circumstance, functions as it does today, as a single monolithic entity but with sources of supply at both ends. However, during emergency situations, it has the ability to split itself into multiple microgrids depending upon where the source of supply is and continue operation in an independent manner until everything is back to normal.

Instead of going from microgrids to distribution grids, one should look at distribution grids as a large network of microgrids. These microgrids could be defined along predefined lines or dynamically configured based on system conditions. Changing the paradigm from a bottom-up mode into a top-down mode allows one to get much more expansive into how we view the problem, moving away from niche solutions to system-wide solutions focusing on overall problems like planning, reliability, economics, and operations.

The advent of the smart grid has brought along with several new technologies including sensors/controls, smart meters, renewable sources of power, and new systems such as the DMS and OMS, all of which have the potential to drive new control mechanisms that can react in real time, bringing about new levels of flexibility and scalability in how we manage the grid.

The introduction of a new concept called the next generation microgrid, which among other things possesses the characteristics of: (1) being able to accommodate different sources of energy, (2) being self-sufficient in terms of energy balance, at least for short periods of time, (3) having several advanced self-healing capabilities, and (4) using demand and supply in a similar fashion and can automatically optimize between them. We have identified and named this as the next generation microgrid. With the next generation microgrid, the entire distribution grid can be considered as if it were made up of a whole set of microgrids, which could be configured in a variety of ways to resolve any system reliability problems that could be afflict the distribution grid. We have also created a road map that can have the potential to move us from today's situation (niche microgrids) through static (predetermined) microgrids all the way to dynamic microgrids.

Specifically, the self-healing capabilities associated with the next generation microgrids include features like the ability to decouple itself from the main grid automatically under certain system conditions, the ability to reconfigure and reroute power through different feeders upon incidence of a faulted condition, the ability to drop one or more loads depending upon their criticality at that time, and so on.

Given the impacts of Hurricane Sandy in the New York/New Jersey area resulting in loss of power for several thousands of people over extended periods of time, this chapter also compares today's storm restoration mechanisms against the hypothetical future next generation microgrid mechanism and points out the benefits of how it could work better in the future under the microgrid scenario.

For these characteristics to become real, the design and implementation of the distribution grid may need to change in a dramatic manner. For this happen, the grid needs to evolve in this new direction. The chapter has also presented an implementable roadmap for evolving from today's distribution grid to future in which it is made up of multiple microgrids stitched together into the distribution grid along with potential challenges that would need to be faced.

References

[1] Galvin Electricity Initiative, "What Are Smart Microgrids?" 2017, http://www.galvinpower.org/microgrids.

[2] Microgrids at Berkeley Lab, "About Microgrids," 2017, https://building-microgrid.lbl.gov/about-microgrids.

[3] Lasseter, R., "CERTS Microgrid Concept," CERTS, 2017, https://certs.lbl.gov/initiatives/certs-microgrid-concept.

[4] CERTS, "Distributed Energy Resource Microgrids, 2017, https://certs.lbl.gov/research-areas/distributed-energy-resource-0.

[5] Wang, X., et al., "A Review of Power-Based Electronics Microgrids," *International Journal of Power Electronics*, 2012, pre-print, http://vbn.aau.dk/files/170816843/JPE_11076_1.pdf.

[6] Seal, B., et al., "Smart-Grid Ready PV Inverters with Utility Communication," *EPRI*, 2012, https://energy.gov/sites/prod/files/sssummit2012_poster_si_epri_seal%20segis.pdf.

[7] NEMA, "Energy Reliability with Microgrids," 2017, https://www.nema.org/Storm-Disaster-Recovery/Microgrids-and-Energy-Storage/Pages/Energy-Reliability-with-Microgrids.aspx.

[8] Salam, A. A., A. Mohamed, and M. A. Hannan, "Technical Challenges on Microgrids," *ARPN Journal of Engineering and Applied Sciences*, Vol. 3, No. 6, December 6, 2008, https://pdfs.semanticscholar.org/a27f/aa261358df3bd803c14b8e290df448fbd1d7.pdf.

[9] Shahidehpour, M., "Operation and Control Microgrid and Distributed Generation," *e2rg*, 2012, http://e2rg.com/microgrid-2012/IIT_Shahidehpour.pdf.

[10] Wilson, A., "The Growing Role of Energy Storage in Microgrids," *Microgrid Knowledge*, May 25, 2016, https://microgridknowledge.com/energy-storage-in-microgrids/.

[11] Asgeirsson, H., "DTE Energy Advanced Implementation of Energy Storage Technologies: Technology Performance Report," *DTE Energy*, 2015, https://www.smartgrid.gov/files/OE0000229_DTE_FinalRep_2016_03_16.pdf.

7

Data Analytics: Bringing Intelligence to the Grid

Smart grid data analytics is the process of converting the plethora of data collected from meters, sensors, switches, and other devices deployed in the field into actionable intelligence for use by the utility. It provides the utility with insights into performance of the power grid, consumer energy use, peak demand, and business risks.

Introduction

A smart grid optimizes the generation, distribution and consumption of electricity through the extensive implementation of information and communication technology (ICT) on the electricity grid. Examples of these include new systems such as advanced distribution management systems (ADMS), outage management systems (OMS), and distributed energy resource management systems (DERMS); decentralized energy producers such as residential solar photovoltaic cells; electric vehicles; smart home and buildings; and new communicating equipment such as smart meters, sensors, and remote-control points. All this will cause a deluge of data that energy companies will have to face.

Smart grids bring to the surface the cost of storing and processing the huge quantity of data used to manage the grid. Unfortunately, much of the new data collected is not exploited because of the lack of infrastructure and/or data analysis skills to deal with it. The unused data could provide opportunities for utilities and their customers to improve their abilities to control their energy consumption, avoiding peak loads, better integrate with DERs and other benefits.

Big data can be defined as a huge quantity of datasets, but it includes other features. In addition to the volume, big data is based on the variety to present various data formats (structured, semistructured, or unstructured), the frequency to provide timeliness requirements, the value to give the ability to extract the meaning from the collected datasets, the variability that injects inconsistency in the data, and veracity to work on the trustworthiness of the data.

Data analytics is not new to the utility industry. Utilities have always analyzed data to improve the operation of the electric grid and the processes. However, for the longest period (as illustrated in Figure 7.1), there was not much data to analyze. The main sources of data are listed next along with the changes happening to the sources of data, frequency of data gathering, and new sources of data[1] being implemented.

- *Transmission data:* For the longest period, this was data coming through systems such as SCADA being brought into systems (mostly energy management systems [EMS]) [1] at a general rate of once every 2, 4, 6, 8, or 10 seconds (and sometimes even slower than this). The data is used to support actions of transmission and/or regional transmission operators (RTO) but also feeds various analytic packages such as contingency analysis, optimal power flow, and others within the EMS, but also extracted from the EMS for other analysis internal or external to the transmission control center.

  In addition to adding more sensing points, newer technologies have arrived such as phasor measurement units (PMUs) (see Chapter 3), bringing more data into the control center, very often at speeds of 30 to 60 times a second. In general, utilities have not yet decided how to use this data in a broad scale.

- *Distribution data:* Like transmission data, much distribution data used to come through the SCADA system, very often through the same system servicing the transmission system.

  This is one area where much is changing. New sources of data from distribution automation devices are growing rapidly bringing large amounts of data into the utility. Added to this, the emergence of dis-

1. Relay data was not included here on purpose, because, for the most part, it was not used for anything else other than relay action analysis. I recognize that this is also changing with the broader movement toward digital relays and the ability to remotely access the relay data. When this reaches critical mass, it is reasonable to expect that this category of data would also enter mainstream data analytics.

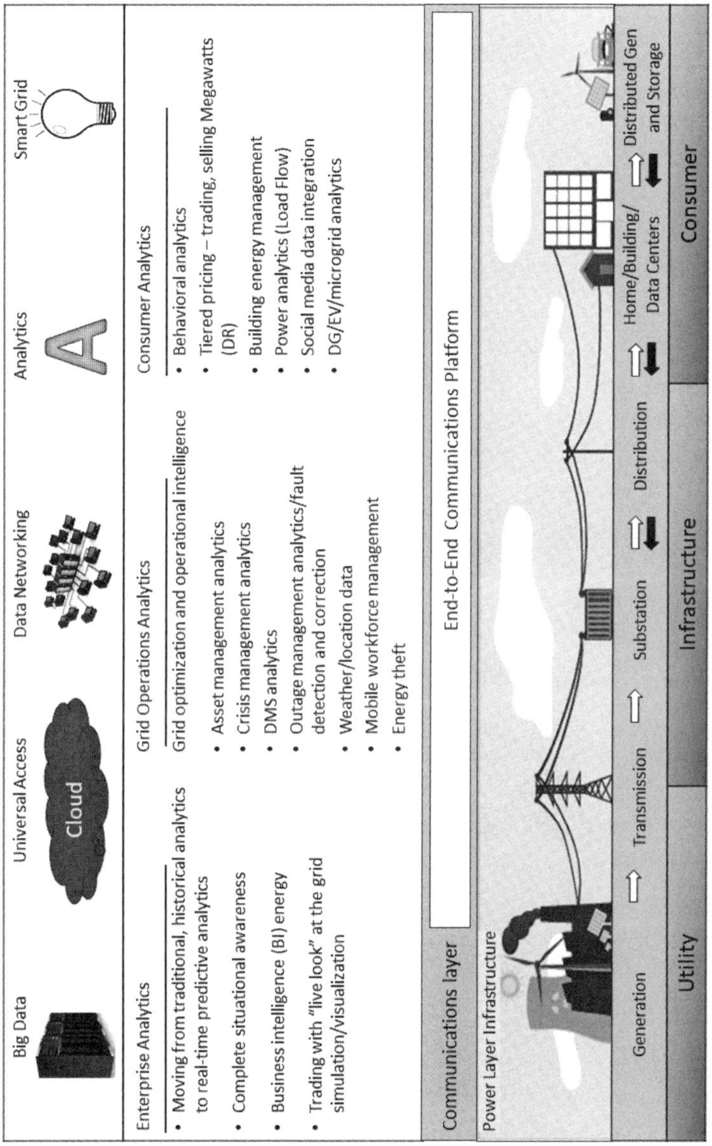

Figure 7.1 Transition of utility data types over time. (© GTM Research Grid Edge Service.)

tribution PMUs could lead to an explosion of data coming in from the distribution system in the field.

- *Meter (customer consumption) data:* The advent of AMI and smart meter systems is allowing the movement of meter data from one data point read per month per customer to several data points being read remotely every 15 minutes per customer. This immediate and tremendous data increase has left utilities just trying to decide what to do with it and take advantage of it.

  Much of the work being done with meter data is relegated to simple, specific value-added pilot applications at some utilities. However, utilities are beginning to purchase data analytics software packages from vendors using meter data as a starting point.
- *Asset data:* Utilities have always installed sensors at critical assets in the field. This was done to protect them from failure, especially when the failure of that component could potentially cause a cascading, widespread blackout or other forms of damage. Any analytics performed with this data was specific to that asset and generally not applicable to others, either due to a lack of sensors at the other components or due to different configuration/loading or other aspects.

  As the cost of sensors and communication is coming down, utilities are putting more sensors on key assets in the field and bringing data into core systems such as asset health centers [2].
- *GIS data:* Geographic information systems (GIS) are designed to capture, store, manipulate, analyze, manage, and present all types of geographically referenced data. Utilities use GIS systems to store and correlate their asset characteristics and location (GPS coordinates), along with connectivity and geospatial rendering.

  For data analytics to deliver good insights, it must be based on an accurate representation of various power system components in the field and their connectivity.

Data analytics is slowly being recognized as a critical tool in the utility's arsenal. However, for the utility industry, it is still in a nascent stage. As a result, many of the decisions taken are very often counterproductive to the utility and even if it delivers value and ends up delivering small amounts of value at a large cost.

A good data analytics implementation does not silo individual components such as meter analytics or customer analytics, but encompasses the entire utility to enable greater impact.

It is about defining what the utility wants to be known for, where the sources of value are, and what services it wants to provide for its customers, employees, regulators, and shareholders. Only then can it collect the right kinds of data and right quantities of data, at the right frequency, store them appropriately, and then draw value from it.

In short, this requires the utility to transform itself. Let's start by looking at other industries and how they have grown and learned from data analytics.

Lessons Learned from Data Analytics in Other Industries

Several other industries have been taking full advantage of the capability to drive a broad range of decisions based on data. Here are some examples of how other industries are using data analytics [3].

- *Hotel and airline frequent flyer programs:* The data gathered from these programs allows the entities to better understand their customers and better develop the capability to offer customized products and services to them.
- *Airline industry's route planning, ticket selling, seat management, and so on:* The airline industry is also one of the more advanced users of data analytics to perform specific activities such as the ones list above and more. For example, when a storm hits a specific location in the country, analytics are used to reroute the passengers in the best way possible to get them to their destination with minimum extra cost to the airline.
- *Internet search engine optimization:* Internet search engines such as Google and Bing use advanced data analytics to deliver the most optimum search response based on users' queries. These responses not only scan the entire internet for the right response but also prioritize it based on their knowledge of the user and his or her interests, online behavior, and other similar characteristics. The better search engines also use data analytics to manage data storage within their servers, which, in turn, allows them to respond to user queries so quickly.
- *Internet advertisements:* It sometimes seems uncanny how online advertisements appear to be targeted to the reader's/user's needs and either related to an online conversation they had or a recent search on the internet. In fact, every piece of data moving across the internet is stored by these companies and used to deliver targeted advertisements to its users.
- *Online purchasing:* Online retailers such as Amazon use the same data and search capabilities indicated above to better understand their customers and make specific products available to them on their main pag-

es. In addition, these online stores also crunch their data to use advanced artificial intelligence (AI) techniques to perform tasks such as delivery logistics, how much product to store in which warehouse, what to price a specific product, and how to bundle products for a better sale, and several other such services.

For most of these industries, this is not something done as an afterthought. It is core to their business and core to their survival in the fierce competition existing in their respective industries. To make it work, these companies embrace data sciences and analytics as a core capability. Every system implementation within their environments is designed to pass data between systems in such a way as to "slice and dice" the data in different ways to meet their business objectives, some internally focused and some externally focused.

So how can utilities take advantage of their increasing access to data and lessons learned from these other industries to become better at operating and servicing their customers?

Let us start with defining the kinds of data available at a utility.

Defining Data Analytics

Each chapter in this book identifies specific technologies such as DA, DERs, storage, and microgrids. Each of these technologies deliver data, some locally and some to centralized locations within the utility, where it is stored and very often not much else is done with it. It is also important to note that this section is not meant to define the entire set of utility data in an exhaustive manner, but rather offer a comprehensive, high-level list and characteristics as illustrated in Figure 7.2. As competition between electricity providers increases, this information will become increasingly important to utility marketing and operations.

Smart Meter Data

The U.S. government and private industry are investing billions of dollars to build the smart meter network infrastructure to save energy, reduce cost, and increase reliability. At 15-minute data sampling rates, 96 data points are collected for one variable in each household daily. If data collected from each household are sent to the utility and stored in a meter data management system (MDMS), the bandwidth of the communication network and the storage capability at the utility need to be significantly higher than what exists right now [4].

Traditionally, energy meters record household electricity usage by real power consumption only. A number is read at the end of each month to record the energy usage. Because the meter may not be read at the same time interval,

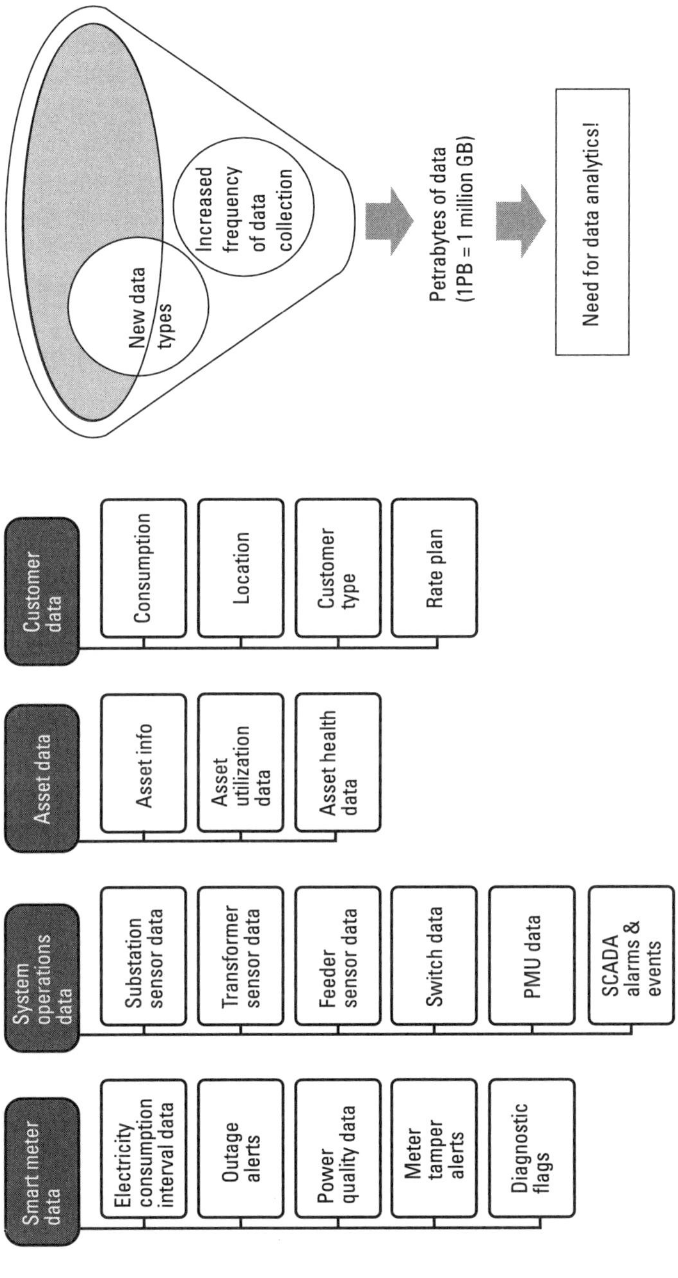

Figure 7.2 Utility data types and specifics. (© Modern Grid Solutions.)

the monthly electricity consumption data provide few details on consumption. A smart meter can record voltage (V), real (watt), and apparent (VAR) power consumptions at 15-minute intervals. The monthly data can be collected and reported from selected start and end dates.

In addition, just knowing whether the meter is on or not provides the system dispatcher with critical information on various households without power. This kind of near real-time information is critical to have during an outage either storm-related or not. Smart meters are designed to respond to a meter ping in which the operator can ping either 1 meter or a set of meters to understand whether a house or neighborhood is out of power, or if power has been restored.

These kinds of data are stored in systems such as MDMS, AMI head-end, CIS, and/or and outage management system (OMS) both for real-time use and for future use in other capabilities.

System Operations Data

With the deployment of a broad variety of intelligent sensors and devices in the field, utilities are inundated with voluminous data in the order of millions of gigabytes. This includes the transmission and distribution data discussed earlier. In addition to SCADA data, it also includes alarms from system operations. Specifically, system operations data comprises data from sensors deployed in substations, transformers, feeders, and switches.

As mentioned earlier, it also includes data from PMUs. A PMU is an electronic device that samples AC voltages and currents on the grid synchronized to a GPS clock at the rate of 48 samples per cycle. It provides time-synchronized, subsecond data of typically 20, 30, and 60 samples/second compared to traditional SCADA measurements every 2 to 10 seconds. The greater the phase angle differences between grid locations, the greater the power flow, indicating stress on the grid and proximity to instability.

EMS, ADMS, SCADA, DERMS, OMS, and related systems take in this data and perform their own set of analyses as well. These systems process the data, extract intelligence from it, and apply it to support utility operational decisions. As more sensors are being installed in the field, utilities are trying to find newer ways to extract intelligence from the data.

Asset Data

Aging assets, an aging workforce, the introduction of networked smart grids and a proliferation of intelligent devices on the power grid are challenging utilities to find more effective and efficient ways to maintain and monitor their critical assets and to do so with high availability and reliability.

Asset data comes from a broad variety of sources; some are hosted in systems such as GIS, asset management systems (AMS), work management systems (WMS), asset health systems, and others. However, a lot of asset data also comes paper-based. This is because much of asset inspections and maintenance work is done by field personnel who may or may not have access to the asset management systems at the utility in their mobile terminals. Their data recording is collected on paper and then is reentered into a system at the service center.

The ultimate objective of traditional or smart asset management is to help reduce, minimize, and optimize asset life-cycle costs across all phases, from asset investment planning, network design, procurement, installation and commissioning, operation, and maintenance through decommissioning and disposal or replacement.

Optimizing the costs associated with each of these life-cycle phases remains among the key objectives of an asset-intensive utility organization.

These objectives cannot be met without solid data on assets. Per utility innovative and visionary requirements, comprehensive and consistent sets of measurements are required to be taken for all assets when visiting the substations. These include:

- *Visual inspections:* The visits include a systematic review of all key control points, such as oil leakages, gas pressures, status of gauges, environment status, corrosion, civil condition, abnormal noise, and so forth.
- *Thermal analysis:* Via temperature measurements and infrared analysis, field inspection teams collect a thermal image of assets. This is used to detect any anomalous behavior or hotspots that could generate damage to the equipment or are the sign of a degradation of the asset itself.
- *Partial discharge:* Evaluation of partial discharge (PD) around each piece of equipment helps to detect abnormal situations, locate anomalies, and define corrective actions.
- *Oil analysis for oil immersed transformers:* This is done in a systematic manner, providing dissolved gas analysis (DGA), oil quality, and other analysis and helps in the overall evaluation of the aging asset. For dry transformers, vibration analysis replaces oil analysis [5, 6].

Customer Data

Customer data such as consumption data, location, customer type, and rate class is generally collected in the customer information system (CIS). This data goes beyond consumption data and includes other information such as payment history, credit rating, connection details, rate or tariff, and other similar information. The past historical data on the customer is almost always brought forward as these CISs get upgraded or replaced, making it very important and critical to a utility's financial backing.

For a utility to better understand its customers and predict customer behavior, it first needs to gain a 360° understanding of its customers. At most utilities, customer transactional data is isolated in multiple different systems. Interactive voice response (IVR) data is separated from web and mobile transactions and most utilities have little understanding of why their customers call into their call centers.

Understanding individual customer segments and their payment behavior can help to identify customers at risk for late or no payment, develop counter initiatives, and thus reduce the number of customers slipping into nonpayment of bills. This in return can help reduce bad debt write-offs and better predict balances at risk once a program is put in place to assist customers with payment difficulties. Lessons learned from these programs' successes can then be applied to the overall credit collection process. Customer analytics based on customer data enables this process by quantifying results and predicting success [7].

Using advanced data and analytics gives utilities insights into its customers, enabling them to improve customer service and better plan for future customer needs and streamline operations.

Key Drivers for Data Analytics

There are several drivers for grid analytics for a utility. Given the large diversity of utilities worldwide and even in the United States, where they change from large investor-owned utilities to smaller municipalities and even smaller coops. Even within a utility, the drivers vary from those impacting utility operations to those impacting the customer experience. As Figure 7.3 shows, the widespread adoption of DERs and their potential impact on all aspects of utility operations is having a profound impact on the need for better access to data analytics.

Let's look at each data analytics drivers in more detail.

Seeking Improved Return on Investments

Distribution utilities invest in AMI technology for one of two basic reasons: To achieve operational cost savings (e.g., reduce meter-reading costs), or in response to regulatory sanctions to implement systems capable of supporting mechanisms such as dynamic time of use (TOU) tariffs, reduce peak energy demand through enhanced load control, and/or implement demand response capabilities. Similar situations exist with the implementation of DA necessary to support new requirements such as increased reliability, increased penetration of DERs, and others.

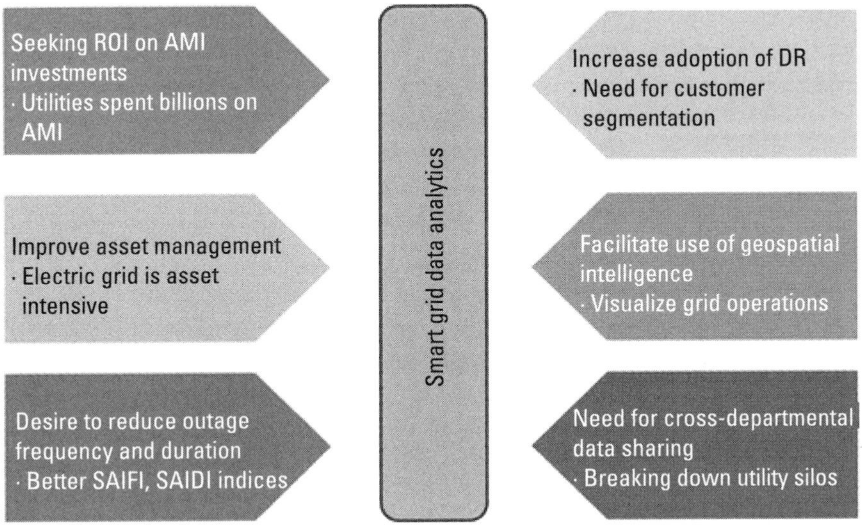

Figure 7.3 Key drivers for data analytics. (© Modern Grid Solutions.)

Utilities are responsible to their shareholders to carefully consider risks associated with the decision to deploy these capabilities. However, regulators have a responsibility to review the socioeconomics of these implementations to evaluate the potential benefits of the investment for all stakeholders.

Data analytics are a way to increase this return on their investments. Reduction of meter-reading costs is not adequate to recover the costs of the AMI implementation. Added implementations such as outage support, asset utilization enhancements, and others improve the ROI.

Utilities seek to recover their investment in the shortest period to decrease risk and improve the rate of return. When making the decision to commit to AMI deployments, utilities consider the regulatory rulings on rate recovery as fundamental. At the heart of these rulings are decisions on the treatment of depreciation, which are founded in studies and conclusions on technological obsolescence [8].

Data analytics presents one of the most important mechanisms to improve the ROI. This is done by using data delivered by sensors and meters and converted to useful information that can be used to improve the operations of the utility.

Improve Asset Management

Traditional transmission and distribution (T&D) asset management has mostly been a matter of maintaining and replacing assets according to fixed schedules. For example, it is common for T&D companies to check the quality of the oil in their power transformers every 2 to 3 years, regardless of the actual rate at which oil quality worsens. However, managing assets following traditional mechanisms has drawbacks; not all assets depreciate in the same manner and so maintaining them all in the same manner results in increased costs to the utility.

Maintenance programs tend to vary across companies. The lack of standards for asset management and the variety of asset-management plans make it harder for regulators to determine whether T&D companies are doing what they should to ensure their networks will perform properly.

Sensors, communications devices, and other hardware allowing objects to be tracked and controlled remotely have become increasingly affordable and reliable. Shifting to an asset-management model based on analytics makes it possible to improve productivity in different ways, which companies should prioritize according to local regulations, the quality of their assets, and other factors. Companies endeavoring to lower costs are beginning to use analytics to identify routine procedures that can be eliminated, particularly for less valuable assets, and to keep their assets in service for longer. Other companies might consider it more important to increase the reliability of their services. For these companies, analytics can aid in planning additional maintenance work for assets that can disable their networks if they fail [9].

Desire to Reduce Outage Duration and Frequency

Consumers' growing appetite for digital products, combined with more industrial and commercial use of computerized machinery, is putting performance pressure on electric utilities. Momentary outages and power quality issues, which have been on the rise, are frustrating and increasing costs for users of sensitive digital technology. This means the present state of electric grid reliability is unacceptable to today's consumers.

Regardless of how brief, a power outage generates broad complaints. Residential customers get distressed and complain when their power is interrupted due to the disruption to their lives. Retail businesses are upset at the hassle, costs, and lost sales occurring when customers leave rather than wait for cash registers to reboot. Manufacturing plants also incur significant costs because of lost production and idle workers while product assembly line controls are reset.

Momentary outages resulting in electric blinks are an inevitable consequence of utility system designs that for 100 years have focused on minimizing sustained outages.

Voltage sags (voltage that drops below 70% of nominal) can be troublesome to customers and affect their electronic equipment. Voltage sags are discussed less, but they are a well-known problem resulting from typical substation breaker-protection schemes [10].

Data analytics allows utility personnel to remotely assess such problems and resolve them more quickly, bypassing traditional approaches of installing special sensors after the customer has complained.

Increase Adoption of DERs

The nature of electricity generation, distribution and consumption are changing rapidly to prevent power disruptions occurring from storms, flooding, cyberattacks, and other threats. Customers also want to become more involved in controlling their energy usage to better manage their electric bills. As a result, utilities are facing a new market landscape where DERs will play an increasing role. Compared to just a decade ago, distributed energy resources are more affordable and efficient. Utilities are finding ways to financially reward customers for the deployment of such resources instead of outdated traditional investments [11].

DERs could provide cheaper alternatives than traditional grid investments because they are small and do not involve major costs such as those related to constructing high-voltage transmission lines. Distributed resources could also be built quickly with less risk than larger and costlier investments such as power plants. In addition, some DERs can provide more than one service.

Most DERs deliver electricity to the grid through an interface such as an inverter (some of them are smart inverters, a standard that is getting more acceptance across the United States and the world) [12]. Improved access to sensor data supported by advanced analytics allows for improved command and control of the DERs both in real time for operations and in advance for planning.

Facilitate Use of Geospatial Intelligence

Currently geospatial data and technology are used tactically by electric power utilities for many purposes including outage management, vegetation management, disaster management, renewable energy facility siting, universal electrification planning, asset management, and energy density mapping, to name just a few. However, smart grid implementations are fundamentally transforming the role of geospatial data and technology in electric power utilities. These implementations provide improved situation awareness and effective anticipation of and response to events that might disrupt the performance of the grid.

Spatial data underlies almost everything an electric utility does, and the GIS provides a foundational view potentially linking every operational and planning activity at an electric utility including design and construction, asset management, workforce management, and outage management as well as SCADA, ADMS, renewables, and strategy planning. As a result, geospatial data is extremely strategic in nature. Location is the basis on which data is organized and spatial analytics is how information is extracted from the data. Spatial analytics is becoming a key technique for electric utilities because everything that a utility impacts (its customers, assets, and operations) involves location [13].

Overlaying data and analytics from sensors on top of geospatial data has the potential to completely change the ability of the utility to be proactive in their approach to drive value to all its stakeholders.

Need for Cross-Departmental Data Sharing

In most companies, data is maintained by various departments and in different systems. This is no different in a utility. There is value in sharing this data between departments to craft efficient enterprise-level and operational processes.

For example, T&D asset data is shared across multiple utility business systems such as GIS enterprise asset management (EAM), CIS, OMS, EMS, and ADMS. For each system, the data must reside in the applications native database to support its functionality.

It is not uncommon for these systems to be integrated at some level so data can be shared between systems. For data to be shared across multiple business systems, well-defined data governance policies and maintenance practices are required. Often, discrepancies between data shared by these systems occur. When this happens, data integrity is compromised and the organization's ability to make informed decisions is negatively impacted. To prevent data integrity issues, it is important for utilities to develop a data integration process. This process should involve alignment of data models, create data governance policies across departments, and use an integration platform (also called democratization of data) to keep the data consistent and help quickly flag any discrepancies that show up so once the data is cleaned up, it stays that way [14].

A Conceptual Architecture for Data Analytics

For the most part, data analytics implementations are also siloed, following their sources of data. For example:

- *AMI meter data:* Meter data analytics focuses on value-added applications such as outage detection (and nested outages), advanced customer care and billing, load research and forecasting, asset utilization, and so on.
- *Operational data:* Operational analytics focuses on applications such as contingency analysis, optimal power flow, asset performance analysis, volt-VAR optimization, and so on. Very often, the data from the operational systems is also extracted from the systems into historical database

(e.g., PI Historian) and then offline and online analyses are performed on it to provide more relevant information to both operational and planning personnel.

- *Asset data:* Asset-based analytics focuses on applications such as asset health diagnostics and other related diagnostics. Sometimes these analytics are augmented by addition of operational and meter-based data.
- *Customer data:* Customer analytics generally include applications such as load profiling, customer profitability, customer satisfaction, customer quality of power delivered, and so on. Consumer analytics also looks at customer segmentation activities that help tailor effective demand response programs, electric vehicle usage, and deployment of distributed generation on customer premises.

Core Components of the Conceptual Architecture

Figure 7.4 represents a suggested approach for developing a conceptual architecture at a utility. It has the following components.

- *Data sources:* The source systems for the data have been identified earlier in this section.

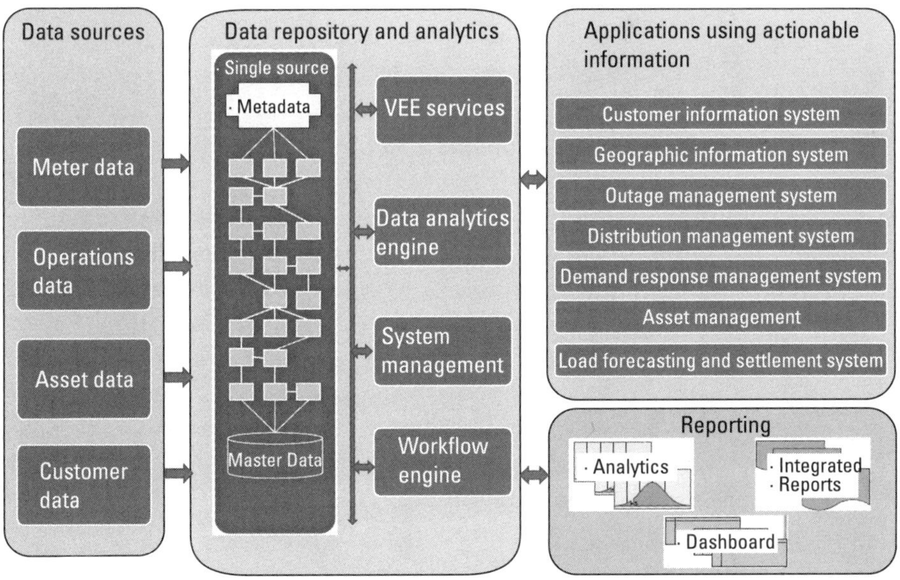

Figure 7.4 A conceptual architecture for utility data analytics. (© Modern Grid Solutions.)

- *Metadata:* The knowledge of the original source of data, where it is stored and other information about the types of data.
- *Integration architecture:* The integration architecture allows the multiple disparate systems to work with each other seamlessly.
- *Validation and error checking:* The data needs to be scrubbed to check for errors to ensure it is clean and safe. The old phrase "garbage in, garbage out" stays true and is even more important here. The decisions made at the utility depend on the validity of the input data and need them to be clean.
- *Performing analytics:* There are various kinds of analytics such as utility enterprise, grid operations, and consumer operations analytics. The key to these analytics is that they are executed on the original source of data based on the metadata information.
- *Reporting and visualization:* Special analytics in the form of reports, dashboards, and advanced visualization allow for a deeper view into the data showing trends and other analyses. It is of high value for enterprise operations, regulatory conformance, intradepartmental and interdepartmental communication, and so forth. It can be in the form of integrated reports or dashboard presentations. Data visualization techniques can greatly enhance reporting. There are several visualization techniques based on multivariate high-dimensional visualization, which provide the ability to use two-dimensional and three-dimensional visualization [15, 16].

Guiding Principles

Key guiding principles governing architecture are listed here:

- *Guiding Principle #1:* Do not try to bring all data from different systems into one single system. Given the broad variety of the systems and their data, bringing them into a single location would cause both redundancies in data storage as well as unnecessary inefficiencies. One cannot control the specific data sources. They come with specific systems procured for the systems and data management. For example, the meter data management system (MDMS) is designed to manage, validate, and store meter data efficiently. Similarly, operational systems such as ADMS and OMS are designed to process and store real-time operational data efficiently.
- *Guiding Principle #2:* Create a metadata layer that maps various data elements to their specific data sources and their characteristics. Know

exactly the contents of each data source and how they are structured. This should include both raw and processed data.

- *Guiding Principle #3:* Ensure that the power system model is accurate and up-to-date and available to all utility applications. Smart grid analytics need to be based on a strong foundation of the power system model. This includes the power system components and their connectivity, location, and characteristics. There are three types of model: as-designed model, as-built model, and as-operated model. The first two models are hosted in a system like the GIS and the third one is hosted in a system like the ADMS.
- *Guiding Principle #4:* Keep data in its original location to the extent possible so that it is maintained by the source system that creates it and kept accurate. Keeping the same data in multiple locations creates data maintenance headaches and raises the possibility of data being corrupted. Data analytics systems tend to bring data from their source systems into a single location and then apply their analytics. It is not uncommon to see two sets of analytic engines performing their analysis on the same data and delivering different results.
- *Guiding Principle #5:* Use a data integration approach to keep these disparate sources of data and their systems to work with each other seamlessly. There are several technologies and approaches to integrate the various data sources:
  - *Service-oriented architecture (SOA):* Software design where services are provided to the other components by application components, through a communication protocol over a network. The basic principles of service-oriented architecture are independent of vendors, products, and technologies.
  - *Enterprise service bus (ESB):* Implements a communication system between mutually interacting software applications such as GIS, OMS, and CIS in a service-oriented architecture (SOA).
  - *Common information models (CIM):* Industry standard for defining device and application characteristics so that system administrators and management programs will be able to control devices and applications from different manufacturers or sources in the same way. Used for smart grid persistence and for the integrated data architecture and are critical, especially in the success or failure of data management.
  - *Messaging:* Represents communication systems based on exchanging messages. These messages include data and other information from different applications managed by messaging server.

Key Challenges

There are significant reasons why implementing data analytics at utilities has either not been implemented as successfully or only in small specific pilot installations. Some of the reasons are identified in Figure 7.5 and explained next.

Dealing with Legacy Data

Utilities have a more than 100-year legacy of accumulating and storing data of various forms, in various locations and of various types. Given this legacy, the data is stored in forms such as old-file-based systems, on paper, legacy database systems, and others. The amount of data is also extremely large.

An example of this data is weather data. It is a lot of data and very critical to utility operations and planning. Keeping even a few months of this data in memory can become prohibitively expensive. Utilities may end up with a monstrous solution that far outweighs the need. Often, what a utility requires is a simple snapshot, which could be around a specific weather meter, specific location, or a specific period. Relying purely on in-memory computing means that there is a limitation on how dynamically the historical analysis can be integrated.

Even when memory and storage problems are solved, utilities using more traditional database approaches to work with historical data often struggle with speed. Rather than perform intraday analytics, they resort to overnight batch processing. It is not uncommon to have to extract millions of records from a database and move it to a separate process, which may even be on a separate machine, for further analysis. This worked successfully when dealing with a few hundred thousand or a few million records, but now companies are looking at billions of records. Extracting this and analyzing it in yet another program does not scale.

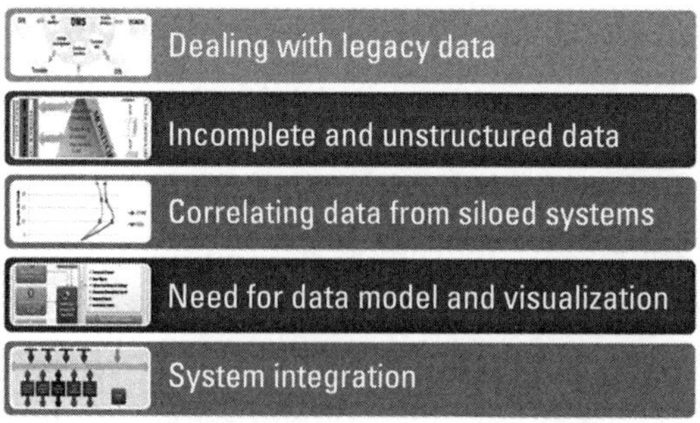

Figure 7.5 Key challenges for implementing a data analytics solution. (© Modern Grid Solutions.)

Utility data from legacy systems often need custom applications to integrate into mainstream analytics tools.

Incomplete and Unstructured Data

Upstream from the consumers' meters, operational data from newly deployed smart grid technologies, synchrophasors, combined with sophisticated generation monitoring and control systems, and diverse and sometimes unstructured sources such as weather information, facility surveillance systems, and even social media (such as Twitter) and e-mails should be considered as part of a consolidated data store. In doing so, utilities can gain a holistic view of their business environment and employ predictive analytics tools, enabling those companies to better anticipate outcomes, not merely react to events.

While utilities gather a lot of data, much of it can, at times, be incomplete and unstructured. An estimated 75% to 90% of all new data being generated is either incomplete or unstructured. No existing platform in place for unstructured data. Most utilities have not explored or tested big data platforms in a meaningful way [16–19]. Examples of these include:

- *Incomplete data:* As various systems execute their tasks at a utility, there are instances of incomplete data. Examples of these include SCADA data, meter data, and others, which can lose data when situations such as loss of communications, bad sensors, and others.
- *Unstructured data:* This includes such as many of the logs maintained at a utility. Examples of these include operator logs, field personnel logs, field personnel readings of transformer data, and other similar information. Unstructured data is highlighted as being too difficult to interpret. Utilities are concerned the volume of unstructured data has increased over the last few years. Apart from complexity, data access and privacy issues also pose implementation challenges.
- *Still on-paper data:* Asset health information is very often recorded on paper by field staff, brought to the service center and then sometimes entered into the computer system. It is not uncommon for some of this data to stay on paper for either extended periods of time or permanently.
- *Data accuracy and validation:* A key aspect of managing data in a large enterprise such as a utility is the need to maintain accurate record of things that utilities have and a process to maintain and validate them and to keep it clean. It is critical for analytics because the accuracy of the insights is dependent on the quality of the inputs (garbage in, garbage out).

In addition, the accuracy of the power system model is also an important attribute impacting the accuracy of any results that come from existing analytics solutions, especially the ones relying on the model. There are numerous instances of data inaccuracy in the distribution power system model at utilities.

Correlating Data from Siloed Systems

Utilities currently make decisions based on extensive experience and information living in siloes in different parts of the organization. Much of that data is of questionable quality and requires a high degree of manual manipulation, often leading to debate among leaders about the relevancy of data, rather than the actual significant decisions. Siloed data leads to difficulty in understanding cross-organizational impact. Decisions made using operational data can have unexpected consequences if there is little correlation between these separate types of data—consequences that are sometimes only realized after decisions are made and implemented.

Many smart grid applications are composite applications drawing on data and functions from multiple systems in the field such as SCADA RTUs, IEDs, and smart meters. Siloed systems prohibit easy data sharing. Over the past 5 years, utilities' chief data concern has been ensuring smart meter data flowed reliably into their CIS/billing systems, so the utility could ensure payment. There have been some gestures and claims made in the industry about integrating siloed departments and building intelligent IT enterprises, but in truth, maintaining accurate and efficient billing standards has been the leading concern.

Data access issues include silos preventing data from being pooled for the benefit of the entire organization, while data privacy concerns security and confidentiality of sensitive data. Failing to see the implications of data remaining siloed within utility departments can cause inefficiencies and lead to bad decisions based on inadequate information.

Need for Data Model and Visualization

Utilities need the GIS to collect, organize, maintain, and manage geospatial data from the grid. The GIS is foundational for utilities because it helps them have visualization of maps and points of interests to manage spatial data and present it. This system can be considered as visualization technology of the grid to have a global vision of consumers, generators and power lines position, and so forth.

GIS has been widely used by utilities for years for automated mapping/facilities management, back-office records management, asset management, transmission line siting, and more recently for design and construction, energy conservation, vegetation management, mobile workforce management

(MWFM), and outage management (OMS). Now utilities are integrating GIS with advanced meter infrastructure (AMI) and supervisory control and data acquisition (SCADA) systems. Intelligent design has crossed over from the office to the field in utilities, also enabled by the capabilities of GIS. Geospatial-related analytics (spatial analytics) is one of the key aspects of success for electric utility operations in the smart grid era. Looking for patterns and correlations between different land, weather, terrain, assets, and other types of geodata will be increasingly important for utilities. Power-related analytics with geospatial components include network fault tracing, load flow analysis, volt/VAR analysis, real-time disaster situational awareness, condition-based maintenance, and vegetation management.

Real-time visualization tools can then integrate with GIS and IT/OT systems to correlate and analyze data for system operations, business owners, and so forth.

Systems Integration

Systems integration is no small task. Added to this is the challenge of creating the underlying architectures allowing easy data access, sharing, and collaboration between systems. It is particularly difficult to upgrade architectures serving as the foundation for electric grids on which millions of customers depend.

No single platform is going to be able to handle all needs. Companies like Facebook and Twitter have had to constantly rebuild and update their architectures to meet their ever-evolving, rapidly expanding needs. This experience likely will be applicable to the utility space as well. Utilities must look to hybrid architectures to integrate the totality of their smart grid systems, as well as their emerging big data needs. Further, massive data warehouses are difficult to support over the long term; often, the best data architecture designs are those keeping the data close to the processing engine and analytics or vice versa [16–19].

Enabling Technologies

Several new and innovative technologies have emerged over the last few years offering new opportunities to the utility industry. They address different challenges and contribute to creating capabilities and value twenty-first-century utilities will generate from smart grid big data through:

- *Integrating larger quantities of data faster:* Efficiently managing large sets of miscellaneous data through database technologies;
- Mining complex, varied and large quantities of data efficiently;

- Business intelligence, portals, mobile device support, dynamic data exploration, geospatial;
- Faster and easier adoption and deployment benefiting from engineered solutions and cloud services.
- AI approaches to managing and analyzing patterns in data.

Some of the key technology enablers include the following.

Data Modeling and Metadata Management

This technology enables flexible metadata and data model definitions on data sets generally stored in distributed file systems and data stores. It is used any time that enterprises need to act on data stored in an unstructured file format. Specific actions that are implemented include techniques such as a Structured Query Language (SQL) query against data in a file format such as in the Hadoop Distributed File System (HDFS); data manipulation scripts (PigScripts); data governance and lineage; element-level and attribute-level security policy; reporting; and visualization.

Data Preparation

Data preparation software eases the burden of sourcing, shaping, cleansing, and sharing diverse and messy data sets to accelerate data's usefulness for analytics. These tools are generally used by data scientists and business users employing self-service business intelligence and customer analytics tools.

Data Quality

These products conduct data cleansing and enrichment on large, high-velocity data sets, using parallel operations on distributed data stores and databases. The volume and velocity of data processes by such systems require new technologies, even though many data quality vendors are evolving their tools to support larger data sets.

Big data tools and techniques are moving to the front office in the form of next best actions for customer engagement, fraud detection, and process automation. As this happens, the risks from dirty or untrusted data can increase. Enterprises use data quality to ensure business decision-makers trust an analysis performed on big data.

Data Virtualization

Data virtualization for big data is a technology delivering information from various data sources, including big data sources such as Hadoop and distributed data stores in real time and near real time. It includes several integrated components, such as data access, portions of data management, data discovery, a federation engine, and integration with Hadoop, NoSQL, and other big data platforms.

Big data virtualization integrates data across big data platforms. It supports many use cases, including a single version of the truth, real-time data integration, enterprise search, real-time business intelligence and analytics, and other federation requirements.

Distributed File Stores

This is a computer network where data is stored on more than one node, often in a replicated fashion, for redundancy and better performance. Key-value stores[2] are the simplest example; another example uses a key-value architecture where data is stored as flat files and accessed via a file system API.

These products are used for storing diverse and large data sets and exposing them to distributed analytics jobs, primarily for batch processing, but performance improvement via in-memory and solid-state drive caching enables some interactive jobs to execute.

Insight Platforms

Insight platforms combine big data management, preparation, and different combinations of SQL, advanced analytics, search and knowledge discovery, insights testing, and delivery into a single offering or integrated suite. When designed for general-purpose SQL analytics against Hadoop, they may be called BI-on-Hadoop platforms. When they use advanced databases and analytics for customer insight, vendors may refer to them as customer, digital, or marketing analytics platforms. Besides Hadoop, these technologies often include a variety of other big data management, integration, and execution technologies such as massively parallel processing (MPP), NoSQL, or Spark analytics databases.

Firms can use insight platforms to perform advanced analytics, customer and product analytics, mobile analytics, location analytics, SQL analytics, data visualization, data integration for search and knowledge discovery, data prepa-

2. A key-value store, or key-value database, is a data storage paradigm designed for storing, retrieving, and managing associative arrays, a data structure more commonly known today as a dictionary or hash (https://en.wikipedia.org/wiki/Key-value_database).

ration for advanced analytics, and predictive model building and deployment as well as insight to software interfaces.

MPP Data Warehouse

MPP data warehouse technology delivers row-oriented or column-oriented parallel database appliances optimized for the write-once/read-many operations typical of business intelligence (BI) applications. These appliances typically have large libraries of in-database analytics functions, data compression algorithms, and advanced query optimizations.

MPP data warehouse solutions offer many use cases, including operational reporting, analytics, and predictive analytics. They are the foundation for BI to support timely reports, ad hoc queries, and dashboards and to supply trusted, integrated data to other analytics applications.

NoSQL Database

NoSQL provides more sophisticated functionality associated with databases, such as data sharding[3], complex query, complex schema definition, and record-level or object-level security and locking. There are three types of NoSQL databases: key-value, document, and graph databases. Key-value databases are ideal for low-latency, fast key lookups; document databases for managing and querying richly structured and variable data; and graph databases to simplify access to connected data.

NoSQL offers new opportunities for enterprises to support next-generation business applications. NoSQL use cases include social network applications, IoT, mobile, recommendation engines, pattern analysis for detecting fraud, heat map dashboards and understanding consumer behavior, and analysis of communication networks.

Predictive Analytics

Predictive analytics supports the identification of meaningful patterns and correlations among variables in complex, structured and unstructured, historical, and potential future data sets for the purposes of predicting events and assessing the attractiveness of various courses of action. Big data predictive analytics technology encompasses the software and/or hardware solutions allowing firms to discover, evaluate, optimize, and deploy predictive models by analyzing big data sources to improve business performance or mitigate risk.

---

3. Sharding is a type of database partitioning that separates very large databases into smaller, faster, more easily managed parts called data shards. The word shard means a small part of a whole.

Usage scenarios encompass all types of customer segmentation: behavior and propensity prediction, supply chain and operational process optimization, logistics, risk management, fraud, technology management security and persistent threat analysis; pricing analytics; call center routing, and help-desk knowledge management [20].

AI-Based Data Analytics

Millions of people are now using AI (Siri, Alexa, and Google Assistant). The biggest advances have been in two broad areas: perception and cognition. Although AI is already in use in thousands of companies around the world, most big opportunities have not yet been tapped [21]. In addition, machine learning systems are not only replacing older algorithms in many applications, but are now superior at many tasks once best done by humans.

AI systems are already being piloted, tested, and implemented in utility systems and examples are showing up in almost all areas of smart grid data analytics [22].

Benefits

Figure 7.6 represents a view of data analytics opportunities all along the electric value chain from generation to the customer. The examples provided in the figure are a snapshot of the realm of possibilities. Nevertheless, as indicated earlier, utilities have barely scratched the surface of possibilities with data analytics.

However, there are several areas where utilities are investing in implementing data analytics. These implementations are also delivering a tremendous amount of value to utilities and their customers. In many instances, these have already reached production-level implementations to the extent that even vendors of the systems are embedding the functionality into their applications.

Let's look at four specific benefit areas in more detail (as illustrated in Figure 7.7).

Load/DER Forecasting

Utilities are using the data they collect from smart meters and other smart grid devices to better understand customers, design DR programs, make buying and selling decisions on the energy market, and increase the reliability of the grid. Forecasting plays a key role in each of these areas from modeling future load growth to predicting the impact of DR.

Forecasting is also becoming more critical to the operations of a utility because of the increasing penetration of DERs, EVs, and energy-efficient appliances. Previously, when forecasting electricity demand, utilities did not have to

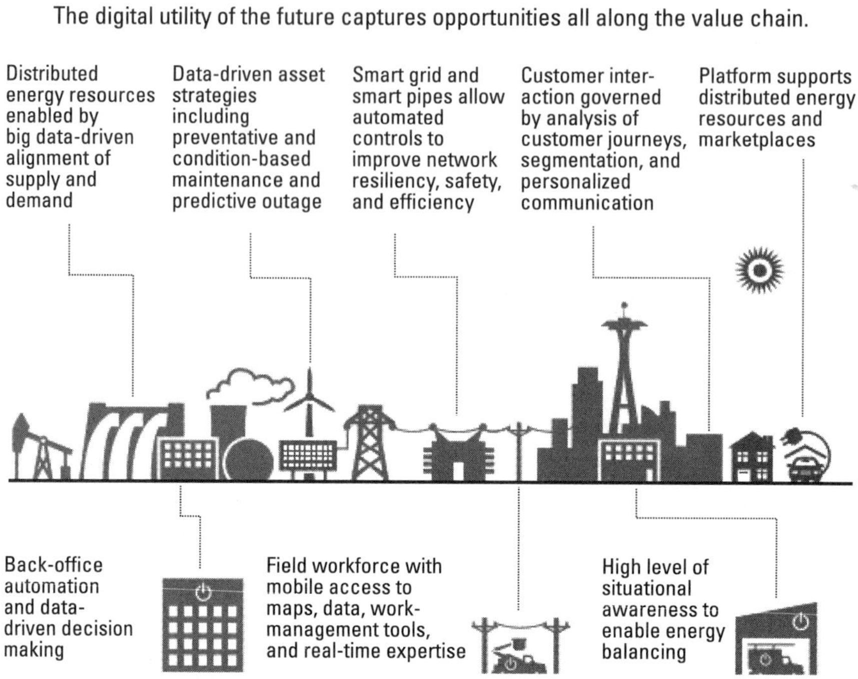

Figure 7.6 Data analytics presents opportunities all along the electric value chain. (Exhibit from "The Digital Utility: New Opportunities and Challenges", © 2017 McKinsey and Company, May 2016, www.mckinsey.com. All rights reserved. Reprinted with permission.)

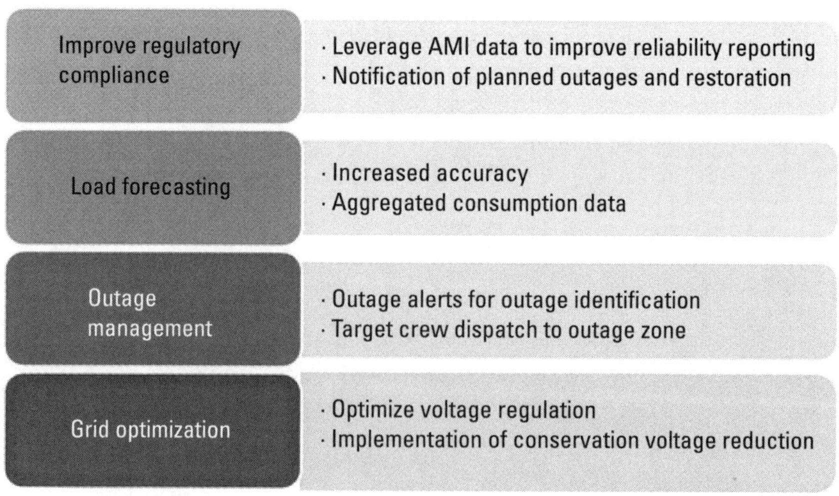

Figure 7.7 Key benefit areas of data analytics at a utility. (© Modern Grid Solutions.)

worry about electric vehicles or solar panels on rooftops or wind farms because these technologies were not present in significant enough numbers to have any real effect. Now, however, they are increasing in prevalence and therefore increasing the challenge of accurately forecasting electricity demand.

AMI is the primary technology offering forecasters more timely and granular data for load analysis and forecasting. With AMI, the utility has two-way communication with the meter (electricity, water, or gas) and gets readings back in an automated fashion in near real time, which means that energy consumption information, down to the meter level, can now be available is smaller periods of time than ever before [23].

Outage Management

Utilities use data analytics to resolve power outages more quickly, which, in turn, reduces costs, improves reliability, and promotes customer engagement. Utilities typically know the average duration of power outages, but are not sure why some outages last longer than others or how they can restore power sooner. By mining the data in an OMS and adding data sources such as telematics, asset-location data, and weather, utilities can produce a finer-grained view of their outage processes and the length of time taken for each step to restore power. Field supervisors and managers can now filter the view by time of day, type of crew, and location to review operations and find ways to increase reliability and reduce costs.

Using advanced analytical techniques, utilities can normalize differences across work locations and determine what processes and business rules affect the duration of each step. This effort can help prioritize process-improvement initiatives, based on their value potential.

Utilities can also use this data to build predictive models delivering better forecasts of restoration times and give customers more accurate information—a critical component of utility customer satisfaction. By adding variables to estimation models and using more advanced statistical methods, utilities can double the accuracy of their outage-duration forecasts and find new ways to tell customers when power is likely to be restored, further strengthening customer engagement [24].

Grid Optimization

Factors such as changing weather patterns, aging infrastructure, and increased adoption of electric vehicles and other new technologies are creating new challenges for utilities. While not a panacea for all challenges, transformer load management analytics can utilize smart meter data and actual weather models to continuously monitor and analyze distribution transformer loading levels and report on asset health, helping utilities make informed decisions to balance loads.

These analytics also include scenario analysis capabilities allowing operators to predict the impact of new loads such as EVs will have on transformers. In addition to transformer load management, voltage monitoring assesses voltage at every delivery point in the distribution network, allowing analysts to evaluate and understand the impacts of customer loads, variable energy sources, or distributed energy. Trends from voltage monitoring can be synthesized to allow for system improvements with a holistic approach using measured data, rather than relying on individual customer complaints or system models.

In more real-time utility operations, optimization of grid operations increasingly calls for predictive analytics accounting for weather patterns, variable resources, utility programs, and emerging technologies, creating a dynamic load environment and impact customer loads [25].

Improve Regulatory Compliance

Regulatory requirements drive the need for consistent, complete and accurate reports on operational key performance indicators (KPIs). Noncompliance could result in penalties.

Utilizing a big data strategy, companies do not have to perform data triage, discarding much of the historical information previously too expensive (due to storage costs) to capture and maintain. These solutions provide a singular repository in which companies can keep all data, both structured and unstructured, and as these solutions utilize commoditized equipment, the cost of doing so is significantly reduced versus specialized high-speed, high-availability equipment that would otherwise be required. This ability to quickly recall and reconstruct the entire circumstance of a trade or trades is a prerequisite for being able to answer regulatory inquiries or internal audits.

While these types of solutions do work with traditional relational databases, when utilized in conjunction with big datasets, additional value can be derived, including:

- Full view of all historical trading activities, including highly granular power transactions down to the smallest time interval;
- Ability to quickly recall, review, and analyze historical data from high-volume algorithmic trading programs, permitting better views of trading patterns to uncover anomalies;
- Positions the company as a proactive, compliance-oriented organization in the eyes of regulators, rating agencies, and shareholders.

Evolving to the Next Generation of Data Analytics: A Road Map

Currently, much of the work done on data analytics at utilities is done in a targeted manner, not as a company-wide strategic initiative. As a result, utilities are having islands of analytics, and some examples provided earlier in this chapter of this are meter data analytics, operational data analytics, and asset data analytics.

However, as pointed out earlier, this is an inefficient approach given that it restricts the user to a subset of analytics applications and, in addition, constrains them from going across the analytics systems to provide greater value to each other.

I believes that a change in approach is warranted to ensure that, instead of working in siloes, the various data models work together to deliver increased value-added information to the utility.

It is my core premise that, over time, the utility must move away from these islands of analytics to develop master data management. This requires a road map and a strategy. There are four basic steps to the development of a big data road map (Figure 7.8):

1. *Develop a big data strategy:* More information on this is provided later.
2. *Assess readiness:* Assess the utility's readiness from a technological and organizational perspective based on the current state of different ways data is procured and stored, both manually and in computer systems.
3. *Follow key guiding principles:* Every organization needs to follow a set of guiding principles. Having these principles published sets the stage for the entire organization to follow them consistently ensuring the cleanliness of the data being managed and thereby in return, the accuracy of the results of the analytics.
4. *Develop the road map:* Define how this strategy will be achieved over time based on critical constraints such as budgets, resource availability,

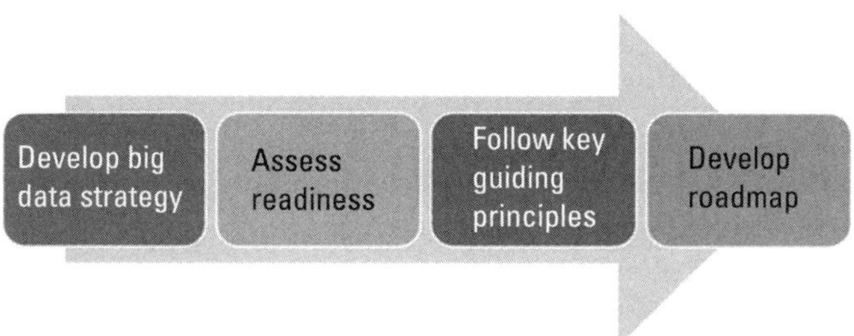

Figure 7.8 Key aspects to a big data road map. (© Modern Grid Solutions.)

future changes to the business framework, and other similar considerations.

Analytics presents a level of disruption and many organizations fail the first time that they attempt a major analytics initiative. Extracting value from data to drive intelligent business decisions requires a cultural shift within a business to institutionalize analytics-based decision management.

Developing a successful strategy requires a thorough understanding of technological and organizational capabilities, limitations, and opportunities and then charting a path toward building a common vision for analytics throughout the organization.

The analytics strategy should assess the unique business challenges for an organization, matching those challenges with relevant data and resources and establishing processes that grow capabilities and institutionalize analytics to ensure that key decision-makers have access to actionable results. Making the results accessible to business decision-makers is vital for adoption, so as a part of the strategy, we work with our clients to build visualization and deployment solutions tailored for their production environment.

Before developing an analytics strategy, an organization should assess their current capabilities and aspirational goals for analytics. A strong analytics strategy relates business goals and use cases with how analytics will support employees and the business. To ensure a cohesive and sustainable strategy, goals need to be identified and set. Goals set out the purpose and vision for analytics within the firm. This may include sustained competitive advantage, incremental revenue opportunities, or cost reduction.

The most important step to developing the roadmap is the strategy. Figure 7.9 presents a proposed approach to developing the big data strategy. The key steps are the following.

Identify Data Sources

The first step in developing a data analytics strategy is to perform a detailed analysis of all the data sources including all their source systems. A description of some of the more important data and their source systems have been provided earlier in this chapter.

Confirm the Use Cases

Use cases identify the potential short- and long-term uses of analytics to drive achievement toward goals. They will be used to:

- Understand the overall business case and show different types of value derived from analytics.

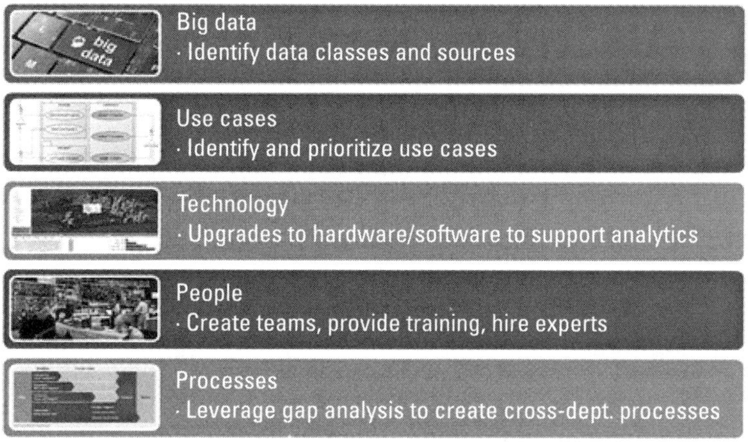

Figure 7.9 Key factors required to develop a big data strategy. (© Modern Grid Solutions.)

- Define holistic requirements within the firm for the types of analytics required and associated information management to support them.
- Understand the user groups and business processes that may be impacted and whether the usage will be a one-time process or an ongoing operational process.

Use cases are the areas of business operations where analytics could be used within a company. They help to define the depth and breadth of how analytics will be used in the business and how that will help the firm. The use cases will be specific to the type of company and the strategy they are following.

Identifying the potential use cases within the firm allows for a deeper understanding of the business needs, the requirements for the technologies and quantitative methods required, and any commonality that may be leveraged across the different use cases. Additionally, these may be used as the foundation for the overall business case to develop the firm's analytics capabilities.

Get the Right Technology

Quantitative methods and related technologies are the heart of how an analytics strategy is different from other strategies. This is what gives analytics its power and utility. The primary focus is to model real-world systems. The four key ways models may be represented are as an operational exercise, a game, a simulation, or an optimization model.

- Operational exercises represent a model by executing experiments with the real-world system and leveraging the results to make decisions about how to operate in the future. It leverages the least information technology (e.g., historical reports) and is slower and costlier.

- Gaming represents a model by creating a simplified response to various scenarios or strategies. Gaming is generally used as a management learning tool to highlight complexities in the decision-making process. This model representation is less realistic, but also does not leverage significant information technology, making it faster and less costly.
- Simulation is like gaming, but decision-makers are augmented with quantitative models. This can include creating a simulation of a real-world system that has not yet been run, or it could be a simulation of various scenarios as a diagnostic to understand why something has happened. The model evaluates the performance of the alternatives. This model representation is more abstract, uses significant quantitative methods, and requires information technology; therefore, it is faster and less costly than using the real system.
- Optimization represents a model completely in mathematical terms, usually by setting an objective to be maximized or minimized under different constraints. The model finds the best possible value of the objective function that also satisfies all the constraints. Like simulation, this model representation is more abstract, uses significant quantitative methods, and leverages information technology; therefore, it is faster and less costly.

Ensure That People Are Aligned

Most organizations now know they need data scientists, technology architects and software developers to make an analytics project successful. These are the people who bring sophisticated statistical data management and technological skills to the table.

What they do not always realize is that these skills alone are not enough. Becoming an analytics-based organization is a cultural change. Bringing about this cultural change requires change managers, political navigators, and senior executive influencers on the job, people with skills such as communication, business acumen, and political know-how. It requires a massive shift from the leadership level down. People must be ready to move beyond gut feelings and rely more heavily on analytics insights when plotting a course for the future, and this commitment must begin at the senior leadership level. Yet, despite the effort, the benefit is clear: with the proper culture in place, the organization must turn analytics into an actionable advantage rather than relegating it to remaining a static report.

The organization needs to build a team of people with requisite skills complementing each other. Further, the analytics team needs access to expertise in IT infrastructure, data storage, data transformation, statistics, and data visualization.

Confirm That Processes Exist (or Need to be Developed) for This Effort to Succeed

When it comes to analytics, existing processes tend to focus on capturing, certifying the accuracy of and distributing the right data, but that is only the beginning when it comes to process. The other process to address is the process to turn data into insight and to act upon that insight.

Some organizations have quite robust processes for turning data into insight, but they are generally confined to generating hindsight, or their use is limited to one individual, one team, or one operational function.

As soon as organizations try to get more sophisticated by, for instance, generating predictive and prescriptive insights that guide a company's decision-making without the benefit of human involvement, their processes are not defined and are inconsistent across the organization.

Processes that guide organizations on how to act upon the insights they generate are even more elusive. That is likely because they often leave out key ingredients. To build this capacity, the organization should include a governance framework that ensures that:

- Analytics priorities align with corporate vision.
- Tools and processes are centrally managed.
- Successes and best practices are shared across the organization.
- The operating model matures through learning events, sharing of successes and rewarding innovative and collaborative behaviors [26, 27].

Case Studies

Case Study #1: Oklahoma Gas & Electric—Customer Segmentation Analytics

Oklahoma Gas & Electric (OGE), Oklahoma's largest regulated electric utility, serves 843,914 customers over a territory of 30,000 square miles in Oklahoma and western Arkansas with 23,000 miles of overhead distribution lines and 500 substations.

OGE's 2020 Initiative prohibits it from building any new fossil-based generation plants until 2020. OGE's 2020 Initiative will require the utility to shed substantial load.

OGE implemented AMI and gets about 52 million meter reads per day, a figure that is expected to double in the years ahead. In addition, the utility expects to receive approximately 2 million event messages per day from AMI, data networks, meter alarms, and outage management systems (see Figure 7.10). Other areas in which OGE is driving big data adoption include the planned deployment of a new distribution management system (DMS), as well as an OMS and an integrated volt/VAR control program.

To achieve the goal of 2020 initiative, the utility has developed a strategy that relies on segmentation analytics, which will allow it to gain visibility into individual customers' responses to price signals, and as well as to identify the best customers to target with specific marketing campaigns. It will also allow the utility to perform the measurement and verification tasks necessary to develop and offer the most optimal rate structures.

OGE believes that it is important to break down organizational silos because correlations of time-synched data from multiple inputs (MDM, CRM,

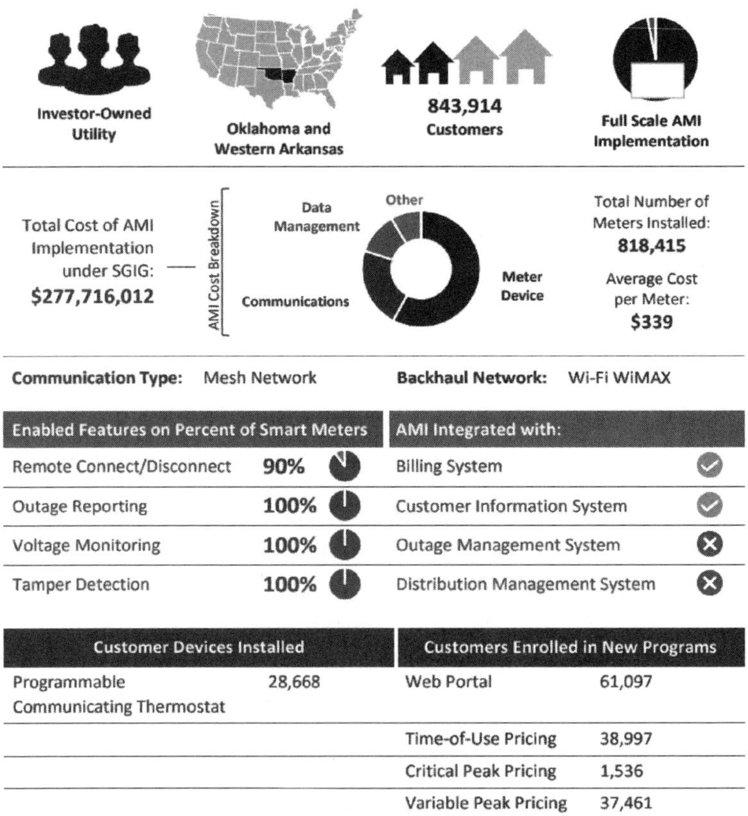

Figure 7.10 Oklahoma Gas & Electric case study. (Source: Energy.gov https://energy.gov/sites/prod/files/2016/12/f34/AMI%20Summary%20Report_09-26-16.pdf.)

billing/CIS, asset management, and outage management) all provide valuable data that can contribute to the process of developing effective demand response programs and rate structures.

Its analytics areas include:

- Consumer analytics and customer segmentation;
- Peak load management/load shed (via demand-side management analytics);
- Grid optimization (voltage control and conservation);
- Geospatial and visual analytics for centralized view of multiple systems.

To deal with this influx of data and the analytics needed, OGE is implementing an integrated operations center, which is an enterprise reporting and analytics solution. The platform provides the backbone for the data and helps the utility gain a clearer understanding of customer behaviors and preferences from an enterprise perspective.

OGE's information architecture includes a data warehouse, improved and expanded data integration and data management, and new analytics and presentation capabilities.

The utility's data warehouse is engineered to expand to accommodate hundreds of terabytes of data. Its data integration and management layer will be improved by real-time messaging. Outputs from the new system will be configured so that employees with access will get a live look at a near-real-time version of the data; this solves the dreaded versioning problems that are widespread in the utility industry. While the analytics layer is still relatively new, the utility reports that the experience thus far has been positive. OGE is particularly impressed with the capabilities of its geospatial and visual analytics, which has granted the utility a new look at its assets [28, 29].

Case Study #2: Potomac Electric Power Company—Grid Analytics

Potomic Electric Power Company (Pepco), an Exelon utility, supplies electric power to energy to more than 842,000 customers in the District of Columbia and Maryland.

The Pepco–District of Columbia Smart Grid Project in Washington, D.C., involved distribution automation (DA), AMI, and demand response programs that involved load control devices and time-based rates (see Figure 7.11). DA included deployment of smart substation devices, automated distribution circuit reclosers or switches, and network and substation transformer monitors.

The AMI deployment was designed to provide Pepco and its customers with detailed electricity usage information that, when combined with the

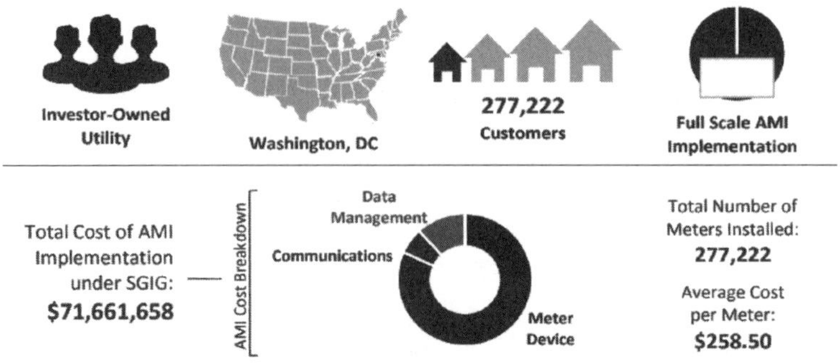

Communication Type: Wireless Mesh **Backhaul Network:** Cellular

Enabled Features on Percent of Smart Meters	
Remote Connect/Disconnect	96%
Outage Reporting	97%
Voltage Monitoring	97%
Tamper Detection	97%

AMI Integrated with:	
Billing System	✗
Customer Information System	✓
Outage Management System	✓
Distribution Management System	✗

Customer Devices Installed	
Direct Load Control	16,010
Programmable Communicating Thermostat	11,383

Customers Enrolled in New Programs	
Web Portal	14,093

Figure 7.11 PEPCO case study. (Source: Energy.gov, https://energy.gov/sites/prod/files/2016/12/f34/AMI%20Summary%20Report_09-26-16.pdf.)

demand response programs, helps customers reduce electricity usage and peak demand on the system. DA equipment was installed to improve system reliability while decreasing operations and maintenance costs.

During 2013, AMI technologies helped Pepco prevent over 6,000 customer outages. Pepco has also begun to use AMI data for transformer load management. The accuracy of transformer loading data allows for a more proactive replacement under a planned outage strategy (as opposed to the prior practice of waiting for transformer overloading, leading to asset failure). Transformer replacement under a planned outage results in a lower outage duration than that associated with emergency repairs.

Pepco, until recently, had limited visibility and awareness beyond substations and SCADA. Deployment of sensors, AMI and smart meters, and distribution automation devices with two-way communication capabilities at the edge of the grid has provided it with enhanced visibility and control into distribution operations.

Pepco deployed a comprehensive physics-based grid analytics package that includes a suite of applications aimed at enhancing asset life, reliability, and distribution system optimization. The analytics solution includes a broad swath of applications for distribution grid modeling capability, including outage management and voltage optimization, asset management and revenue protection, and load planning and demand response management.

The term physics-based alludes to the heavy math, which makes its modeling more flexible and accurate than the model-based DMS of today, which essentially use technology scaled down from transmission grid modeling. The solution combines its physics-based models with statistical analysis common to other analytics platforms, which can help find relationships between different parameters where physical modeling cannot.

The result is a real-time power flow model of thousands of miles of low-voltage grid circuits, able to adapt to changes coming in and update the other utility systems it's communicating with. Pepco's analytics started with a solution, which feeds grid modeling data into how utilities plan for investments in grid infrastructure, something that states like California, New York, and Hawaii are requiring their utilities to do for future years. Other use cases include real-time operations like conservation voltage reduction, which is on Pepco's road map [30–32].

How Data Analytics Will Impact Utility Transformation

At Stanford's first Women in Data Science Conference, Stanford Engineering Dean Persis Drell said in her opening remarks, "Almost anywhere we turn, evidence of a data revolution abounds. Solutions to the challenges of our future increasingly link back to data and data science. Finding those solutions, will require diversity of thought, approaches, and styles, and, ultimately, of teams" [33, 34].

While the article delved into areas such as how data and data science are impacting personalized medicine, personalized viewing habits, market design, cybersecurity, and several other topics, the electric utility industry should have been at the top of the discussions, given the supreme importance over time and its even greater importance going forward.

In this book, there have been several examples of how data and analytics are creating an ability for utilities to both improve their own operations but also in how they are able to improve the services provided to their customers (or prosumers). In this section, the author provides a few direct and tangible examples of how data analytics will impact utility transformation.

- *Enhanced predictive asset management:* It is a well-known fact within electric utilities that most major substation equipment outages occur at 3 AM on a Saturday. The problem with a major outage then is that it takes longer to fix (finding the right crew and materials takes more time in the middle of the night) and costs more (overtime costs that need to be paid to the crew).

Data on the usage of various equipment is available from a combination of AMI, SCADA, and other DA sources. Taking advantage of this information and augmenting with the support of the right types of sensors and analytics can prevent a much higher percentage of equipment failures at a significantly reduced cost and reduced number of customers impacted. This is done by maintaining the equipment when needed based on its condition instead of maintaining it on a schedule.

- *Asset utilization:* Smart meter data can be aggregated to reflect the transformers to which they are connected, and then utilization can be compared to the capacity of the transformer to build detailed capacity utilization trend analysis. Examples of questions this type of analysis can answer include:
  - What percentage of the time is a transformer operating within 10% of its peak rating?
  - Are there certain times of day or times of year when transformers are nearing overload?
  - What is the minimum size transformer that could be used to replace an aging transformer?

  Utilization patterns can also be compared against pre-failure data for similar transformers to begin building proactive asset maintenance and failure prevention.

- *Power quality issues:* The advent of DERs at the residential level makes power quality (PQ) a more important issue than it used to be earlier; a customer may have PQ issues at his or her home even if he or she does not install PV on the roof, but his or her neighbor did. PQ issues generally start with the customer first complaining about it, the utility placing PQ monitors at the customer site, performing an analysis, and then determining the appropriate course of action.

  Analysis of the data on residential DER installation and/or generation supported by AMI data should allow the utility to become proactive about the residential PQ situation and respond proactively instead of waiting or the customer to complain. Improved power quality results in improved customer satisfaction leading to regulatory reception leading to better rate discussion relationships.

- *Microgrid and macrogrid integration:* Chapter 6 discusses the concept of using the microgrid as a building block for the macrogrid. So when a storm such as Superstorm Sandy or Hurricanes Irma, Harvey, or Maria come in, the macrogrid could devolve into an interconnected set of microgrids that could either disconnect from each other or stay in a connected mode but still able to deliver power to the customers in a reliable manner.

For this to work, as described earlier, there needs to be access to a lot of sensor and analytics that provide the grid with the ability to identify the points of failure, the points of supply, whether centralized or distributed and get them connected in a near-real-time situation so that power can be restored much faster to the individual microgrids that still have access to sources of power. Once the storm blows over, the microgrids can be stitched back together to get the macrogrid back up and in operation.

Conclusions

The utility environment is changing in ways that even they are not able to comprehend. Over the next 10 years, they will face a triple-whammy of significant hits that will require them to respond in a strong manner.

1. DERs and associated technologies such as distributed generation, renewables, demand response, storage, PEVs, and others are here and changing the way power flows over the distribution grid. This specific change will impact all areas of the utility's operation ranging from system operations, protection, and planning, all the way to customer services and field operations.

   As an example, if one is aware of how complicated system planning is now with having to define and assess various future scenarios and their impacts as planning personnel prepare for a rate case, one can only imagine future scenarios in which they also need to evaluate the impact of nonwire solutions in which they may need to depend on the customers (mostly consumers at this time) becoming generators and having to rely on this segment to deliver on their mandate of a reliable and resilient grid.
2. The movement of customers from being consumers to prosumers (producer and consumer) is creating new business models with the introduction of entities such as aggregators who appear to be positioning themselves to take customers away from the utility. Lastly, as New York state is discussing retail markets within the REV initiative, a concept that could grow to other states, the utility's own business model

could undergo a dramatic change somewhat like the change seen with the passage of FERC orders 888/889 but at a wholesale level.

Microgrids bring in a combination of the business and technical changes that have the potential to take customers completely away (disconnected) from a utility thereby leading to an increased potential for stranded assets for the utility.

Imagine a storm scenario in which the utility is trying to restore power with a broad variety of business and technical challenges playing a role in the recovery process and in parallel, dealing with the field crews (both internal and external) handling safety issues that come with having reverse power flow on to the grid.

3. A 2006 article in *Public Utilities Fortnightly* [35] stated that "almost 40 percent of utility workers will become eligible for retirement in the next five years. Assuming only nominal growth, by 2010 the industry will need to hire 10,000 new skilled workers each year." Utilities depend on institutional knowledge to solve a lot of problems. However, with every retirement, much of that knowledge is "walking out the door." These statistics are now all too familiar among utility human resources professionals.

   This is not the kind of experience that can be easily replaced by newer recruits. However, the newer recruits come with a higher level of theoretical knowledge, ability and interest to work with computers and make decisions based on data mainly amplified by their lack of background experience that comes with long years of service in the utility.

For utilities to respond to this trilogy, they need to be more open to the intelligence gleaned from data. The key is for the utility to convert all its data to useful information so the new utility worker, operator and others will have the right tools to make the right decisions, at the right time, within the right deadlines.

References

[1] Vadari, M., *Electric System Operations: Evolving to the Modern Grid*, Norwood, MA: Artech House, 2013.

[2] Fleeman, J., "AEP Launches Asset Health Center," *T&D World*, February 1, 2013, http://www.tdworld.com/asset-management-service/aep-launches-asset-health-center.

[3] Analytics Vidhya Content Team, "13 Amazing Applications/Uses of Data Science Today," September 21, 2015, https://www.analyticsvidhya.com/blog/2015/09/applications-data-science/.

[4] "Smart Meter Data Analysis," https://www.researchgate.net/publication/261118785_Smart_meter_data_analysis.

[5] Nguyen, Q., S. Kumar, and K. G. Girish, "Using Predictive Analytics to Optimize Asset Maintenance in the Utilities Industry," *Cognizant 20-20 Insights*, December 2014, https://www.cognizant.com/whitepapers/using-predictive-analytics-to-optimize-asset-maintenance-in-the-utilities-industry-codex964.pdf.

[6] Coullon, J. -L., "Advanced Asset Health Management for Effective Maintenance and Asset Replacement Planning," *CIRED Workshop*, June 11–12, 2014, http://www.cired.net/publications/workshop2014/papers/CIRED2014WS_0023_final.pdf.

[7] Kaestner, M., "The Power of Customer Analytics to Transform the Utilities Industry," *Electric Light & Power*, June 21, 2016, http://www.elp.com/Electric-Light-Power-Newsletter/articles/2016/06/the-power-of-customer-analytics-to-transform-the-utilities-industry.html.

[8] Reilly, J., "The AMI Investment Decision," *Metering & Smart Energy International*, June 2, 2008, https://www.metering.com/the-ami-investment-decision-12636/.

[9] de la Peña, C., "How Analytics Can Improve Asset Management in Electric-Power Networks," McKinsey & Company, October 2016, http://www.mckinsey.com/industries/electric-power-and-natural-gas/our-insights/how-analytics-can-improve-asset-management-in-electric-power-networks.

[10] Qualheim, T., "Utility Industry Targets Growing Concern: Momentary Outages," *Electric Light & Power*, June 12, 2015, http://www.elp.com/articles/powergrid_international/print/volume-20/issue-6/features/utility-industry-targets-growing-concern-momentary-outages.html.

[11] Douris, C., "Third Party Energy Resources Are the Future of the Electric Grid," Lexington Institute, February 13, 2017, http://www.lexingtoninstitute.org/third-party-energy-resources-future-electric-grid/.

[12] Smart Inverter Working Group, http://www.cpuc.ca.gov/General.aspx?id=4154.

[13] Zeiss, G., "Future of Utilities: Utility Asset Management Will Be Based on Shared Geospatial Platform," *Between the Poles: All About Infrastructure*, July 5, 2016, http://geospatial.blogs.com/geospatial/2016/07/future-of-utilities-future-of-utility-asset-management-is-map-based-platform.html.

[14] "How to Achieve T&D Asset Data Integrity When Information Is Shared by Multiple Enterprise Systems," GeoNexus Technologies, April 19, 2017, http://www.geo-nexus.com/lessons-learned-from-an-electric-utility-td-asset-data-integrity-review/.

[15] Daki, H., et al., "Big Data Management in Smart Grid: Concepts, Requirements and Implementation," *Journal of Big Data*, April 28, 2017, https://journalofbigdata.springeropen.com/articles/10.1186/s40537-017-0070-y.

[16] "The Soft Grid 2013-2020: Big Data & Utility Analytics for Smart Grid: Research Excerpt," GTM Research, December 2012, https://pdfs.semanticscholar.org/b660/651e4439f103598cb9a5b39275d5867fbbe1.pdf.

[17] "Big Data BlackOut: Are Utilities Powering Up Their Data Analytics?" Capgemini Consulting, May 5, 2015, https://www.capgemini.com/consulting/resources/bigdata-blackout/.

[18] "Big Data—Challenges and Opportunities for the Energy Industry," Sungard, 2013, https://www.sungard.com/~/media/fs/energy/resources/white-papers/Big-Data-Challenges-Opportunities-Energy-Industry.ashx.

[19] "Unlocking the Value of Analytics," Accenture, 2014, https://www.accenture.com/t20150520T210556__w__/us-en/_acnmedia/Accenture/Conversion-Assets/Microsites/Documents6/Accenture-Grid-Analytics-Report-Digitally-Enabled-Grid-2.pdf.

[20] Yuhanna, N., and B. Hopkins, "TechRadar: Big Data, Q1 2016, Big Data Is Critical Technology for Insights-Driven Businesses," Forrester, March 16, 2016, http://www.dbvisit.com/EE/content/pdfs/TechRadar_Big_Data_Q1.pdf.

[21] Brynjolfsson, E., and A. McAfee, "The Business of Artificial Intelligence: What It Can—and Cannot—Do for Your Organization," *Harvard Business Review*, July 2017, https://hbr.org/cover-story/2017/07/the-business-of-artificial-intelligence.

[22] Popovic, T., M. Kezunovic, and B. Krstajic, "Smart Grid Data Analytics for Digital Protective Relay Event Recordings," *Information Systems Frontiers*, June 2015, https://link.springer.com/article/10.1007/s10796-013-9434-9.

[23] Hong, T., and A. Farrell, "Utilities Dust Off the Forecasting Playbook-Smart Grid Data Brings Challenges and Opportunities for Power Industry," *Analytics*, July/August 2013, http://analytics-magazine.org/utilities-dust-off-the-forecasting-playbook/.

[24] Guille, C., and S. Zech, "How Utilities Are Deploying Data Analytics Now," Bain, August 31, 2016, http://www.bain.com/publications/articles/how-utilities-are-deploying-data-analytics-now.aspx.

[25] McCracken, J., "Using Analytics to Optimize the Grid," *Greentech Media*, September 8, 2014, https://www.greentechmedia.com/articles/read/using-analytics-to-optimize-the-grid.

[26] Haq, R., "Analytics Strategy: Creating a Roadmap for Success," *Crossings*, 2014, http://www.sapient.com/content/dam/sapient/sapientglobalmarkets/pdf/thought-leadership/SGM_Analytics_Strategy_2014.pdf.

[27] Deloitte, "Becoming an Insight Driven Organization: Realize Return on Your Analytics Investment Sooner," 2016, https://www2.deloitte.com/content/dam/Deloitte/nl/Documents/deloitte-analytics/deloitte-nl-data-analytics-point-of-view-becoming-an-insight-driven-organisation.pdf.

[28] U.S. Department of Energy, "Advanced Metering Infrastructure and Customer Systems: Results from the Smart Grid Investment Grant Program," September 2016, https://energy.gov/sites/prod/files/2016/12/f34/AMI%20Summary%20Report_09-26-16.pdf.

[29] "Utility Case Study: OGE's Data Analytics Deployment," *Greentech Media*, January 7, 2013, https://www.greentechmedia.com/articles/read/utility-case-study-oge-data-analytics-deployment.

[30] U.S. Department of Energy, "Distribution Automation: Results from the Smart Grid Investment Grant Program," September 2016, https://energy.gov/sites/prod/files/2016/11/f34/Distribution%20Automation%20Summary%20Report_09-29-16.pdf.

[31] U.S. Department of Energy, "Pepco–District of Columbia Smart Grid Project," September 2015, https://www.smartgrid.gov/files/Pepco-District-Columbia-Smart-Grid-Project-2015.pdf.

[32] St. John, J., "Pepco to Put Landis+Gyr's GRIDiant Analytics to System-Wide Test," *Greentech Media*, October 27, 2014, https://www.greentechmedia.com/articles/read/pepco-to-put-landisgyrs-gridiant-analytics-to-system-wide-test.

[33] Chipman, I., "How Data Analytics Is Going to Transform All Industries," Stanford Engineering, February 23, 2016, https://engineering.stanford.edu/magazine/article/how-data-analytics-going-transform-all-industries.

[34] Roggendorf, M., "Making Data Analytics Work: Three Key Challenges," McKinsey & Company, March 2013, http://www.mckinsey.com/business-functions/digital-mckinsey/our-insights/making-data-analytics-work.

[35] Burr, M. T., "Baby Boom Blues: A Series of Articles, Reviews, and Strategies for the Anticipated Utility Workforce Shortage," *Fortnightly*, July 2006, https://www.fortnightly.com/fortnightly/2006/07/baby-boom-blues?page=0%2C2.

8

Electric Transportation: First Mover to a Mobile Carbon-Free Future

Electrification of transportation is the use of hybrid electric and all-electric vehicles instead of all-petroleum vehicles. It also includes the implementation of infrastructure to charge the electric vehicles. An electric vehicle is a vehicle where the propulsion system contains one or more electric motors that contribute, partly or entirely, to powering the vehicle.

Introduction

If I had asked people what they wanted, they would have said faster horses.

—Henry Ford, American industrialist, founder of the Ford Motor Company, and sponsor of the development of the assembly-line technique of mass production

We will not stop until every car on the road is electric... I think there are more politicians in favor of electric cars than against.... You have to match the convenience of the gasoline car for people to buy an electric car.... In order to have clean air in cities, you have to go electric.... You should not show somebody something very cool and then not do it. At Tesla, any prototype that is shown to customers, the production must be better.

—Elon Musk, cofounder, investor, CEO, and product architect of Tesla Inc., founder, CEO, and CTO of SpaceX, cofounder and former chairman of SolarCity, Canadian-American business magnate, investor, engineer, and inventor

The quotes from Henry Ford and Elon Musk illuminate the shift from where the mass market of automobiles began to where it will travel in the future. However, this chapter is not about cars or automobile engines. It is about the electrification of transportation, what it means to the electric utility, and, more importantly, the transformation of the electric utility.

Electric cars are not new [1]. They were introduced more than 100 years ago and have resulted in a broad variety of cars. Technologies currently in the market include hybrid electrci vehicle (HEV), plug-in hybrid electric vehicle (PHEV), or plug-in electric vehicle (PEV). In the mass market, we have seen various makes of cars such as the Toyota Prius, Nissan Leaf, Tesla Model S/X/3, Chevy Volt/Bolt, and the BMW i3. In addition, several automakers have announced new electric models to be released over the next few years. According to a report by Navigant Research, electric vehicles currently constitute about 3% of new vehicle sales and the sales could grow to nearly 7% or 6.6 million per year by 2020 worldwide [2].

Shifting over to electric vehicles requires more than simply changing from gasoline engines to electric motors. It is about moving from using gasoline as a propulsion mechanism to electricity. The resultant mechanism delivers a quieter car with no tailpipe emissions, reduced greenhouse gas emissions, and a reduced use of gasoline. It also creates a new and potential consumer of electricity and a new customer for the utility, one that is not solidly located in one location, is mobile, and potentially delivers power to the grid from different locations.

This chapter starts by defining the impetus for moving to electric cars. Exploring the different types of electric vehicles reveals the larger picture of electrifying transportation. The electrification of transportation has gone far beyond personal transportation. It is possible that other forms of electric transportation may become more prevalent than personal cars. The chapter then defines the infrastructure requirements of electric vehicles (EVs) and a description of the impacts of EVs on the grid. The chapter introduces the concept of Vehicle to Grid (V2G) and its potential to make a significant difference in the operation of the grid. The chapter presents two case studies about the implementation of electric transportation: one about V2G and the other about the policy impacts of EV penetration in a state. The chapter concludes with the impact of electric transportation on the transformation of the electric utility and the grid.

Motivations for Electrification of Transportation

Electrification of transportation is being driven by several factors (see Figure 8.1), which range from reduced dependence on the use of fossil fuels to cutting emissions and significantly contributing to reducing the carbon footprint of the

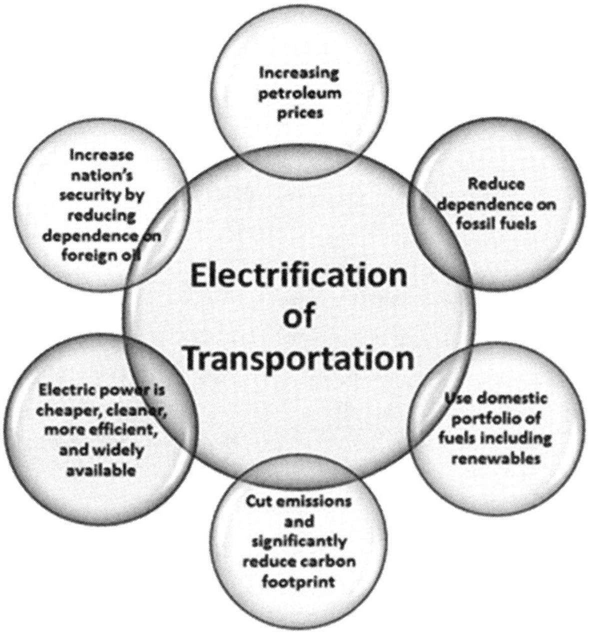

Figure 8.1 Motivations for electric transportation. (© Modern Grid Solutions.)

world. These motivators will become more important over time as the impacts of greenhouse gas emissions become more obvious.

Use Electric Power Instead of Gasoline

The biggest benefit of electric cars is that they no longer need gasoline[1]. Although electricity is not free, an electric car is far cheaper to run. The U.S. Department of Energy has calculated that a typical EV can run for 43 miles on a dollar's worth of electricity. Only a substantial drop in the cost of gasoline could give gas-powered cars anywhere near such a low cost per mile.

A plug-in hybrid eliminates a major portion of the gas bill as well, but it still uses a gasoline engine as a range extender.

Fueling with electricity offers advantages not available in conventional internal combustion engine vehicles. Because electric motors react quickly, EVs are responsive and have good torque. EVs are often more digitally connected than conventional vehicles, with many EV charging stations providing the option to control charging from a smartphone app [3].

1. An exception to this rule is the hybrid car. Some have engines that run on gasoline and some use the gasoline to power an electric motor.

Reduce Dependence on Fossil Fuels

On average, American drivers spend $2,000 to $4,000 on gasoline each year. Electric cars are entirely charged by electricity, thus eliminating the cost of gasoline for the owner. EVs can also help the nation have a greater diversity of fuel choices available for transportation. For example, the United States used nearly 9 billion barrels of petroleum last year, two-thirds of which went toward transportation. Reliance on imported petroleum makes the United States vulnerable to price spikes and supply disruptions. EVs help to reduce this threat because the United States produces almost all its electricity from domestic sources, including coal, nuclear, natural gas, and renewables.

Beyond the fuel-saving benefit, EVs offer another major cost savings: maintenance. Because an EV is fully electric, it no longer uses oil to lubricate the engine. The same is true for a lot of other expensive engine repairs and maintenance, which could afflict a gas-powered car.

Cut Emissions and Reduce Carbon Footprint

EVs can also reduce emissions that contribute to climate change and smog. This will improve public health and reduce ecological damage. There are two categories of vehicle emissions: direct and life cycle.

Direct emissions are emitted through the tailpipe, through evaporation from the fuel system, and during the fueling process. Direct emissions include smog-forming pollutants such as nitrogen oxides, other pollutants harmful to human health, and greenhouse gases, primarily carbon dioxide. All-electric vehicles produce zero direct emissions, which improves air quality in urban areas. Plug-in hybrid electric vehicles (PHEVs), which have a gasoline engine in addition to an electric motor, produce evaporative emissions from the fuel system as well as tailpipe emissions when operating on gasoline. PHEVs are more efficient than comparable conventional vehicles, however, so they still produce fewer tailpipe emissions even when relying on gasoline.

Life-cycle emissions include all emissions related to fuel and vehicle production, processing, distribution, use, and recycling/disposal. EVs typically produce fewer life-cycle emissions than conventional vehicles because most emissions are lower for electricity generation than burning gasoline or diesel [4].

Use Domestic Portfolio or Fuels, Including Renewables

Charging an EV on renewable energy, such as solar or wind, minimizes dependence on fossil fuels and emissions even more.

Falling costs of electric vehicles and solar panels could further reduce worldwide growth in demand for oil and coal by 2020. The cost of solar has fallen 85% in 7 years. Experts predict that solar panels could supply 23% of global power generation by 2040 and 29% by 2050, entirely phasing coal out and leaving natural gas with just a 1% share. By 2035, electric vehicles could make up 35% of the road transport market, and two-thirds by 2050, when it could displace 25 million barrels of oil per day [5].

Solar-cell prices have fallen 85% over the past 7 years, while electric-car battery costs have fallen 73% in the same time. The average battery price is currently $268 per kilowatt-hour, and could decrease to $100 per kWh by 2020 [6].

Increasing Petroleum Prices

The U.S. Energy Information Administration (EIA) forecasts an increase in gasoline prices in the near term. U.S. regular gasoline retail prices currently average $2.35 per gallon (gal), and are forecast to average $2.38/gal, 15 cents/gal higher than last summer. U.S. regular gasoline retail prices are forecast to average $2.32/gal in 2017 and $2.33/gal in 2018. Although this is not a steep increase, some upward price pressures could emerge in the second half of 2018 if global oil inventories decline during that period [7, 8].

Electricity is cheaper than petroleum in many parts of the world, which is why buying an electric car makes sense as insurance against rising gasoline prices. Even in places where there is only a small gap between the costs of electricity and petroleum fuels, EVs can still make a huge difference in operating expenses as they have a lower cost per mile. Specifically, EV owners spend one-third to one-quarter less per mile than those who drive traditional cars. Vehicles fueled by electricity are far more efficient than conventional vehicles, sometimes achieving efficiency equivalents of 100 miles per gallon (mpg). When these vehicles run on batteries alone, the cost of driving is significantly lower than with conventional vehicles. Fuel costs for EVs are typically 2 to 3 cents a mile, whereas owners of conventional vehicles with average fuel efficiency pay more than 13 cents a mile when gasoline prices are $3 a gallon. At current prices, an EV owner will save up to $10,000 over the vehicle's lifetime, compared to a conventional vehicle [9].

Reduce Dependence on Foreign Fuels

National policies that promote vehicle electrification are critical to reduce a country's dependence on foreign oil, reinvigorate domestic manufacturing, and minimize environmental impacts, while enhancing the nation's competitiveness in the global clean energy economy. If the United States committed to deploying 10 million charging stations and making 25% of new vehicles electric by 2020, it would yield benefits that could help strengthen economic, national, and environmental security far into the twenty-first century.

In the United States alone, 94% of cars, trucks, ships, and planes depend on oil. In 2009, the United States imported 11.7 million barrels of crude oil and refined petroleum products per day [5]. At $100 a barrel, this amounted to sending foreign countries—some of them hostile to U.S. interests—more than $1.1 billion to meet the country's daily energy needs [9].

EVs can be divided into two major segments: on-road and off-road.

EV Set #1: On-Road Vehicles

While EVs have been in existence since the late 1800s, it was the introduction of the Toyota Prius that created the renaissance of the hybrid electric vehicle (HEV). The HEV is just one type of the electric vehicle; each car company, some established and some new, is developing its own version of the EV. However, on-road EVs can be subclassified into a clearly defined set of EV technologies.

Hybrid Electric Vehicles

A HEV (illustrated in Figure 8.2) combines a battery-powered electric motor with an internal combustion engine (ICE) system or other propulsion source that can run on conventional or alternative fuel. The internal combustion engine is designed to operate at or near its maximum fuel efficiency during highway cruising. In steady state highway cruising, the internal combustion engine alone drives the vehicle and is designed to operate at or near its maximum fuel

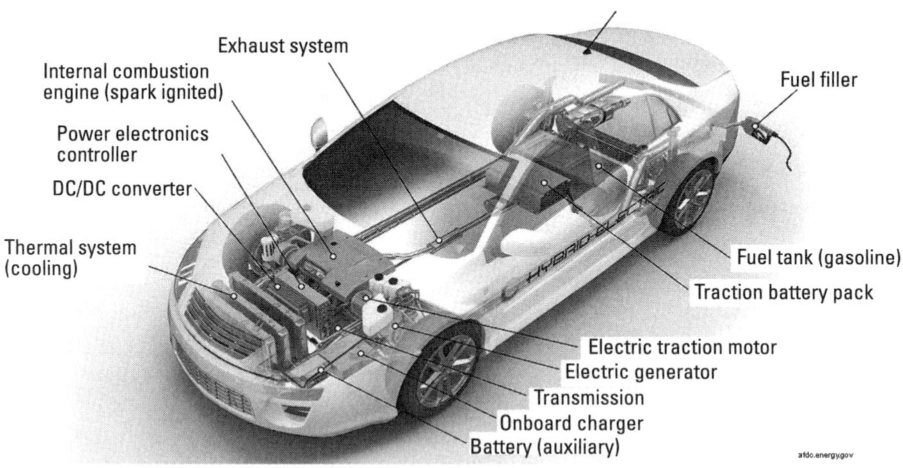

Figure 8.2 Hybrid electric vehicle mechanism. (Source: energy.gov https://www.afdc.energy.gov/vehicles/how-do-hybrid-electric-cars-work.)

efficiency. At low speeds, the electric motor uses the power stored in batteries to drive the vehicle [10].

Plug-In Electric Vehicles

A plug-in electric vehicle (PEV), shown in Figure 8.3, can be charged from an external source of electricity by plugging into an electrical outlet or charging station such as wall sockets. The electricity stored in the rechargeable battery packs powers or contributes to drive the wheels. Plug-in cars have lower operating and maintenance costs than both gasoline-powered cars and HEVs. In addition, they emit no air pollutants when running in all-electric modes. Plug-in electric vehicles (PEVs) can curtail our dependence on oil, cut carbon emissions, and reduce energy costs to fuel our vehicles. However, they do rely on power plants to charge their batteries. By December 2016, PEVs were more expensive than both conventional vehicles and hybrid electric vehicles due to the additional cost of their lithium-ion battery packs. Other factors discouraging the adoption of plug-in electric cars are the lack of availability of public and private recharging infrastructure [11].

As defined in Figure 8.4, there are three subcategories of PEVs: battery electric vehicles, plug-in hybrid vehicles, and a further subcategory, extended-range electric vehicle. Each of these vehicle-types will be defined in more detail next.

Figure 8.3 Example of a PEV by the Oregon Department of Transportation (Fast Charge Exit 33-1 Uploaded by Smallman12q [CC BY 2.0 {http://creativecommons.org/licenses/by/2.0} via Wikimedia Commons https://commons.wikimedia.org/wiki/File%3AFast_Charge_Exit_33-1_(6939232224).jpg].)

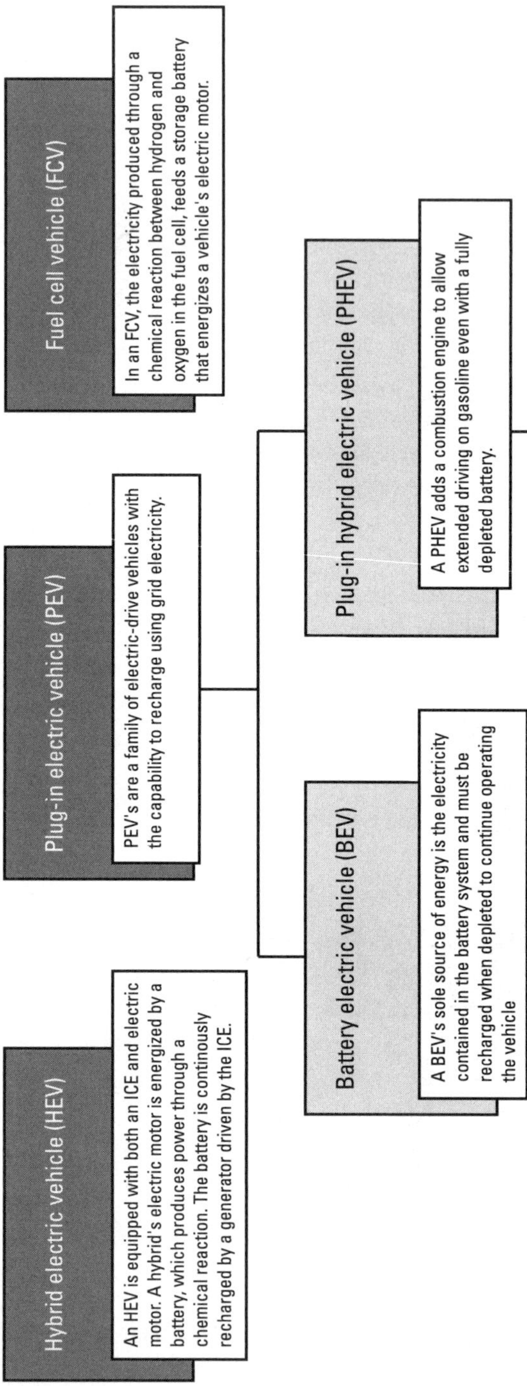

Figure 8.4 Classification of on-road vehicles. (© Modern Grid Solutions.)

Battery-Electric Vehicle

A battery electric vehicle (BEV), shown in Figure 8.5, is an electric vehicle that utilizes electricity stored in a battery pack to power an electric motor and turn the wheels. When depleted, the batteries are recharged using grid electricity, either from a wall socket or a dedicated charging unit. Because the automobiles do not run on gasoline or diesel, but are powered entirely by electricity, battery electric cars and trucks are considered all-electric vehicles. Battery electric cars are significantly cheaper to fuel than conventional vehicles [12].

The cost of replacing the batteries dominates the operating costs of these vehicles. BEVs do not produce pollution; however, the electricity that they use may produce heat-trapping gases and other pollution at the source of its generation or in the extraction of fossil fuels. The amount of pollution produced depends on how the electricity is generated. BEVs powered by renewable energy sources like wind or solar are virtually emission-free. Like other electric and hybrid-electric vehicles, BEVs minimize wasted energy by turning the car off when stopped and by charging the battery when braking. Electric motors are also inherently more energy-efficient than gasoline or diesel engines.

Battery electric cars have the added benefit of home recharging.

Figure 8.5 Battery electric vehicle mechanism. (By Jakob Härter [Bettermann Ladebox B3200] [CC BY-SA 2.0 (https://creativecommons.org/licenses/by-sa/2.0)], via Wikimedia Commons https://commons.wikimedia.org/wiki/File%3ABettermann_Ladebox_B3200_(36434535451).jpg.)

Plug-in Hybrid Electric Vehicle

Plug-in hybrid electric vehicles (PHEVs), shown in Figure 8.6, use batteries to power an electric motor and use another fuel, such as gasoline or diesel, to power an internal combustion engine or other propulsion source. PHEVs can be plugged-in and recharged from an outlet, allowing them to drive extended distances using just electricity. When the battery is exhausted, the conventional engine turns on and the vehicle operates as a conventional gasoline-powered car. PHEV batteries can be charged by an outside electric power source, by the internal combustion engine, or through regenerative braking. During braking, the electric motor acts as a generator, using the energy to charge the battery. PHEVs combine the fuel-savings benefits of hybrids with the all-electric capabilities of a battery-electric vehicle [13].

Extended Range Electric Vehicle

Extended-range electric vehicles (EREV), shown in Figure 8.7, have an internal combustion engine as well as a plug-in battery pack and electric motor. The electric motor always drives the wheels. When the battery is depleted, the internal combustion engine acts as a generator to recharge it, extending the range. Typically, these vehicles have a pure electric battery range of roughly 40 miles, before the vehicle switches to the range-extender mode. The most commonly used range extenders are internal combustion engines, but fuel cells or other engine types can be used. The range-extending vehicle design uses battery power as a primary fuel, thereby reducing the consumption of the secondary fuel, such as gasoline. The secondary system, which kicks in once the batteries have run

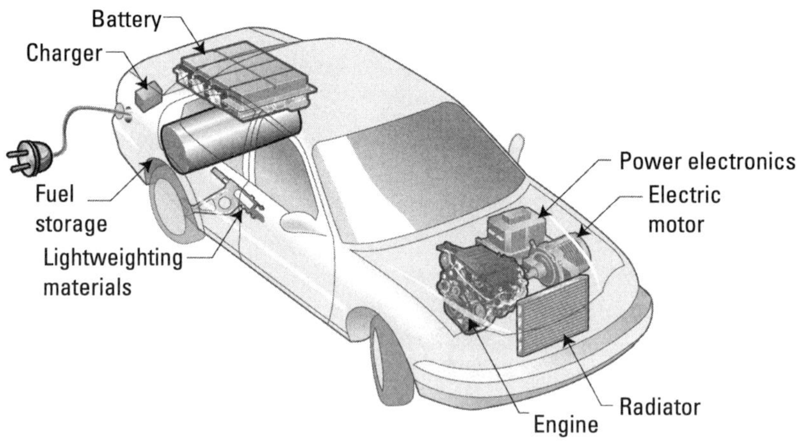

Figure 8.6 Example of a plug-in hybrid electric vehicle. (Source: Matt Howard [CC BY-SA 2.0 (https://creativecommons.org/licenses/by-sa/2.0)], via Wikimedia Commons https://commons.wikimedia.org/wiki/File%3APlug-in_hybrid_electric_vehicle_(PHEV)_diagram.jpg.)

Figure 8.7 Example of extended range electric vehicle. (Source: Raysonho at Open Grid Scheduler/Grid Engine [Own work] [CC0], via Wikimedia Commons, https://commons.wikimedia.org/wiki/File%3AChevroletVoltChargingStation.jpg/.)

down, provides a driving range comparable to a single-fuel vehicle. Because the electric portion is both less expensive and more environmentally friendly, the vehicle uses up the primary fuel source first [14].

Fuel-Cell Vehicle

A fuel-cell vehicle (FCV), shown in Figure 8.8, or fuel-cell electric vehicle (FCEV) is a type of electric vehicle that uses a fuel cell instead of a battery. It may also be used in combination with a battery or super capacitor to power its on-board electric motor. Unlike conventional vehicles, fuel-cell cars and trucks combine compressed hydrogen and oxygen to produce electricity, which runs a motor. Since they are powered entirely by electricity, fuel-cell vehicles are considered EVs—but unlike other EVs, their range and refueling processes are comparable to conventional cars and trucks [15].

Converting hydrogen gas into electricity produces only water and heat as a byproduct, so FCVs do not create tailpipe pollution.

EV Set #2: Off-Road Vehicles

Electric off-road vehicles are an alternative to traditional off-road vehicles. This new generation of vehicles combines the best characteristics of vehicles which

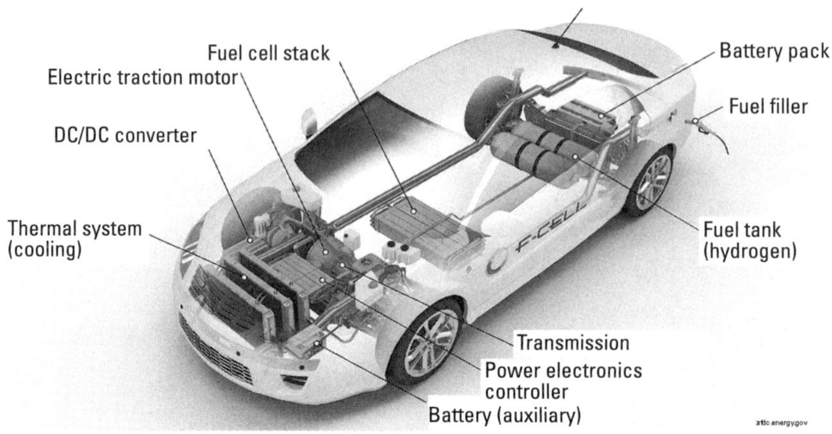

Figure 8.8 Fuel-cell vehicle mechanism. (Source: Energy.gov https://www.afdc.energy.gov/vehicles/how-do-fuel-cell-electric-cars-work.)

are used in nontraditional sectors such as factories, people transportation and road use. The result is a new generation of vehicles with outstanding capabilities, especially useful in sectors where a more standardized electric vehicle is not enough [16, 17].

Figure 8.9 classifies the off-road EVs based on their specific use. Examples of these include airport ground power and long-haul trucking.

Railroad

Electric railways use electric locomotives to carry passengers or haul freight in separate cars or electric multiple units (EMU). Electricity is typically generated in large and relatively efficient generating stations, which is then transmitted to the railway network and distributed to the trains. A nearly continuous conductor running along the track supplies power to the moving trains.

Electric railways offer substantially better energy efficiency, lower emissions, and lower operating costs than conventional diesel trains. Some electric traction systems provide regenerative braking, which turns the train's kinetic energy back into electricity and returns it to the supply system. Disadvantages of electric traction include high capital costs and a vulnerability to power interruptions.

Seaport

In seaports, heavy equipment, which used to be powered by diesel, are now capable of being powered by electricity. This is accomplished either by being retrofit or as new equipment. Electric options for harbor craft are new to the mar-

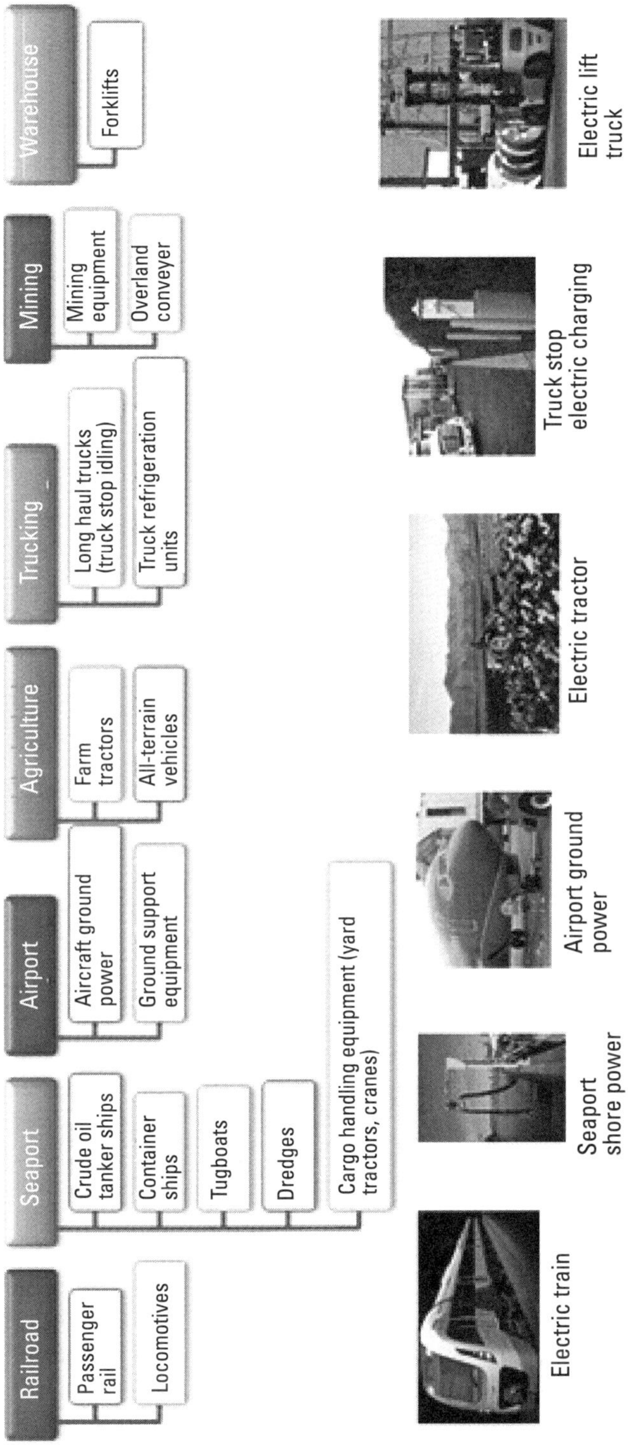

Figure 8.9 Classification of off-road vehicle. (© Modern Grid Solutions.)

ket; however, hybrid tugboats and electric ferries are becoming more common. An electric dredge uses power delivered from shore via a large cable, and can be beneficial for long-term projects. Other electric options for seaports include forklifts and automated guided vehicles. Short-distance electric yard hostlers at ports and distribution centers are also being demonstrated by the manufacturers. The most common drawbacks of electric vehicles in this application are the reduced distance range achievable and the long charging time of the battery. However, a harbor area is often a closed system with relatively short driving distances and relatively predictable usage. Therefore, the utilization of electrical vehicles within harbors could be optimized for each specific harbor's needs.

Airport

Airport fleets operate thousands of vehicles every day. These fleets include both ground service equipment (GSE) such as tugs, tows, and baggage tractors and landside vehicles including buses, shuttles, taxis, and limousines. Airports are uniquely suited for the use of electric vehicles. They present excellent opportunities for alternative fuel vehicles (AFV) as the fleets need to travel many miles a day and use large quantities of fuel. They also tend to require long idle times and frequent stops. Use of AFVs can help these fleets reduce engine wear, pollution, and fuel costs. They can also reduce the ground level emissions and exposure of travelers and workers to air pollution.

Agriculture

Agricultural vehicles pose unique challenges. Requiring a long operating life and minimum down time, these vehicles are exposed to shock and vibration, which is not a factor for other electric and hybrid electric vehicles. Advantages of electric farming machinery range from less fuel consumption to reduced emissions. Use of electric, rather than hydraulic, power in agricultural auxiliary machines would lead to increased productivity by enabling higher speeds and accuracy. Auxiliary machines include saws, balers, and mowers. Low-speed provides high torque, making them ideal for heavy-duty applications common in agricultural situations.

Trucking

Utility and delivery fleet managers are successfully adding medium-duty and sometimes heavy-duty plug-in hybrid work trucks and vans to their fleets. Plug-in hybrid line, bucket, crane, and other utility work trucks are available today. Manufacturers are integrating the electronic components with different chassis and engines. Since a typical utility bucket truck idles 3 to 4 hours a day, the emissions reductions and fuel savings are significant. Some configurations also operate in hybrid-drive mode, providing even more savings. While parked, long-haul truck drivers can plug into the grid instead of idling their truck or auxiliary engines to power heating, air conditioning, and accessories. Prime electrification candidates are delivery vehicles. They typically travel fixed daily routes under 80 miles, stop and start often.

Mining

Underground and surface mining utilizes electricity for much of their heavy equipment. Examples are the continuous miner and long wall systems that remove coal and other minerals, roof bolters that secure the overhead rock underground, and most draglines, shovels, and conveyors. Electric equipment enhances worker safety in underground mines because it runs quietly. In the prevailing darkness, where it is not easy to see coworkers, this aids in communication. Zero emissions improve air quality. Some underground mines with rail systems use mining locomotives to haul people, equipment, and materials inside the mine. Electric locomotives can replace diesel locomotives in many of these applications. Opportunities for further electrification are numerous, especially underground.

Warehouse

In warehouses and industrial parks, on the nation's highways, and on city streets, electricity powers the equipment that supports the transfer of consumer and bulk goods. Studies have shown that electrification of forklifts can result in huge savings. Today, multiple batteries can be charged simultaneously, eliminating the need for extra batteries. These vehicles also save space, are more energy-efficient, and can reduce pollution. Electric transport refrigerators displace the conventional diesel-powered transport refrigeration units (TRUs), which are a significant source of air pollution in transport hubs. TRUs with electric standby can now plug into shore power at home base or during stops.

Infrastructure Needs and Capabilities

Standard cars with internal combustion engines (ICE) have an extensive network of gas stations around the world. These cars can replenish their energy needs with a 5-minute stop. This network was developed over more than 100 years, which grew to match the number of cars sold worldwide. EVs do not have this infrastructure. The infrastructure needs of EVs are very different.

Plug-In Electric Vehicle Components

Compared to the propulsion system in an internal combustion engine vehicle, which has hundreds of moving parts, electric vehicles are simple and have far fewer components (shown in Figure 8.10). The main components of an electric vehicle are described below.

For battery technology:

- The lithium-ion battery is currently the dominant battery technology in the PEV market.
- The cost of a battery pack is expected to drop significantly with greater production volumes.
- Battery performance has also been steadily increasing.

Power electronics:

- Control the flow of energy into and out of the PEV battery.
- Could include onboard or off-board chargers and onboard power conditioning (e.g., power inversion for use by the electric motor).
- Represent a significant upfront cost for PEV development in the near term, but is expected to decrease considerably with volume and with electrification trends in general.

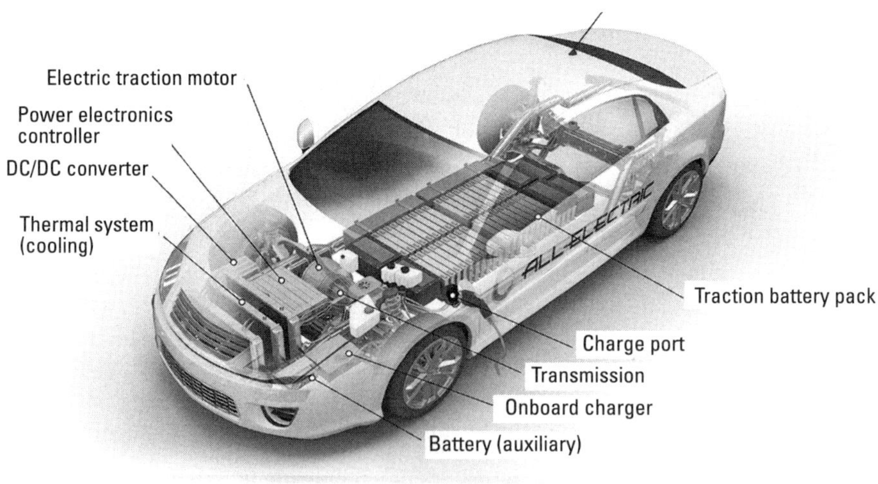

Figure 8.10 Plug-in electric vehicle components. (Source: Energy.gov https://www.afdc.energy.gov/vehicles/how-do-all-electric-cars-work.)

Electric motors:

- Utilize an electric motor to turn the wheels of the vehicles. There are several different drive system designs in use today, for example, vehicles with a single large electric motor coupled to the rear wheels through a differential housing. Other designs utilize two smaller motors to power each wheel separately through independent drive shafts.
- Convert electrical energy into mechanical energy. Two types of electric motors are used in electric vehicles to provide power to the wheels. The direct current (DC) motor and the alternating current (AC) motor [18].

Batteries and power electronics are the two components likely to have a significant effect on the penetration rate of PEV technology.

PEV Charging Infrastructure

Charging infrastructure is a crucial aspect of PEV operation, as illustrated in Figure 8.11. All PEVs require at least one readily available electric vehicle supply equipment (EVSE) at their home parking location, such as at a residence, parking facility, or a fleet yard. National travel survey data shows that vehicles spend 66% of their time parked at home, making residential EVSE the highest priority in the industry. Employer-provided EVSE at the workplace is also important, as vehicles spend another 14% of their time parked at work. As the penetration of batter electric vehicles increase, a public infrastructure network will also be required to provide for the safe recharge and reliable operation of

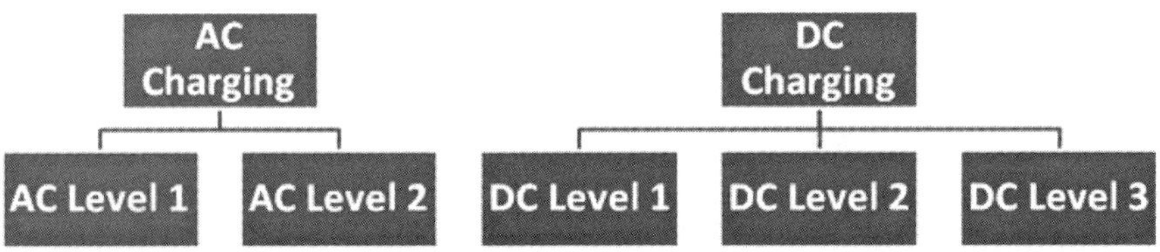

Charging Levels	AC Level 1	AC Level 2	DC Level 1	DC Level 2	DC Level 3
Voltage (Volt)	120	208-240	200-450	200-450	200-600
Current (Amps)	12-16	12-80	<=80	<=200	250, 350, 400
Demand (kW)	1.44-1.92	2.5-19.2	<=19.2	<=90	<=240
Charge Time (Average)	3-10 hours	3-8 hours	N/A	30 minutes	N/A
Supply Connection	NEMA 5-15R	Hardwired	J1772 AC connector	Combo connector, CHAdeMO charger	TBD

Figure 8.11 Characteristics of PEV charging. (© Modern Grid Solutions.)

battery electric vehicles. Public infrastructure also increases the electric utility of plug-in hybrids, allowing them to travel greater distances on electricity because they can charge.

- Charging is grouped into two classifications based whether the electricity delivered to the charge port on the vehicle is alternating current (AC) or direct current (DC).
- With AC charging, an on-board charger (an AC-DC converter) transforms the supply into DC electricity for storage in the battery.
- Devices called electric vehicle supply equipment (EVSE) are used to safely control the delivery of AC electricity to the vehicle. EVSE comprises all the hardware from the electricity source to the vehicle skin.
- The EVSE may also provide other functionality such as informational displays, fee collection capabilities, power metering circuitry, and other features appropriate to specific applications.
- DC charging, often referred to as fast charging, uses an off-board charging station to convert AC electricity to DC and directly charge the vehicle battery without the need for an onboard charger.
- The primary purpose of DC charging is to enable the rapid recharge of battery electric vehicles.

PEV Meets Grid: Factors Impacting PEVs Connecting to the Grid

The electric power grid is the single most important infrastructure required for the EV. The impact of connecting EVs to the grid can vary, based on a broad variety of factors (illustrated in Figure 8.12), which depend on characteristics such as the type of EV or type of charger.

Level of Penetration

Utilities closely monitor sales for plug-in hybrid and all-electric vehicles because of their potential impacts on electricity demand and the opportunity for new business. There has been a steady increase in monthly sales of plug-in hybrids and all-electric vehicles since 2011, along with a growing number of manufacturers and models to increase customer choice. In the United States, electric vehicles are projected to grow from nearly 300,000 in 2014 to more than 2.7 million in 2023. The pace and geographic distribution of PEVs have a huge impact on demand of feeders on distribution network.

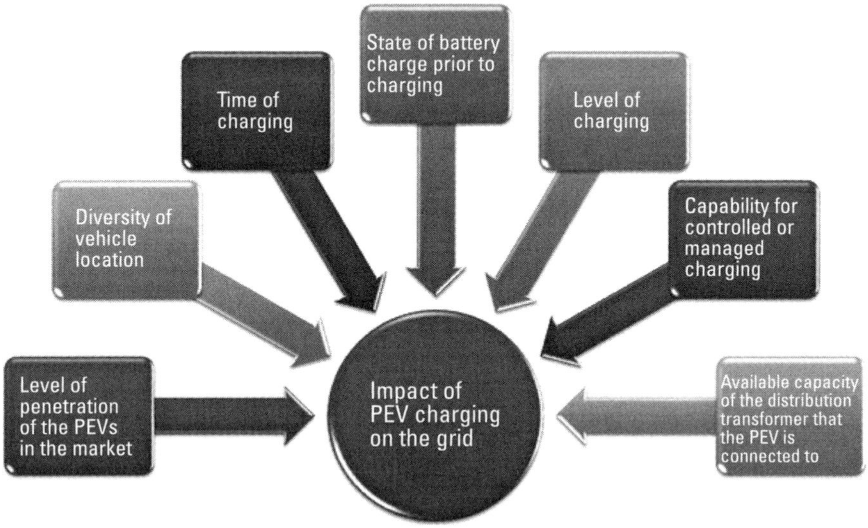

Figure 8.12 Factors impacting PEV connecting to the grid. (© Modern Grid Solutions.)

Diversity of Vehicle Location

To estimate the system-wide PEV charging demands, utilities need to know the travel patterns and estimates for the daily travel distance of PEVs, together with where and when PEVs will be charged. That information can be represented by a time-varying distribution of the number of simultaneously charging PEVs at each distribution node. Usually, national travel surveys or database obtained by GPS recording devices installed in PEVs obtain the information. The data can be analyzed to determine optimal locations for deployment of commercial and public electric vehicle charging stations.

Time, State, and Level of Charging

The standard charging levels for PEVs include: AC level 1, AC level 2, and DC fast charging. AC level 1 charger is for use at home, which requires a small power level (1.4 kW to 1.9 kW) over a long charging time (17 hours to 7 hours). However, AC level 2 and DC fast charger require a huge amount of power (19.2 kW to 40 kW) to charge over a short period (3 hours to 20 minutes). PEVs are suitable for charging in public and commercial areas. Customers strongly prefer shorter charging sessions, but this creates a greater demand on the distribution feeder. Daily charging sessions use an average 7.1 kW/hr, which would add about 2,500 kW/hr to a distribution feeder. This could have an adverse impact on the grid during peak demand times.

Controlled and Managed Charging

Charging scenarios can include both controlled and uncontrolled charging. In uncontrolled charging, a PEV starts charging immediately after it is plugged into the power grid, and the charging rate is fixed. Controlled charging, can be price-based control or direct control (smart charging), with some control over the charging process. Advanced knowledge of PEVs' charging profiles help utilities better estimate system loads. Controlled and managed charging helps to optimize planning of the charging infrastructure to mitigate any negative impacts from PEVs on the power system.

Available Capacity on the Distribution Grid

Utility surveys show that commercial and public stations were used mostly during business hours, and, as such, there was a higher likelihood for overlap with peak time periods for utilities. For example, while this is usually the late afternoon in the summertime, exact peak times can vary. Utility planning departments have found that the clear majority of charging sessions occur between 9 A.M. and 7 P.M. This data has significant implications for utilities to determine the available capacity of transformers and feeders to accommodate this charging pattern.

Heating and Cooling Cycles for the Grid Components

Charging profiles, especially those related to time of use for charging equipment, provide utilities important information on whether the grid assets, such as transformers, have sufficient time to cool down before they are loaded up again. This can prevent early asset failure and reduce maintenance costs for the utility. Time-of-use rates for PEV charging can help mitigate the negative effects of cluster charging.

Grid Meets PV—How the Grid Is Impacted

The future anticipated increase of EVs will impact the low-voltage distribution networks (see Figure 8.13) and can have other unintended consequences. It is anticipated that the majority of PHEVs/EVs will be charged at home, putting additional stress on the existing local low-voltage distribution network, which must then cater to the increased demand in energy. Uncontrolled charging of multiple PHEVs/EVs can raise the daily peak power demand, which leads to increased transmission line losses, higher voltage drops, equipment overload, damage, and failure [19].

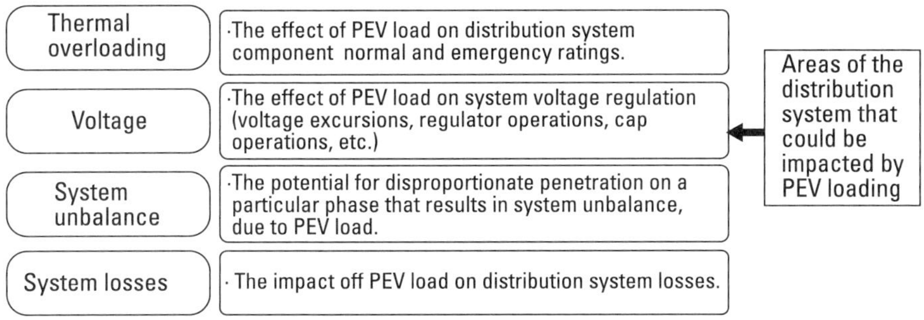

Figure 8.13 Factors impacting the grid when PEVs connect to it. (© Modern Grid Solutions.)

Thermal Loading

Utilities design distribution networks for a long-time horizon, owing to the long service life of its grid assets such as power transformers. Uncontrolled charging of EVs could produce unforeseen and sudden load patterns on a part of the distribution grid, which would have an adverse impact on grid assets such as causing power transformers and feeders to exceed their thermal limits and overheat. If there is insufficient time for the assets to cool down, as in the case of cluster charging during peak hours, it can lead to premature failure of transformers and lines.

Voltage Impacts

As the number of EVs rise and their charging requirements increase, the quality of voltage on distribution feeders will be impacted. A study on uncontrolled charging in a neighborhood of approximately 8,500 households, showed an additional load of only 10%. The EVs were found to be sustainable before voltages dropped below acceptable levels.

At EV penetrations of 30% and in the worst case of uncontrolled charging, losses were calculated to be 6% and voltage deviation calculated to be 10.3% [20]. Other adverse impacts could be voltage fluctuations, which could lead to wear and tear on voltage regulators and capacitor banks due to frequent operations.

System Unbalance

In three-phase power systems the generated voltages are sinusoidal and equal in magnitude, with individual phases 120° apart. However, voltage on the distribution system can be unbalanced due to asymmetrical end-user load. This causes the magnitude and phase angle of voltage to change.

Typically, lighting load is one of the distribution system's most unbalanced loads. Single-phase plug-in electric vehicle (PEV) charging on the distribution feeder is another cause. Voltage unbalance in the low voltage distribution network one of the main power quality issues caused by electric vehicles [21].

System Losses

A residential EV charging station in North America provides a 120-V (level-1) or a 240-V (level-2) voltage supply to the connected EV through either a normal wall outlet or a dedicated charging circuit. EV owners typically charge their EVs overnight at residential charging stations using level-2 chargers. Unfortunately, the increasing number of residential EV chargers may cause several challenges for the distribution system, one of which is an increase in system losses. The increased load demand due to the EVs can overload service transformers, deteriorate the transformers' life, and increase system losses [22].

Vehicle to Grid

On average, an electric car only operates 1 hour a day. The car could be used for the hours they are not being driven, as temporary stationary power storage devices. The energy stored in the vehicle battery could be returned to the power grid at times of increased energy demand.

Electric-drive vehicles (EDVs), whether powered by batteries, fuel cells, or gasoline hybrids, have within them the energy source and some of the power electronics capable of producing the same 60-Hz AC electricity that powers homes and offices. When appropriate interconnections are made available to allow this electricity to flow from cars to power lines, it is called vehicle-to-grid power (V2G), as illustrated in Figure 8.14. These interconnections could either be at the home (installed by the homeowner) or at other locations where they could be installed by the utility or a commercial business. For battery and plug-in hybrid vehicles also, much of the capabilities for the power connection is already available, as they must plug in to the grid to recharge. For fuel-cell and fuel-only hybrids, an electrical connection must be added to make them V2G-capable.

Cars pack a lot of power. One electric-drive vehicle can put out over 10 kW, the average draw of 10 houses. The key to realizing economic value from V2Gs is using grid-integrated vehicle controls that can be used to dispatch according to utility's needs.

- The basic concept of V2G power is that EVs provide power to the grid while parked.

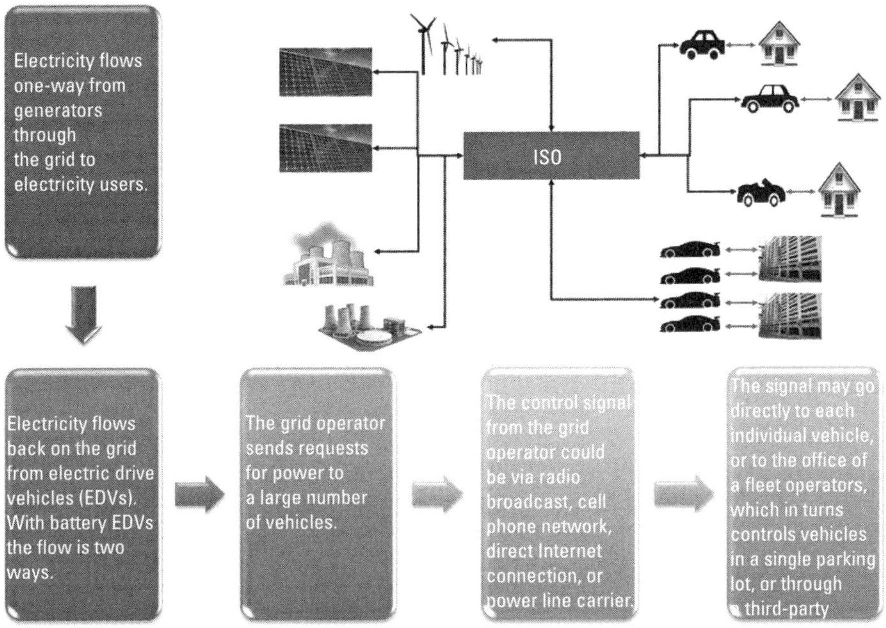

Figure 8.14 V2G mechanisms of working and integration. (© Modern Grid Solutions.)

- The EV can be a battery-electric vehicle, fuel-cell vehicle, or a plug-in hybrid.
- Battery EVs can charge during low demand times and discharge when the grid requires it.
- Fuel cell EVs generate power from liquid or gaseous fuel.
- PHEVs can function in either mode: deliver from their batteries or generate from their built-in gasoline-powered engine.

Infrastructure Required for V2G to Work

V2G requires a vehicle that can be bidirectionally charged. It must also be able to supply energy back into the power grid through an infrastructure, as illustrated in Figure 8.15. This represents an estimated value to the utilities of up to $4,000 per year, per car [23].

Connection to the Grid for Electrical Energy Flow

Whether the vehicle is charging or discharging, it needs to be connected to the grid. This also means the feeder where the vehicle is connected must be capable of handling two-way power flow with all the necessary system protections that go with it.

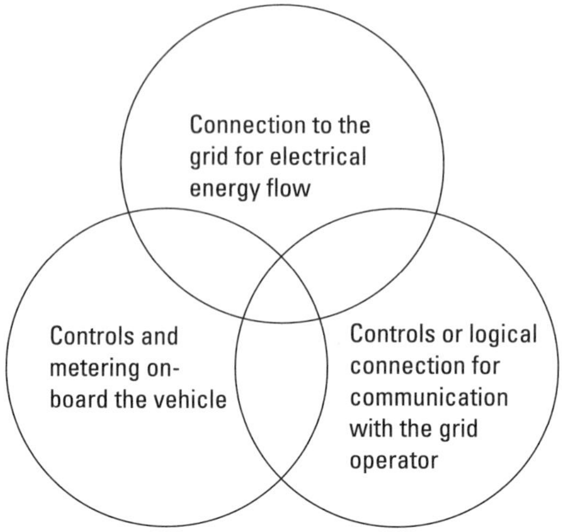

Figure 8.15 Infrastructure required to make V2G work. (© Modern Grid Solutions.)

Controls and Metering Onboard the Vehicle

There are several types of controls that are needed on board the vehicle. More information on these controls is presented in Chapter 2.

- *Smart inverter:* The energy stored in an EV is DC in nature. This means that any connection to the electrical grid will need some form of conversion of the stored energy into constant voltage AC power.
- *Meter:* The meter needs to be able to track the amount of energy stored in the EV's batteries and measure how much energy is consumed or delivered to the grid.
- *Control mechanism:* A local control mechanism must inform the car to charge or discharge the batteries. Based on the capabilities of the smart inverter, the control mechanism may also be asked to deliver either real or reactive power.

Control or Logical Connection for Communication with the Grid Operator

EVs capable of performing V2G connections need to be in telecom contact with the grid operator or equivalent. The grid operator needs to:

- Know where the EVs are located (or connected) on the grid.
- Know how much capacity they can deliver to the grid.
- Control the V2G capable EV with commands to charge or discharge (watts or VARs) based on local need.

Applications of V2G

V2G is still somewhat in a nascent stage from both a technology perspective as well as from a policy perspective. However, there is enough analysis and pilots that have been conducted to define a set of potential applications and value streams [24], as illustrated in Figure 8.16.

Power Markets—Mainly Ancillary Services

Electric vehicles integrated with the grid are suitable for three power markets: peak power, spinning reserves, and regulation services. Some vehicles, based on type and charging capacity, are better suited than others for individual power markets. This indicates that, in terms of their economic value to the market, matching the vehicle type to power market is important.

Based on cost of electricity, studies have found that peak power is the most promising. Typically, battery-powered vehicles serve the peak power market by charging their batteries when demand is off-peak and price is low and selling power back to the grid when demand is at peak and the price is high.

Vehicles can provide ancillary services of a higher quality than currently available, or fast-response, available in small increments, and distributed. Spinning reserves show economic viability for most vehicles. Regulation services

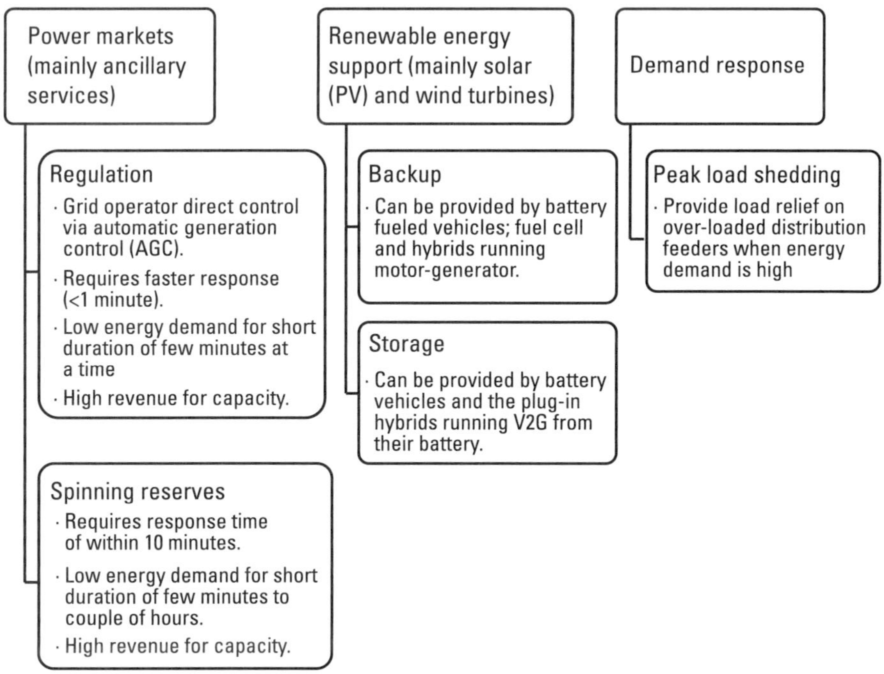

Figure 8.16 Applications of V2G. (© Modern Grid Solutions.)

involve higher numbers, for both revenue and cost, because vehicles can sell regulation more of the time. The battery vehicles appear to be especially suitable for regulation because their shallow cycling causes less battery degradation, and because batteries experience very little discharge when providing both regulation up and regulation down [25].

Renewable Energy Support

Electric vehicles are expected to play a momentous role in helping the electric grid be more flexible. EVs could be deployed to buffer uneven power generation of wind and solar. Owners might, for example, wait to charge their cars until the height of the day, when solar is plentiful and cheap, or they might delay their charging when the grid is under the most stress, during peak hours in the early evening. In addition, excess battery capacity can be utilized to return power to the grid during peak demand times.

Demand Response

Electric vehicles have the potential to provide load relief during times of peak demand either by throttling their charging levels or providing power back to the grid. This is especially feasible with dynamic pricing where electric vehicle owners are encouraged to charge their batteries on the cheap and avoid or minimize the punitive effect of peak power prices. This will incentivize them to provide back to the grid during high price periods. Electric vehicles can therefore serve as demand response devices.

Steps Being Taken by Utilities to Aid in EV Integration

EVs represent a new load on the electric grid. A load that exhibits very different characteristics than those that came before. Some interesting characteristics of how this new load impacts the utility and its operations are listed here:

- Unlike normal customers, who are distributed across the entire footprint, EVs tend to come in clumps. Some feeders may have a large concentration of EVs and some may have none.
- The load is mobile. Depending on the need of the EV's owner, the EV may charge (draw power from the grid) at different locations: the home, the office, and out in the community.
- EVs have a significant load. In many cases, owners install dedicated EV charging circuits (some AC 220V and some DC). When this occurs, charging an electric car is equivalent to adding one more house to the grid. This can be an excessive burden, as the feeder delivering power to

the house may not be sized for it. A typical house is sized at 200 amps, meaning it is designed to draw about 2 kW of power at times of peak demand, but an EV could draw between 5 and 20 kW depending on the car.

Given the low penetration of EVs in the marketplace, much of these issues are not yet a problem. When penetration reaches critical mass, these issues must be addressed, along with any as yet unknown problems these new technologies may bring, as illustrated in Figure 8.17.

To cope with this new load, utilities have begun to implement the following steps, as illustrated in Figure 8.17:

- *Utility tariffs and rates to encourage off-peak charging:* Developing an EV tariff to incentivize owners to charge their EVs at night or other off-peak times, which may be different depending on the type of neighborhood where the chargers are located. This would also include the installation of a separate smart meter to confirm the time and duration of charging.
- *Controlled charging of EVs:* Devices such as smart chargers are being developed and implemented in various utility pilot programs. Dynamic pricing or other market-based pricing techniques are being implemented to assist in the decision process.
- *System asset management and good system design practices:* Improved planning and monitoring of asset usage for the devices on the feeders where an increased concentration of EVs helps utilities define enhanced system design practices.
- *Customer education and outreach programs:* Utilities are educating their customers on the impacts of connecting their EVs to the grid and dem-

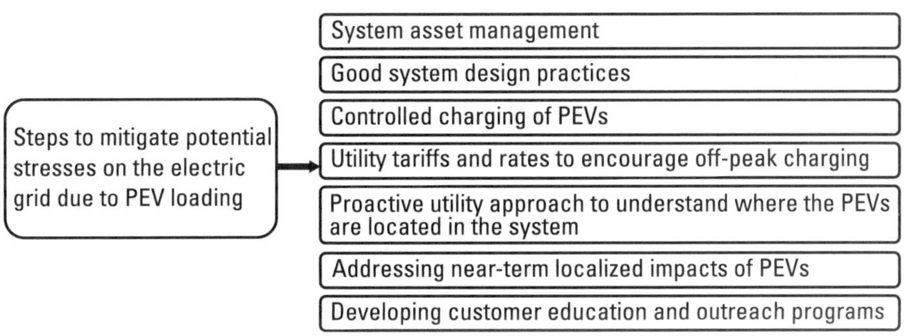

Figure 8.17 Steps taken by utilities to aid in EV integration. (© Modern Grid Solutions.)

onstrate steps such as charge at night to improve usage costs and load problems.

- *Proactive approach to understand where PEVs are located:* Utilities work with EV dealers to understand locations where EV buyers live. This information can be taken into consideration when performing system analysis and design.
- *Addressing near-term localized impacts of PEVs:* In some situations, utilities are also taking some innovative steps to respond to near-term localized impacts of EVs. Some of the steps being taken include installing electric storage supported by solar PV, installing super chargers in communities, and installing residential batteries.

Case Studies

Case Study #1: Demonstration of V2G for Energy Storage and Frequency Regulation in the PJM System

The following case study was a test conducted by a university-industry research partnership between the University of Delaware, Pepco Holdings Inc. (PHI), and PJM Interconnect. This case study is a practical demonstration of vehicle-to-grid power, providing real-time frequency regulation from an electric car. In October 2007, a team of PHI, PJM, and university engineers and officials at the University of Delaware successfully interconnected an AC propulsion "eBox" (a converted Scion xB) to the PJM grid. The team used a direct signal from the PJM control center to dispatch the vehicle as a regulation resource, like traditional generators.

A fully functional, freeway-capable electric vehicle was converted to a V2G capable car. AC propulsion installed an AC induction motor, an AC-150 power electronics unit, and a custom-built battery (355V, 35 kWh battery). At the University of Delaware, an Arcom director, a communications gateway, was installed in the vehicle to receive the PJM signal to control charging and discharging. The test vehicle, responding in real time to PJM's regulation signal, and can provide 19 kW and responding in subseconds. Communications from PJM were translated, directed, and transmitted to the vehicle via a power-line carrier. The command signal was lifted from the power line and decoded onboard the car by the Arcom director.

From PJM's perspective, a new generator was added to the queue for ancillary services called V2Gcar1.

On PJM's SCADA, V2Gcar1 is shown as a generator, plugged in and responding to the encrypted signal used by PJM to control generators' output (see Figure 8.18).

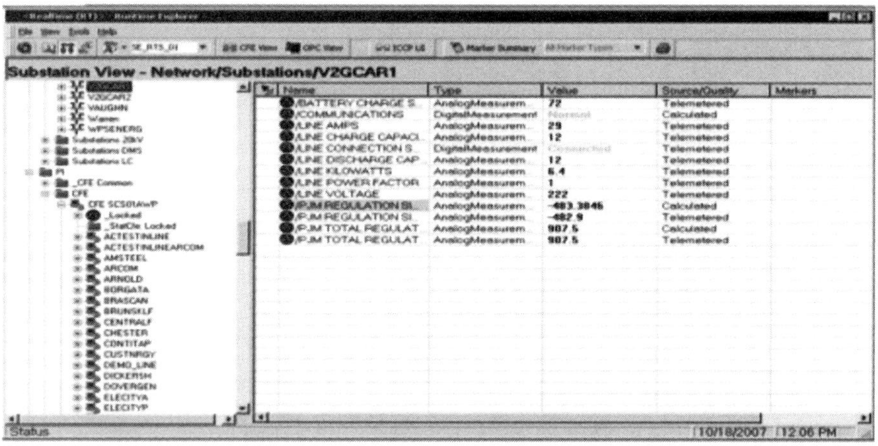

Figure 8.18 V2G Car 1 as a generator. (Source: University of Delaware https://www1.udel.edu/V2G/resources/test-v2g-in-pjm-jan09.pdf [Page 17].)

During the regulation test, when the vehicle was set to respond to the regulation up and down signals, the signal modified the power flow into and out of the battery.

Figure 8.19 shows the regulation service provided by the vehicle.

The data gathered from this test demonstrated that electric vehicles can provide ancillary services and storage. The highest value ancillary service is regulation. In areas with deregulated electricity markets, regulation can have average values of $30 to $45/MW per hour. A second market of interest is spinning reserves, with values in the range of $10/MW per hour, but with much less frequent dispatch. The primary revenue in both markets is for capacity rather than energy, and both markets are well suited for batteries as a storage resource because they require quick response times, yet low total energy demand. Therefore, V2G-capable cars can provide a good revenue stream in ancillary service markets. Additionally, V2G-capable cars can be connected and aggregated in large numbers, and used as dispersed energy storage for intermittent renewable resources such as wind and solar [25, 26].

As V2G moves from pilot to mainstream, it is important to consider manufacturer warranty concerns that tend to be impacted heavily by charge-discharge cycles necessitated by V2G connections.

Case Study #2: Drive Electric Vermont Program

The Drive Electric Vermont (DEV) program was formed in 2012 from a memorandum of understanding (MOU) between three Vermont state agencies, the Vermont Agency of Transportation, the Vermont Agency of Natural Resources, and the Vermont Public Service Department, and the nonprofit Vermont

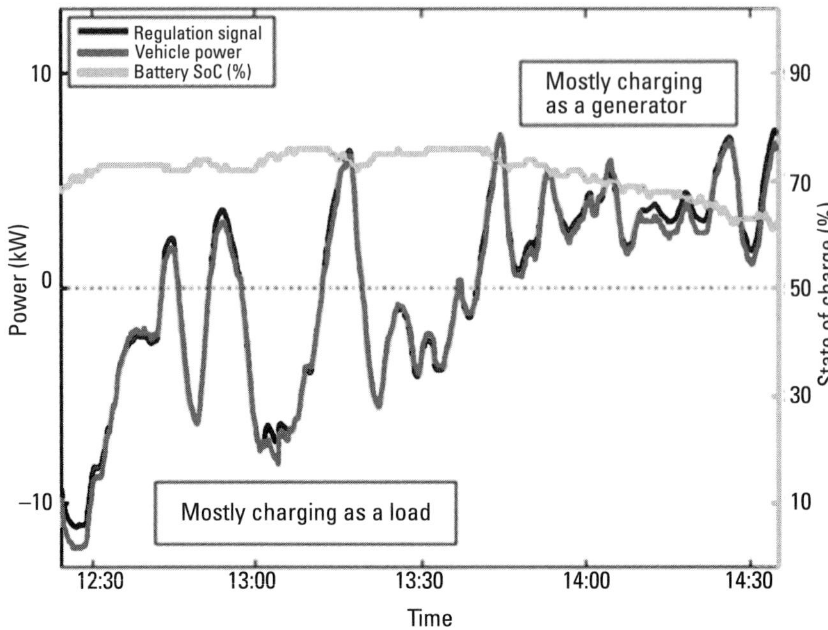

Figure 8.19 Regulation service provided by the vehicle. (Source: University of Delaware https://www1.udel.edu/V2G/resources/test-v2g-in-pjm-jan09.pdf [Page 23].)

Energy Investment Corporation (VEIC). The goal of the DEV Program is to increase use of electrified transportation in Vermont through policy development, education and outreach, and infrastructure development. Measures of success for the DEV program include the number of PEVs registered in the state, the number of people aware of PEV options and considering them for vehicle purchases, the availability of workplace and public charging infrastructure, and state and local policy support (e.g., building codes). Figure 8.20 illustrates steady growth in Vermont PEV registrations since 2012, due in part to the program.

The DEV Program can be broadly broken into four components: strategic planning/leadership, stakeholder/partnership development, education and outreach, and incentives. Early phases of the program focused heavily on strategic planning, and stakeholder and partnership development. As the foundation of the program became increasingly solidified, a transition was made to education and outreach activities and charging infrastructure development. This emphasis has continued with the addition of grant and incentive programs supporting implementation of charging infrastructure and purchasing of PEVs.

Vermont has experienced a significant increase in electric vehicle charging stations over the past three years. The number of electric vehicle charging infrastructure grew from 17 in January 2013 to 111 in January 2016. This growth

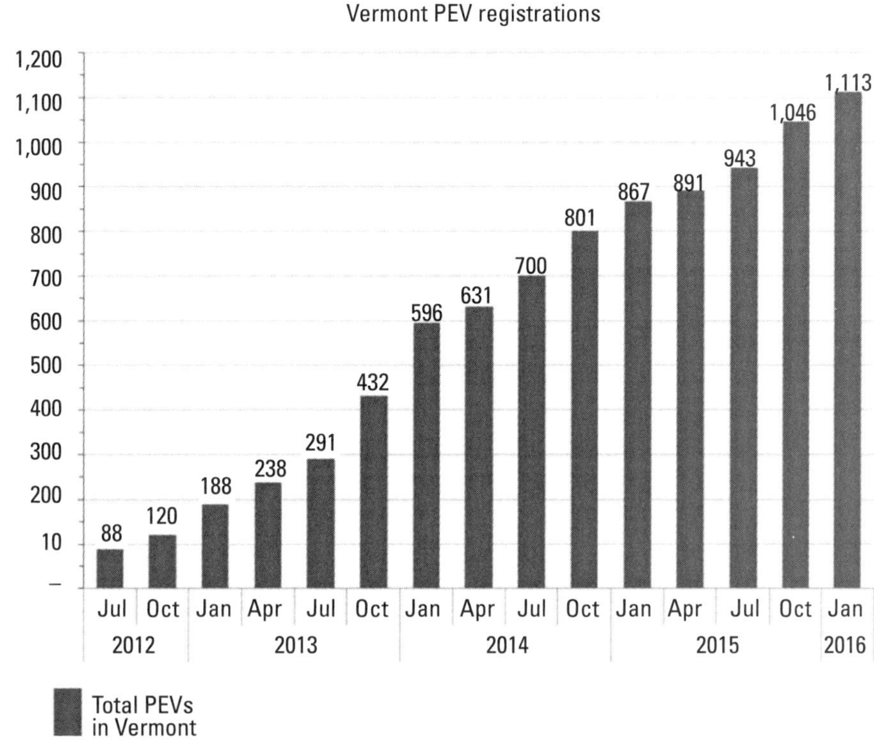

Figure 8.20 Monthly Vermont PEV registrations. (Source: Drive Electric Vermont Case Study [Report ID–INL/EXT-16-38077], https://energy.gov/sites/prod/files/2016/06/f32/Vermont%20 Case%20Study.pdf.)

demonstrates the effectiveness of combined efforts to increase the availability of PEV charging, particularly with the advent of more robust utility and state participation.

As shown in Figure 8.21, the largest segment of the market with EVSE installations is retail, especially standalone businesses where customers purchase goods or services. Most of these are short-term charging opportunities, lasting perhaps 2 or 3 hours (e.g., malls, grocery stores, and markets). Many of these EVSEs are in downtown areas, serving PEV drivers who may be visiting several different places around a town center. The retail and parking (short-term) categories have nearly identical charging level structures, with the clear majority being Level 2 EVSE, followed by DC fast charging (DCFC) and lastly Level 1 EVSE.

Achievements of the DEV program in 2016 and beyond included:

- A new round of consumer/dealer incentives in greater quantities and at higher incentive levels. Incentive forms will be distributed on an as-needed basis to support more rapid use of available funds.

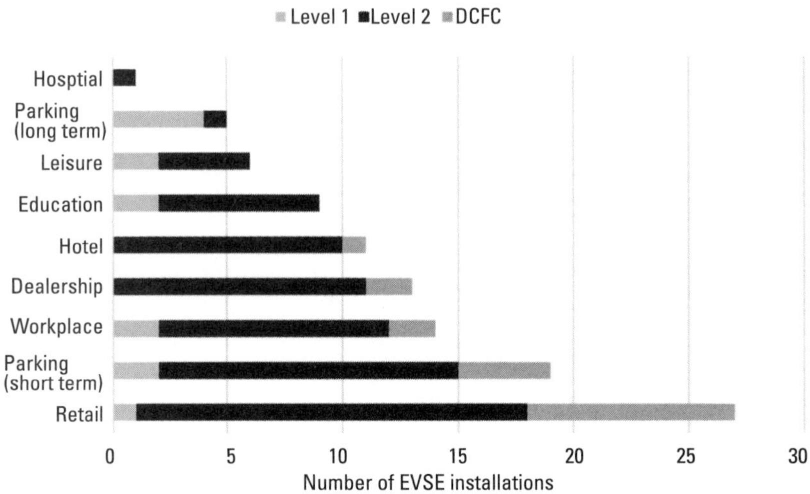

Figure 8.21 EVSE venues and charging levels. (Source: Drive Electric Vermont Case Study [Report ID–INL/EXT-16-38077] https://energy.gov/sites/prod/files/2016/06/f32/Vermont%20Case%20Study.pdf.)

- Continuation of the marketing campaign, placing greater emphasis on search engine optimization techniques to drive organic visits to the DEV website.
- Utilization of a consumer survey to develop new campaign themes and measure the effectiveness of the distribution channels [27, 28].

Conclusions and Impacts to Utility Transformation

The news coming in from around the world is that:

- Tesla has over 500,000 registrations for Model 3, well before any details of the car was known or released to prospective buyers. This is an unheard-of response for any car, let alone one from an automaker barely recognized 10 years ago.
- Indian Power Minister Piyush Goyal announced at the CII Annual Session in 2017, that "India is going to introduce electric vehicles in a very big way. We are going to make electric vehicles self- sufficient. The idea is that by 2030, not a single petrol or diesel car should be sold in the country."
- The Swedish automaker Volvo slammed on the brakes for vehicles powered solely by internal combustion engines. They announced that every

car built from 2019 onward will be solely electric. This move makes the Chinese-owned Volvo the first traditional carmaker to fully embrace electric and hybrid production. "This announcement marks the end of the solely combustion engine-powered car," Volvo's president Håkan Samuelsson said in the same statement.

- Following India, several EU countries are considering a ban on cars with IC engines ranging from as early as 2030 to 2040. Norway, Sweden, Germany, France, Belgium, Switzerland, and the Netherlands are all considering the ban.
- General Motors announced in September 2017 that it will start selling a tiny electric car in China that costs about $5,300 after national and local electric vehicle incentives. The two-seat car's wheelbase, the distance from the center of the front wheels to the center of the rear wheels, is just 63 inches. That is about 10 inches less than Daimler's Smart-For-Two, a car that already delivers remarkable performance despite its short wheelbase.
- New battery technologies and new charging technologies are being developed by various organizations, some in the private sector and some by U.S. Department of Energy research laboratories. These technologies focus on range anxiety, the major source of concern for most people who are not jumping on the EV bandwagon. The newer batteries can hold more charge, thereby allowing the EV to go longer across different types of terrains. The newer charging mechanisms allow the cars to charge faster, aiming to go from near-empty to full in approximately the same amount of time it takes to fuel an IC.

Over the next 3 to 4 years, almost all automakers will have several models in their lineup, which will be one or more of the types of EVs discussed. From an electric utility perspective, this means the following:

- A charging network somewhat parallel to the current gas station network will be required. By 2012, there were approximately 115,000 gas stations in the United States. This means an entirely new network of charging stations with fast chargers must be installed around the country (and around the world).
- The utility industry will see new customer load. Every car will become a new consumer of electricity with a load greater than a typical residential load.

- Utilities will see a new kind of load, one that becomes mobile as the same car can get charged from more than one location, much like the IC engine car that can be fueled from multiple locations.

Much of the discussions above focused on the need for new infrastructure. The transformational aspects of smart transportation as connected to the electric utility come from:

- *Moving to a carbon-free future:* EVs, from their onset, have run on electricity stored in batteries. As a result, every EV on the road potentially takes one car that runs on gasoline off the road and, in turn, reduces the amount of greenhouse gas emissions. It does not reduce greenhouse gas emissions if the electricity supplied to the EV comes from coal or gas-fired generating plants.

  Transformed electric utilities find new ways to provide for the energy needs of the EV from a wide range of different sources:

  - By using renewable energy sources delivered to the grid when charging the EV, every kilowatt hour consumed by the vehicle will be green. For this step to work, utilities need to develop contracts with renewable sources to ensure sufficient capacity to deliver to a rising need of an increasing EV customer base.
  - Incentivize customers to charge during off-cycle times and effectively flatten the consumption curve, thereby reducing the need for peaking generating units, which tend to be less clean than base-loaded generating units.

    For this to work, two major actions are needed: work with the state public utility commission (PUC) to develop a rate case that provides a special rate for EV customers to charge during off-hours and work within the utility's own generation commitment program to deliver to a flattened consumption pattern based on more EVs charging during the off-hours time frame.
- *Vehicle-to-Grid (V2G) and Grid-to-Vehicle (G2V):* The battery within an EV can charge and discharge thereby becoming a consumer or a generator as needed. EVs can charge in off-peak times as identified earlier, but can also charge when there is excess power in a locality, for example, in a primarily residential locality with a lot of PVs but no consumption during the day when most people are out of their homes. If one or more EVs are present in the locality, then a request can be made of the car (or the owner) asking for it to charge up. Similarly, in the same residential

locality, early to mid-evening, when the residential load starts moving towards peak consumption, EVs parked in the various garages will most likely have enough residual charge to provide power locally[2].

For this to work, the transformed utility needs to have the full breadth of V2G capabilities as identified earlier in this chapter. More importantly, the utility must have contracts with the EV owners to charge/discharge on a utility operator's command. The location of the EV must be known. In addition, a critical mass of EVs with the V2G and G2V capabilities must be spread across its service territory to make a difference.

- *Identifying and implementing nonwire alternatives (NWA):* Utilities like ConEd are already assessing NWA to solving normal utility problems. To support EV loads, utilities are looking at implementing newer alternatives such as installing energy storage at the residence level, behind the meter or at the feeder, to support multiple homes with or without EVs.

  For this to work, the transformed utility must work with a combination of the State PUC and the customer to support nonwire alternatives, which would allow the utility not to follow the normal mode of responding to the new loads from EVs with normal business-as-usual approaches such as enhancing substation capacity.

The key is that as the number of EVs increase, the utility must have the ability to handle the power and the grid's capacity must manage this extra load and the profile presented by EVs. There is a risk that this electricity could be delivered from conventional sources of generation such as coal and/or natural gas. If conventional generation is used, it could result in increased costs for not just the EV owners, but also non-EV owners, which could put the overall EV expansion at risk from the perspective of acceptance by the community. As explained earlier in this chapter, the potential benefits of EVs to contribute to a reduced carbon footprint can only come from utilities looking at newer alternatives to deliver to this brave new world.

EVs will also require an analysis of policy adjustments. Increased EV penetration will be accompanied by at minimum, a reduction in cost and the reduced cost of batteries. When this happens, state and federal incentives will need to either decrease or be eliminated. This will also be accompanied by reductions in convention vehicles sales and the implications that come with it. Examples of this backlash would be from automakers who may lose revenue

2. In this scenario, it is assumed that the cars can take advantage of the off-hours (off-peak) capacity and charge early morning.

and consumers who may be left with the stranded asset of already-purchased gasoline-powered car. Additionally, states and regulatory agencies will need to find alternative mechanisms to replace revenues lost from reduced fuel sales. Loss of tax on fuel sales will change how states would pay for the maintenance of roads, highways, and other considerations. All of this requires a new approach to policy, not just to support the utility, but also other aspects of state and federal revenues.

For EVs to succeed, utilities need to be on the forefront of the revolution, starting with the electric infrastructure to facilitate drivers switching from gasoline to electric automobiles. This new mode of thinking can only come from a transformed utility, able to exploit energy from renewable sources or a more efficient use of existing resources. This will reduce the amount of greenhouse gas emissions or at least not contribute to its increase, while still supporting an increased penetration of EVs in the marketplace.

References

[1] Matulka, R., "Timeline: History of the Electric Car," U.S. Department of Energy, September 15, 2014, https://energy.gov/articles/history-electric-car.

[2] Navigant Research, "Electric Vehicle Market Forecasts—Global Forecasts for Light Duty Hybrid, Plug-in Hybrid, and Battery Electric Vehicles: 2013-2020," Second Quarter 2013, http://www.navigantresearch.com/wp-assets/uploads/2013/06/EVMF-13-Executive-Summary.pdf.

[3] EERE, "Electric Vehicle Benefits," Office of Energy Efficiency & Renewable Energy, 2017, https://energy.gov/eere/electricvehicles/electric-vehicle-benefits.

[4] EERE, "Reducing Pollution with Electric Vehicles," Office of Energy Efficiency & Renewable Energy, 2017, https://energy.gov/eere/electricvehicles/reducing-pollution-electric-vehicles.

[5] Press Association, "Electric Cars and Cheap Solar 'Could Halt Fossil Fuel Growth by 2020,'" *The Guardian*, February 2, 2017, https://www.theguardian.com/environment/2017/feb/02/electric-cars-cheap-solar-power-halt-fossil-fuel-growth-2020

[6] Edelstein, S., "Big Energy Hugely Underestimates Electric Cars, Renewable Power," *Green Car Reports*, February 12, 2017, http://www.greencarreports.com/news/1108835_big-energy-hugely-underestimates-electric-cars-renewable-power.

[7] EIA, "Short-Term Energy Outlook," U.S. Energy Information Administration, July 11, 2017, https://www.eia.gov/outlooks/steo/.

[8] EIA, "Short-Term Energy Outlook: Prices," U.S. Energy Information Administration, November 7, 2017, https://www.eia.gov/outlooks/steo/report/prices.cfm.

[9] PEW, "Electric Vehicles: Reducing Foreign Oil Dependence, Enhancing U.S. Competitiveness and Decreasing Pollution," The PEW Environment Group, May 2011, http://www.ehcar.net/library/rapport/rapport095.pdf.

[10] Severinsky, A. J., "Hybrid Electric Vehicle," U.S. Patent 5343970, September 6, 1994, https://www.google.com/patents/US5343970.

[11] Sandalow, D. B., (ed.), *Plug-In Electric Vehicles: What Role for Washington?* 1st. ed., Washington, D.C.: The Brookings Institution, 2009, https://www.brookings.edu/book/plug-in-electric-vehicles/.

[12] California Air Resources Board's DriveClean.ca.gov, https://driveclean.arb.ca.gov/pev/.

[13] "Position Statement: Plug-in-Electric-Hybrid Vehicles," adopted by the IEEE-USA Board of Directors, June 15, 2007, https://web.archive.org/web/20071004203527/http://www.ieeeusa.org/policy/positions/PHEV0607.pdf.

[14] Vortisch, P., et al., "Electric Vehicles with Range Extender as a Suitable Technology (EVREST)," Karlsruher Institut für Technologie, June 2015, http://www.ifv.kit.edu/english/26_EVREST.php.

[15] Department of Energy Fuel Cells Technologies Office, https://energy.gov/eere/fuelcells/fuel-cell-technologies-office.

[16] "Commercial & Industrial Guide to Electric Transportation," Electric Power Research Institute, https://www.firstenergycorp.com/content/dam/customer/get-help/files/PEV/guide-to-electric-transportation.pdf.

[17] Alternative Energy News, http://www.alternative-energy-news.info/off-road-electric-vehicles/.

[18] "Electric Vehicles," The University of Tennessee, Chattanooga, 2017, https://www.utc.edu/college-engineering-computer-science/research-centers/cete/electric.php.

[19] Alonso, M., "Optimal Charging Scheduling of Electric Vehicles in Smart Grids by Heuristic Algorithms," *Energies*, April 17, 2014, www.mdpi.com/1996-1073/7/4/2449/pdf.

[20] de Hoog, J., "Electric Vehicle Charging and Grid Constraints: Comparing Distributed and Centralized Approaches," Research Gate, July 2013, https://www.researchgate.net/publication/255786695_Electric_Vehicle_Charging_and_Grid_Constraints_Comparing_Distributed_and_Centralized_Approaches.

[21] Thongchai Klayklueng, S. D., "Impact Analysis on Voltage Unbalance of Plug-In Electric Vehicle Home Charging in Thailand Distribution System," *Cired*, June 2015, http://cired.net/publications/cired2015/papers/CIRED2015_1201_final.pdf.

[22] Dubey, A., "Electric Vehicle Charging on Residential Distribution Systems: Impacts and Mitigations," *IEEE Access*, September 14, 2015, http://ieeexplore.ieee.org/document/7264982/.

[23] Motavalli, J., "Electric Cars Provide Power Back to the Grid (and Get Paid for It)," *Plug-In Cars*, April 26, 2013, http://www.plugincars.com/payback-v2g-electric-cars-provide-power-grid-and-get-paid-it-127091.html.

[24] Shankleman, J., "Parked Electric Cars Earn $1,530 From Europe's Power Grids," *Bloomberg News*, August 11, 2017, https://www.bloomberg.com/news/articles/2017-08-11/parked-electric-cars-earn-1-530-feeding-power-grids-in-europe.

[25] Kempton, W., "Vehicle-to-Grid Power: Battery, Hybrid, and Fuel Cell Vehicles as Resources for Distributed Electric Power in California," University of Delaware, June 2001, http://www1.udel.edu/V2G/docs/V2G-Cal-ExecSum.html.

[26] Kempton, W., "A Test of Vehicle-to-Grid (V2G) for Energy Storage and Frequency Regulation in the PJM System," University of Delaware, November 2008, https://www1.udel.edu/V2G/resources/test-v2g-in-pjm-jan09.pdf.

[27] Wagner, F., "Drive Electric Vermont Case Study," Office of Energy Efficiency and Renewable Energy, March 2016, https://energy.gov/sites/prod/files/2016/06/f32/Vermont%20Case%20Study.pdf.

[28] Drive Electric Vermont, 2017, http://www.driveelectricvt.com/.

[29] "Global EV Outlook 2017: Two Million and Counting," International Energy Agency, Electric Vehicles Initiative, https://www.iea.org/publications/freepublications/publication/GlobalEVOutlook2017.pdf.

9

Smart Homes and Buildings: The Final Frontier

A smart home or building integrates building energy systems with information and communications technologies. Empowered by its automation system, the building provides actionable information that enables the owner or facility manager to optimize energy usage, space, and services available to the occupants.

Introduction

Homes and buildings are the final frontier for electricity consumption. The entire grid and all its resources—generators, the transmission system, and the distribution system—exist solely to bring electricity from the source to the destination. Homes and buildings consume electric energy daily. Lights turn on and off, refrigerators and air conditioners turn on and off, and individuals perform a variety of tasks requiring electric power.

Ever since the dawn of the electric age, generation followed load through the emergence of entities such as control area and balancing authority. This means that as consumption goes up and down, generation follows.

A review of a typical utility consumption profile at the distribution level shows a consumption pattern that varies broadly over a 24-hour period. In reality, the profile is not as smooth as it looks in the graph. When examined closely, the graph is jagged. The profile is generated from a portfolio of generators with very specific generation profiles, not from individual generators.

Base-loaded generators are at the bottom of this profile and are on all the time and delivering steady generation output hour after hour, day after day.

Nuclear power plants and the larger coal-fired power plants generally fall into this category. At the other end of the profile are the peaking power plants and other plants that provide reserves such as operating reserve. These generators are typically brought on to serve the peak load as identified in the load profile in Figure 9.1. These generators only operate for a few hours a day, a few hours a month, and sometimes a few months a year. These intermittent generators are primarily combustion turbines and other generating units. Their unique ability to start and stop quickly makes them useful. Between these two sets are natural gas and hydro plants. They are generally smaller in size and are also designed to turn on and off. Dramatic reduction in natural gas prices and the identification of large reserves of natural gas ensures that energy from these types of plants will be in use for a long time.

A newer category of generators is the distributed energy resources (DERs) that include generation multiple sources including renewables, storage, and demand response [2]. More information on DERs is covered in Chapter 2.

The list and types of generators above is not an exhaustive list. The list illustrates the various types of generators, all of which follow the consumption profile, which changes based on consumer behavior.

Homes and buildings are where consumers reside, work, play, shop, and in general live their lives. For a long time in the history of electricity consumption, consumers and industry professionals gave little thought to the consumption profile of buildings. The consumption amount was considered important because it had a direct impact to the energy bill. However, little attention was paid to the consumption profile in terms of questions such as:

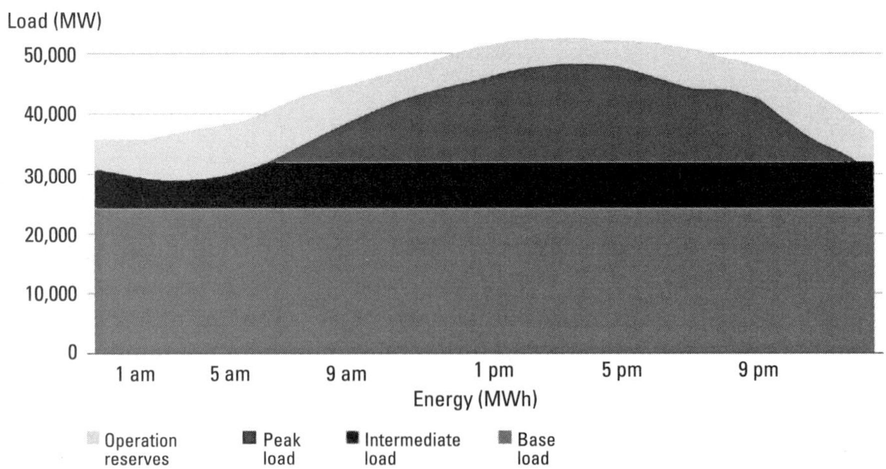

Figure 9.1 Schematic of typical daily load curve. (Source: Energy.gov, https://energy.gov/sites/prod/files/2017/08/f36/Staff%20Report%20on%20Electricity%20Markets%20and%20Reliability_0.pdf [1].)

- How much energy is consumed and when?
- How does this consumption profile align with the utility's generation profile?
- Can consumption be controlled to move from one time to another or be reduced without a negative impact on quality of life?
- Can consumption be controlled in such a way as to be supportive of the needs of a utility and the larger society by utilizing the best generation at the right time? This would require the home or building to reduce or increase consumption on demand.

The answer is yes. The answer is to move to the next generation of construction with smart homes and buildings. To understand the potential of this change in consumption, the concepts, definitions, and importance of smart homes and buildings is paramount.

Define Smart Homes and Buildings

A building or residence is called smart (see Figure 9.2) when it can:

- *Proactively monitor premises energy usage:* The premises have one or more sensors to monitor energy consumption at different levels for different locations and/or appliances. Monitoring can include basics, such as a smart thermostat to control home temperature or a smart meter

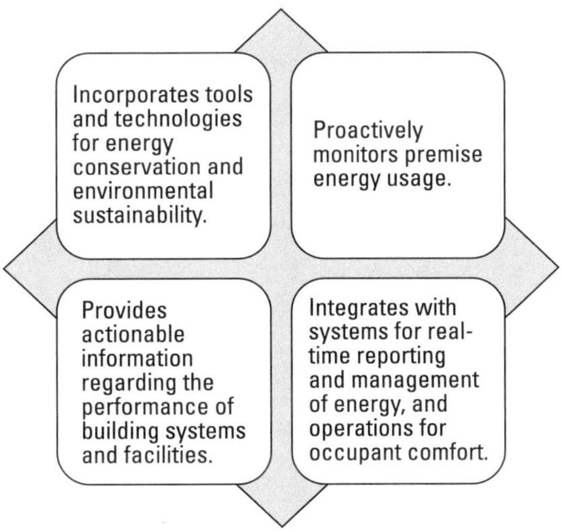

Figure 9.2 Characteristics of a smart home or building. (© Modern Grid Solutions.)

to monitor home energy consumption. More complex monitoring includes smart light-emitting diode (LED) lighting controls, and other more advanced programs. The most important role for sensors is to supply data to a central monitoring system located either on the premises or in the cloud.

- *Incorporate tools and technologies for energy conservation and environmental sustainability:* Smart homes and buildings all have a central monitoring system that records and analyzes data from various sensors monitoring the energy consumption in real time. These systems incorporate a range of tools and technologies using data collected to perform analyses. The kinds of analyses that can be performed cover a broad range of actions from energy conservation to environmental sustainability to focus on reducing the energy bill.
- *Provide actionable information regarding the performance of home/building systems and facilities:* A smart home or building gathers information and performs calculations to deliver actionable information to the homeowner or building manager/operator. The recommendations could be in the form of short-term actions to be taken in real time to modify the energy consumption or long-term actions to modify the energy consumption over a period of time.

  Examples of short-term actions are modifying temperature settings or shutting off one or more lights based on present occupancy. Examples of long-term actions are adding insulation to the home or installing a lighting control system.
- *Integrate with systems for real-time reporting and management of energy and operations for occupant comfort:* Smart homes and buildings very often integrate with external systems. That integration can be either with a building integrator, one that manages a broad portfolio of buildings or residences, or the utility.

This integration allows the utility, the aggregator, or another entity to manage and report on the energy profile of these residences/buildings.

Examples of integration benefits are:

- Managing energy consumption as a portfolio;
- Driving energy savings across multiple buildings;
- Taking advantage of this critical mass of energy consumers to participate in wholesale/retail markets;

- Get bulk discounts on energy and/or technology/conservation implementations.

Comparing Smart Homes and Buildings

As the name implies, smart homes are residential in nature, generally single-family residences, and smart buildings are generally commercial. Apartment complexes also fall into the category of smart buildings. By their very definition, smart buildings are more complex and with more points of control. Another major difference between smart homes and smart buildings is their rate base. Smart homes tend to be on a residential rate, which means, for the most part, they have a standard rate covering the cost of energy consumed independent of the time it is consumed. Smart buildings tend to be on a commercial rate, meaning that they have an energy charge like that of the residential rate but lower. They also tend to have a demand charge calculated on the coincident peak of the building's consumption for last year. The utility typically updates this demand charge annually.

Similarities

There are many similarities between smart homes and smart buildings, including that they allow residents to be more aware of their energy usage.

- *Provide actionable data to owner/operator to optimize energy usage and performance:* Homes and buildings account for nearly 40% of society's energy use. Data from smart meters provides significant opportunities for both utilities and consumers to optimize energy use and minimize waste. That data also provides insight into how modern homes and devices use energy. Meter data is used to analyze user consumption and provide recommendations for future energy savings. In addition, the utility uses meter data to develop a better understanding of their asset utilization, such as for both amount and usage patterns.
- *Need automation systems to control end-user devices:* Smart homes and buildings are equipped with technology that enables occupants to remotely control or program an array of automated home electronic devices by entering a single command. Homeowners can use a smartphone to arm a home security system, control temperature gauges, switch appliances on or off, control lighting, program a home theater or entertainment system, and perform many other tasks [2].

- *Help detect issues with malfunctioning devices and reduce repair/maintenance costs:* Sensors in smart devices help detect issues and alert the user. The user can intervene and take preventive measures to fix issues and save on repair costs. Smart appliances are examples of such devices, which respond to power quality issues such as low voltage or frequency deviations in intelligent ways and prevent unwanted damage. Sensors in Smart meters can detect meter tampering issues, general malfunctions, or any installation issues in the electric system.
- *Contribute to occupant/tenant comfort and convenience while saving money:* In-home/building energy monitors and web portals provide direct interface to the end-user to set energy preferences such as temperature, light intensity levels, on/off for individual appliances. Some of these preferences can also set the upper limit on price of electricity. Energy can be reduced by turning off appliances or car charging when that limit is reached.

Differences

The differences between smart home and buildings are more about the complexity of control and allowing the resident to save money in areas where it may not be as beneficial for the home owner.

- *A building automation system (BAS) typically needs greater investment than for a single-family home:* Buildings, such as those used for office space and commercial purposes, typically have more energy infrastructure than a single-family home, such as complex heating, ventilation, and air conditioning (HVAC), lighting controls, and window shutter controls. In addition, some of the buildings may also have on-site generation in the form of rooftop solar, microturbines, and diesel generator sets. These lead to greater investment.
- *Buildings may have a stronger business case than home automation:* The deployment of various load controls and on-site generation requires an energy management system to monitor and control the devices. In addition, the installation of smart meters at the building premises provides a more granular look at energy usage. A building automation system that ties building load with on-site generation can optimize overall energy usage and provide enhanced savings.
- *Automation systems for commercial buildings are typically more complex than single-family homes:* The building automation system that must

communicate between smart meters, loads, and on-site generation requires more intelligence than simpler systems for private homes. Some of this intelligence may need to be in the form of translating between various communication protocols. To do this, the automation system operates between various vendor devices using different communication protocols. In addition, home automation systems need to be simpler to use than commercial systems. This is because commercial buildings will likely have a dedicated staff, while the single family home does not.

- *End-user devices are different:* The end-user devices such as lighting systems and controls, HVAC systems, appliances, vehicle charging controls, in-building displays, on-site generation controls, and smart meters may all be from different vendors. This may lead to the use of different communication standards. These devices may also have various degrees of automation, ranging from partial to a full set of controls.
- *Load controlled in commercial buildings is typically higher than for homes:* Typically, office and commercial spaces tend to have energy usage 24 hours a day, with load being greatest during high-occupancy times, decreasing drastically during late evenings, nights, and weekends. However, usage during off-peak times is still greater than those for homes, largely due to the operation of sensors, security cameras, some internal lighting, external lighting in parking lots, and other mandatory systems.
- *Commercial buildings may have on-site generation to be coordinated with load:* Given the emergence of various distributed generation options and reducing installation costs for choices such as rooftop solar, businesses are deploying more on-site generation as a way to offset utility costs. This on-site generation is often primary, with any shortfall purchased from the local utility. To determine this shortfall, on-site generation coordinates with load on the premises.
- *Demand response programs offered to commercial buildings may be different than for single-family homes:* Demand response programs are often designed based on the type of consumer load and energy usage. Programs for homes could be incentive-based such as direct load control, as in the case of curtailing air conditioning load during peak summer months. However, commercial building owners may opt for price-based programs such as time of use (TOU) and critical peak pricing (CPP) reducing overall building load or use greater on-site generation during periods of high utility energy prices.

Key Drivers of Smart Homes and Buildings

Several elements drive the move from traditional homes and buildings to smart homes or buildings ranging from conserving energy to managing energy consumption remotely to having a connected home able to provide additional services to the resident. As shown in Figure 9.3, these go beyond energy to a broad range of services.

Energy Efficiency/Conservation

Buildings account for roughly 40% of all U.S. energy use. Of this, residential buildings account for 22% and commercial buildings account for 18% of energy use. The load is typically larger in commercial buildings and office spaces than in homes, due to larger floor space and greater occupancy. To a large extent, building load is attributed to HVAC and lighting systems. In homes, most of the load is due to appliance use, lighting, and heating and cooling systems. Providing users the mechanisms to choose preferences that alter energy use and lead to conservation go a long way in lowering electricity bills. Appliances that are not only energy-efficient, but also contain intelligence that respond to grid needs, are increasing in usage due to this reduction in electricity costs.

Ubiquitous Communications at Home or Work

Homes and buildings are already networked for various electronics such as computers, personal data assistants (PDA), and smartphones. Most living spac-

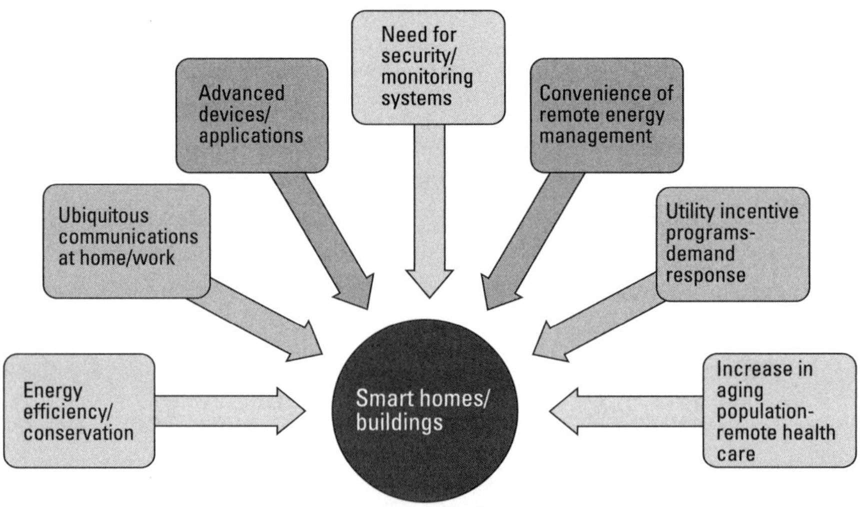

Figure 9.3 Key drivers of smart homes and buildings. (© Modern Grid Solutions.)

es have available Wi-Fi. Many in-home or in-building monitors that allow consumers to set preferences for energy use are now widely available in the market. These devices can easily plug into existing local area networks via Ethernet or Wi-Fi. This makes installation and operation extremely simple. In addition, the availability of energy portals makes it easy for consumers to take an existing laptop or PDA and convert it into an energy monitor.

Advanced Devices and Appliances

We are on the brink of a smart home boom, in which 71% of consumers can expect to see at least one smart-home device in their home by 2025. There are many amazing home automation devices on the market today. Many of these, such as Amazon's Echo, are cloud-connected, voice-activated, and designed to make things easier and more accessible. Others, such as the Nest Thermostat, help with controlling energy consumption, which can lower bills while providing a comfortable environment for occupants. Industry experts anticipate that many of these would become as common as smartphones within the next decade or so.

Need for Security and Monitoring Systems

Home automation gives users access to control devices in their home or building from a mobile device anywhere in the world. From a home security perspective, this also includes alarm systems and all the doors, windows, locks, smoke detectors, surveillance cameras, and any other sensors linked as well. Monitoring apps can provide a wealth of information about the home or building, from status to detailed historical data of previous energy consumption. The user can check the status of the security system, including whether the lights are on, the doors are locked, what the current temperature is in the home, and much more. In addition, with cameras as part of the automation system, the user can view video feeds in real time [3].

Convenience of Remote Energy Management

One of the main characteristics of cutting-edge home automation is remote monitoring and access. While a limited amount of one-way remote monitoring has been possible for some time, only since the rise in smartphones and tablets has the ability to truly connect to home networks increased. With the right home automation system, one can use any internet-connected device to view and control the system and any attached devices.

Utility Incentive Programs—Demand Response

Demand response programs curtail energy use and mitigate rate spikes during peak events seamlessly and automatically, while minimizing disruption to utility customers. These programs can be incentive-based, such as direct load control, or price-based, such as time of use TOU and CPP. These, coupled with automated demand response (auto DR) solutions, allow customers to predetermine customized load shed strategies during a peak period and directly connect the utility to the home or facility for automated delivery of the load shed.

Increase in Aging Population and Remote Health Care

The average age of the population has risen consistently worldwide, both in urban and rural areas. This statistic indicates that an Internet of Things (IoT)-based remote health monitoring system designed specifically for the elderly would be a successful venture, especially if that system does not compromise convenience or the choice to receive care at home. Such systems would generate large amounts of data, which would need to be able to accommodate the bandwidth requirements and the volume of that data, such as an efficient remote health monitoring system that offers health-care providers the ability to constantly monitor the well-being of elderly population or those suffering from chronic or debilitating medical conditions [4].

Why Are Smart Homes and Buildings Important?

As the chapter title reads, smart homes and buildings are the next frontier for the smart grid. Homes and buildings are where much of electricity consumption takes place. With the advent of DERs, this is also where a lot of the new generation comes from. Given these two points, the importance of smart homes and buildings needs to be evaluated from different perspectives: the utility, the customer, businesses, and the regulator.

Utility Perspective

A home or building automation system enables the control of end-user devices or loads. These loads could be anything from water heaters to refrigerators or furnace/home heating to building lighting. The presence of controllable loads offers the utility the ability to modify the consumption of electricity. The controllable loads also allow the utility to pack this capability into specific programs such as demand response opportunities to its customers.

For the utility, these programs are more than just an ability to control load. The control of the load allows the utility to:

- *Request the reduction of consumption during peak hours:* Reducing the load during peak hours allows the utility not to bring online generating units such as peakers that are generally more expensive, more polluting, and typically used only a few hours per day, a few hours a month or a few hours a year. Avoiding the use of these generators presents significant value to the utility.
- *Demand the reduction of load during emergencies:* During emergencies, the utility can call on the same customers participating in demand response programs to immediately curtail their load, thereby allowing the utility to continue to manage the system and its reliability until the emergency is resolved.

As a future capability, the advent of DERs on the grid may also result in the utility requesting the customer increase the consumption of energy at times when there is surplus energy in the grid.

This situation could happen because of excess generation from renewable sources of energy. An example of this could be wind energy. Wind often maxes out in the middle of the night. EV programs could focus charging during these high output times.

Customer Perspective

Automation offers the consumer the convenience of remote operations of home and building devices to optimize energy usage and comfort. It enables the consumer to conserve energy and reduce the cost of energy usage. For the building operator, in addition to the benefits above, it also allows a green branding opportunity that may be attractive to tenants. It offers the customer flexibility in accommodating new devices and appliances. It helps maximize home security by connecting motion detectors, surveillance cameras, automated door locks, and other tangible security measures throughout the home or building into the automation system. Smart homes can also help run appliances better by connecting appliances and other systems with automation technology. An example of this is using automation to turn lights on and off when needed instead of being on all the time. It also improves appliance effectiveness as well as increases comfort levels by allowing home owners, for example, by turning the heat on at home as the residents are driving towards their home. The insights gained from automation may be analyzed to improve daily habits and energy consumption patterns [5].

These types of controls also allow the home or building to participate in utility programs such as demand response. Probably the biggest benefit for the smart home resident or the smart building operator is savings in their energy costs, which can change significantly depending on the type of program offered by the utility.

Business Perspective

There has been a lot of interest in home automation (HA) and connected home devices. The HA market is anticipated to grow to 1.8 billion devices worldwide in 2019. While the HA market is still fragmented and chaotic, the growth coincides with an increase in interest from utilities and retail energy providers in distributed generation, energy storage, and in-home devices and signals. This is a natural intersection for the HA market and energy suppliers. While demand response programs have offered thermostats to customers in small numbers, a relative new phenomenon is being tested by utilities to capitalize on the growth of the consumer market for smart thermostats by connecting consumer-owned smart thermostats to the utility programs. Experts believe that households with smart thermostats interacting with utilities are expected to reach about 1.5 million customers in the United States by 2020 [6].

Home automation is a global market, expanding to bring new and exciting opportunities for technology companies. Industry professionals believe the market will be driven toward energy-efficient, safe, luxurious, and convenient homes or smart homes. Consumers constantly look for automation in day to day activities at home, which creates excellent business opportunities for automation and device vendors in developing programmable, self-learning, sensor-driven, Wi-Fi-enabled thermostats, smoke detectors, and other security systems.

Automation services such as security and monitoring can also be bundled as a service. These services can be combined with existing services such as telephone, internet, and cable, and then offered to the consumer. Load control when participating in a utility demand response program can even be offered as a service by a third party.

Regulatory Perspective

For many years, the utility's generation profile was called load-following. This meant that the consumption of electricity at homes and buildings generally goes unchecked and the generation profile follows the load. Smart homes and buildings create a new paradigm. Automation allows the home and building consumption to be modified up or down based on conditions.

As these devices proliferate and more homes and buildings become smart, regulators will begin to view smart technology as a place where consumption and supply can respond interactively. This means that, with the introduction of DERs, as loads go up, generation goes up, and utilities now have the option to supply this change instead of the traditional method of expanding substation or feeder capacity or both. Demand response is increasingly seen by utilities as a new option.

The regulator's response comes in many forms: newer and more innovative rates such as TOU, rebates for converting lightbulbs to LED, rebates for buying energy-efficient appliances, rebates for installing PV on the roof, and rebates for buying electric cars.

Elements of Smart Homes and Buildings

For a home or building to be smart, it needs to have several components (see Figure 9.4) These components allow the control of and the ability to monitor and visualize consumption as the ability to monitor and visualize it. The components are as follows.

Communications

Communications is key to home and building automation. Communication protocols work to form home and building networks. These networks can either be wired or wireless or a combination of both. An important concept in home networking is the mesh network, where each node or device in the network also acts as a router to move information.

Wired protocols, like UPB and X10, use a home's existing wiring to communicate. Wired protocols tend to be reliable, but can be slow and difficult to encrypt.

Wireless protocols are usually faster and more compatible with other devices. Wireless home automation protocols include: Z-Wave, ZigBee, Wi-Fi, Thread, and Bluetooth. These connect and communicate without relying on power lines, making them easier to secure.

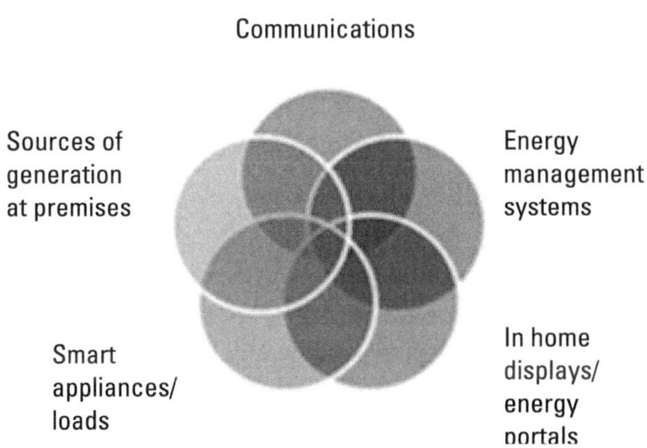

Figure 9.4 Elements of smart homes and buildings. (© Modern Grid Solutions.)

Compatibility with different smart home devices is an important consideration in the move toward smart homes and buildings [7].

Home and Building Energy Management Systems

Home and building energy management systems (HEMS/BEMS) are computer-based systems used to control and monitor mechanical and electrical equipment for energy consumption. The communication between the energy management system (EMS) and the end-user devices is based on the communication protocols identified earlier. The three basic functions of a central, computer-controlled EMS are controlling, monitoring, and optimizing the home or building's facilities and equipment for comfort, safety, and efficiency.

The energy information coming to and from the home or building through the smart meter can be run through the HEMS or BEMS. These systems allow the user to view data in an easy-to-understand format on a computer or hand-held device. A home EMS allows the user to track energy use to better save energy. For instance, one can see the energy impact of various appliances and electronic products simply by monitoring the EMS while switching the devices on and off. An EMS also allows for monitoring real-time information and price signals from the utility and creating settings to automatically use power when prices are lowest [8].

An EMS provides the automated interface between consumer and utility. It enables receiving current and forecasted electricity pricing information. It also enables the customer to participate in utility demand response programs. It uses the electric pricing information from the utility and the consumer preferences set via the in-home display (IHD)/energy portal to autonomously optimize energy usage. A significant reduction in energy can be obtained by linking equipment/appliances to an EMS.

In-Home Displays and Energy Portals

These are installed inside the home or building and serve as the interface between the consumer and the EMS. They serve as the interface for the consumers to configure their usage preferences. IHD units provide energy customers with real-time energy consumption feedback. They serve as the energy usage dashboard and typically display the following data: current energy usage/cost, usage in the past hour(s); projected monthly bill, and daily and hourly energy usage graphs. These interfaces are key to reducing energy usage.

IHD units can acquire consumption information through a sensor with built-in RF and/or power line communications (PLC). However, a more effective solution transmits information from a Smart Meter via a home area network. IHDs can have a range of microcontrollers with embedded liquid crystal display (LCD) display controllers and flexibility to support any application.

The IHDs can have flexible touch solutions, from buttons and wheels to sophisticated touchscreens and provide support for a wide range of user interface features and capabilities. They could be PLC or system-on-a-chip (SoC) solutions, with full digital implementation delivering best-in-class sensitivity, high performance, and high temperature stability [9].

IHD units vary in complexity, from simple wall-mounted segment LCDs to battery-operated products with displays and touchscreens. Advanced IHDs can show not only consumption information, but energy consumption advice from energy providers. They can also support additional functions such as home automation.

With the end of the ARRA funding in the United States, much consumer interest in independent devices such as IHDs went down, primarily due to the associated costs. The need for the capabilities did not disappear. The industry expects this capability to move to a combination of a smartphone app or the web or both.

Web-based or smartphone app-based energy portals are convenient for viewing home and building energy data from any location. These energy portals enable the consumers to configure their usage preferences remotely. They also offer greater functionality than fixed hardware devices since they are web-based and not limited by hardware. Unlike the IHD, they do not need to be replaced or upgraded, which provides cost savings and convenience.

These portals allow users to view data from a smart meter via a computer, smartphone, or tablet. Typical features of a portal include viewing electricity consumption by week, day, and hour, viewing a breakdown of usage by household appliance, comparing energy usage to similar homes in the area, viewing a forecast of the energy bill and see how the resident is tracking, and setting a budget target or goals to help manage costs.

Smart Appliances and Loads

Smart appliances can also respond to signals from an energy provider or the customer to reduce or avoid energy consumption during times of system stress. Very often, these are more complicated than a simple on/off switch. For instance, a smart air conditioner might extend its cycle time to reduce its load on the grid; by itself, this may not be noticeable, but millions of air conditioners acting the same way could significantly reduce the load on the power grid. Likewise, a smart refrigerator could defer its defrost cycle until off-peak hours or a smart dishwasher might defer running until off-peak hours. Examples of smart appliances include:

- *Smart refrigerator:* Delay defrost cycle to off-peak periods; it can also take stock of inventory and expiration dates.

- *Smart thermostat:* Enable heating and AC control remotely.
- *Smart dishwasher:* Delay operation to off-peak periods; Synchronize with renewable wind output.
- *Smart washer and dryer:* Delay operation to off-peak periods; timed back-to-back dryer loads to use residual heat from first load for improved efficiency.
- *Smart water heater:* Preheating of water during off-peak periods.

A smart appliance has an embedded microprocessor enabling it to have two-way communication with the smart meter, typically via the home or building energy management system. In a smart home, many of the appliances are networked together, allowing one to access and operate them through the EMS. An EMS can turn on the heater or air conditioner remotely or keep track of the energy use of specific appliances or equipment, such as tracking the energy use of a pool pump or seeing how much energy was saved with a new EnergyStar dishwasher.

Smart appliances are proliferating in the marketplace with the ability to connect to the internet and remotely control the energy consumption, as well as control of other services in one or more appliances.

Generation at Premises

Smart homes and buildings may have various types DERs, such as solar PVs, microturbines, fuel cells, reciprocating engines, and gensets installed at the premises. They may also have electric energy storage (EES) technologies, such as batteries, installed. The smart home or building, with its system of controls and smart meters, helps to effectively connect all these mini-power generating systems to the grid, to provide data about their operation to utilities and owners, and to know what surplus energy feeds back into the grid versus that used on site.

Smart Building Load Controls

Smart buildings employ control devices to monitor and control end-user loads, using energy when it is required, using the solely the amount of energy actually required, and applying the energy used with the highest possible efficiency. Examples of building load controls include:

- *Heating/cooling control:* Individual room control with communication between controllers; interlock between heating and cooling control.

- *Ventilation/AC control:* Airflow control at room level depending on occupancy; air humidity control.
- *Window shutter control:* Adjust closure angle based on availability of natural light.
- *Lighting control:* Automatic dimming and on-off lighting control at room level depending on occupancy; coordination with window shutters to optimize interior lighting based on availability of natural light.

Home Gateways

Automation systems generally consist of switches and sensors connected to a central hub sometimes called a gateway, from which a user interface controls the system with a wall-mounted terminal, mobile phone software, tablet computer, or a web interface. This is often done via internet cloud services.

A gateway can be an IP-based system with RF, ZigBee, and Z-Wave and optional Wi-Fi capabilities. It can provide comprehensive smart home services, including remote control, home security, live visual monitoring, home automation, energy management, and emergency monitoring. A gateway can report alarm events and sends videos and images. The system can be controlled via a home automation app on a smartphone or online.

Example Architectures

The architectures for smart homes and buildings mainly differ in their complexity mainly due to providing greater granularity of control and the complexity of the devices being monitored and/or controlled. In addition, buildings also tend to have multiple tenants, which leads to a greater need for customization for each tenant and/or his or her unique business requirements.

Home Automation

A home automation network or a home area network (HAN), shown in Figure 9.5, connects devices in the home such as smart appliances, localized (behind-the-meter) DERs, and other devices seamlessly into a single network, which is also integrated with the smart metering system. It also contains the necessary software applications to monitor and control these devices. The HAN is the backbone of interactions between all the components.

In a HAN, multiple components interact to provide a wide range of capabilities. The basic components of a HAN are:

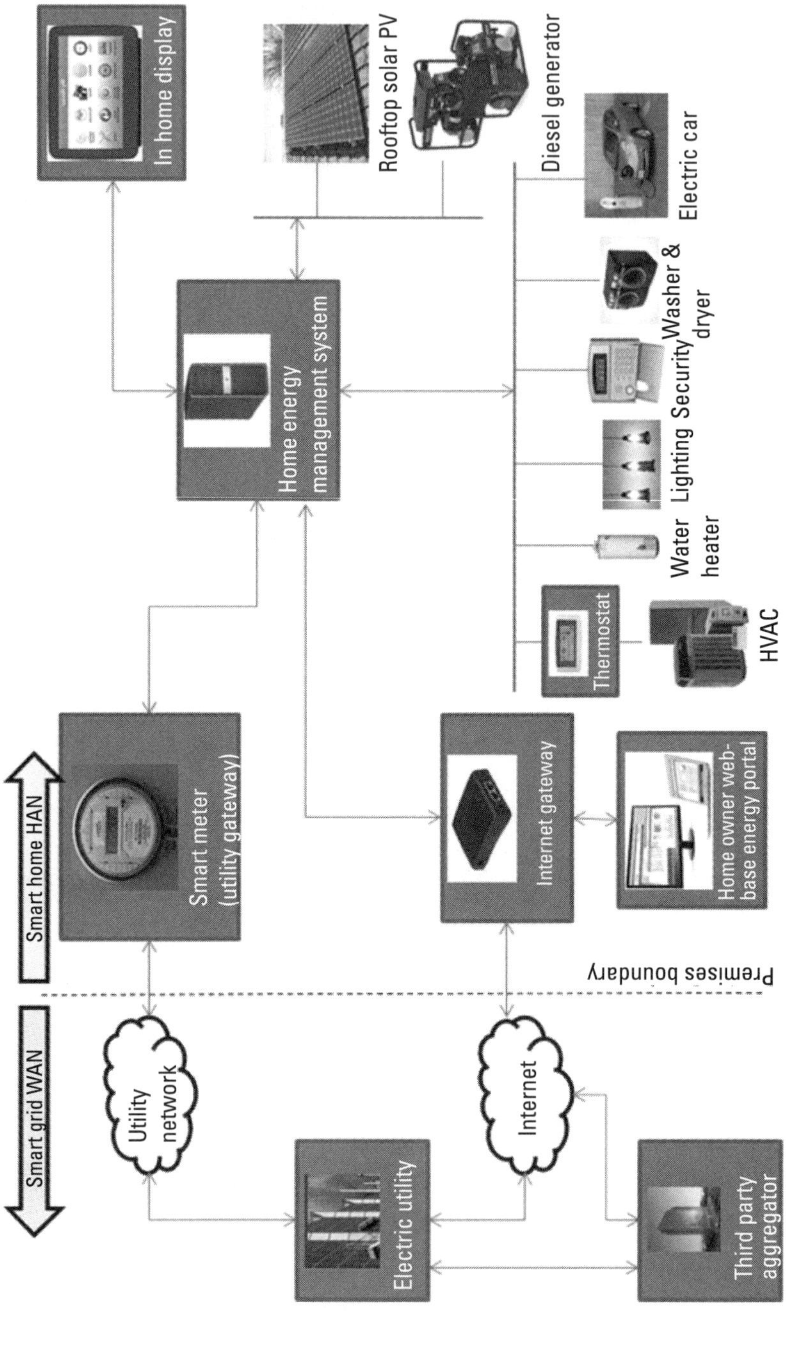

Figure 9.5 Home area network architecture. (© Modern Grid Solutions.)

- The network portal or gateway that connects one or more outside information services to the home area network;
- The access point or network nodes that form the wired or wireless network;
- The network operating system and network management software;
- The end points such as thermostats, meters, in-home display devices, and other appliances.

HAN extends smart grid capabilities into the home using different networking protocols. Some of these are ZigBee, Wi-Fi, Ethernet, Z-Wave, HomePlug, Wireless M-Bus, and Wavenis.

Several aspects influence HAN architecture and its integration into the larger smart grid. These include:

- *Pace of technology innovation:* Emerging standards, such as the Wi-Fi, ZigBee Smart Energy Profile, and OpenHAN, are rapidly evolving, along with demand-side management applications leveraging these standards.
- *Upgradability:* Consumer devices, such as in-home displays and programmable communicating thermostats, which deliver demand-side benefits, are giving way to smartphone apps and online or cloud-based presences. This means any presence at the home needs to be upgradeable over time and flexible to delivering more services.
- *Consumer choice:* Choice is a critical component in consumer adoption. Choice requires simultaneously sustained product innovation and attentiveness to existing technologies.
- *Device ownership:* The increasing variety of products and product types resulting from technology innovation leveraging smart grid communication capabilities are redefining the relationship between consumers and utilities.
- *Market diversity:* Consumers and utilities face a mix of retail, direct-to-consumer, and services markets that influence and are influenced by device ownership aspects of the HAN.
- *Interoperability:* The ability of devices and systems to work together is critical for the applications delivering value to consumers, utilities, retail energy providers, load aggregators, and other stakeholders.
- *Total cost:* A variety of cost models may apply depending on the HAN gateway architecture chosen. The smart grid business case must account

for this choice. Cheaper alternatives such as cloud-based solutions may become apparent in these situations.

- *Performance:* System level performance of applications can be hindered or enhanced by different architectures.

Demand-side management for the home still holds significant promise for electric utilities. The ZigBee Smart Energy Profile is an emerging standard to define device descriptions and standard practices for demand response and load management smart energy applications used in a smart energy-based residential or light commercial environment. The smart energy profile (SEP) provides suitable, core demand-side management functionality to satisfy many of the needs of utility programs for energy efficiency, demand response, and direct load control. It routes message to and from the relevant end points. It may be installed within a meter, thermostat, or in-premise display, or may be a standalone device [10, 11].

However, a major change on the horizon is the move by utilities to distributed resource management. This takes demand response to a broader management of all resources behind the meter: consumption, storage, PV on the roof, storage, and electric cars. Managing all the resources as a portfolio with some resources capable of generating, some of consuming and some capable of both will be key for demand response success. In addition, some of the resources are dispatchable, meaning that they can be turned on or off or modified, through remote control, while others, such as PV on the roof, are not. HANs are evolving into systems resembling a nanogrid that can function either at a single home or a group of homes.

Building Automation

A building automation network or a building area network (BAN) controls an entire building. A BAN can be a single network or a collection of smaller networks.

Like a HAN, building automation and control systems rely on sensors and actuators placed at different locations throughout a building. Reducing the power consumption of a modern building requires continuous monitoring of various environmental parameters both inside and outside the building. To be an efficient monitoring and controlling system, all sensors and actuators must be addressable over the network. This includes a collection of sensors that determine the condition or status of parameters to be controlled, such as lighting, temperature, relative humidity, and pressure. An action based on the sensor data received by the control unit is to devices like electric relays, dampers, or valve actuators via electronic signals that activate physical action to control the devices.

Building automation enables intelligent features such as:

- *Airflow control:* When occupants in the room increase, local sensors and the thermostat will increase the room temperature. AC control unit will open its damper, allowing more air into the room, which will cause a drop in the duct static pressure sensed by the duct static pressure sensor. To maintain the static pressure in the duct, BAN activates the control to increase the fan speed, which builds up the duct pressure back to the desired point.
- *Temperature and fan control:* When the control unit is not functioning, the BAN detects and communicates the off status of the unit, thus shutting it down.
- *Lighting:* The lighting system can be controlled using motion and detection sensors that detect occupancy and motion. On and off switches can be configured based on predefined time schedules. Daylight-linked automated response systems can also be incorporated into the system.
- Security and telephone systems, among others.

Building automation provides several benefits, which include:

- *Increased energy efficiency:* A substantial amount of costs can be saved by effectively controlling equipment use in real time. In addition, it is far easier to monitor aspects of the system for potential problems or provide preventative maintenance.
- *Streamlined operations management:* Smart building automation greatly reduces operational expenses and the hassle of installing and operating multiple autonomous building systems managers. Operators can view data from all over the facility and make quick changes or provide preventative maintenance, lower installation costs, and lower failure rates and downtime.
- *Better services:* Building automation enables management to provide better services to occupants and users. Accessing building systems via remote control makes it easier for facility professionals to assess real-time conditions, detect problems, and monitor building performance from off-site.
- *Effective monitoring:* More accurate data monitoring and control of energy-intensive systems like HVAC and lighting and statistical data report generation help keep costs in check.

- *Increased customer retention:* Smart buildings can demand above-market rents, have lower vacancy rates, and can have reduced turnover through higher customer service, added technologies, and increased efficiency.
- *Environmentally friendly:* Monitoring and control of energy use for reducing consumption define a green building. While it may be possible to have a green building that is not smart, most green buildings will have some form of a building automation system [12].

Communications Mechanisms Used in HANs and BANs

Depending on the type of home or building, there is a broad variety of communications technologies used to enable the automation. Some of them are shown in Figure 9.6.

Home and building area networks can use either wired or wireless technologies to connect end points. Wireless is the predominant option in homes

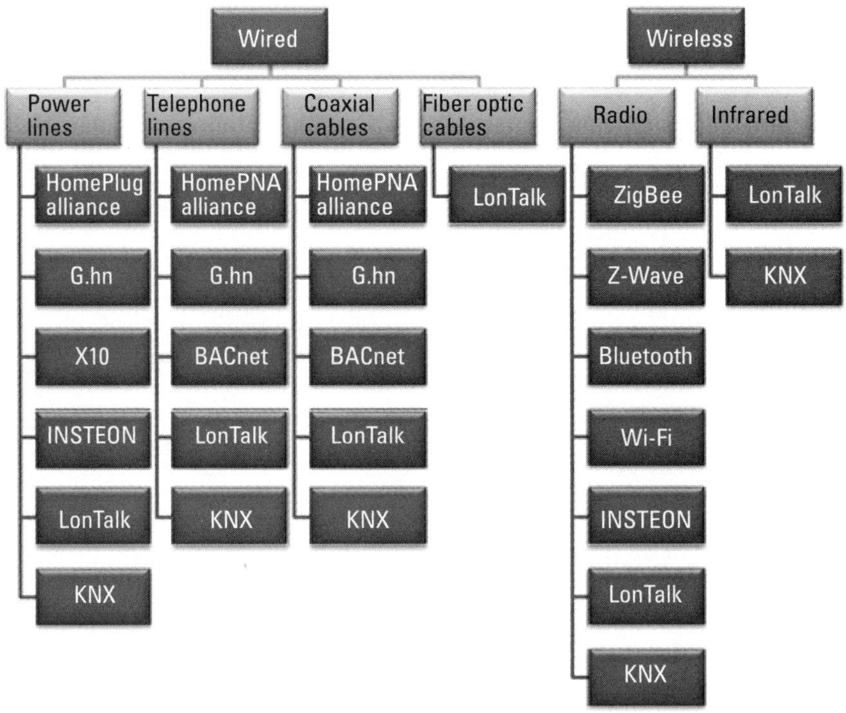

Figure 9.6 Communications mechanisms used in smart homes and buildings. (© Modern Grid Solutions.)

due to the ease of installation, lack of unsightly cables, and network performance characteristics sufficient for residential activities [13].

- *Wired:* Using existing wires within a home offers many benefits over installing a new network with all the inherent costs and frustrations. The two options currently in virtually every home are telephone lines and power cables. There are technologies and standards in existence for both types of wires:
  - *Power lines:* PLC is a communication method using electrical wiring to simultaneously carry both data and electric power. Most PLC technologies limit themselves to one type of wire such as premises wiring within a single building, but some can cross between two types, for example, the distribution network and premises wiring.
  - *Telephone lines:* Minimal use in HANs due to limited coverage of phone lines in new home construction.
  - *Coaxial cables:* Copper cable used by cable TV companies between the community antenna and user homes and businesses; widely installed for use in business and corporation Ethernet and other types of local area networks.
  - *Fiber optic cables:* Like an electrical cable, but containing one or more optical fibers used to carry light. Different types of cables are used for different applications, for example, long-distance telecommunication or providing a high-speed data connection between different parts of a building.
- *Wireless:* The ability to network devices without running wires has tremendous appeal. Wireless technology is spreading from the corporate IT environment into industrial and smart grid arenas. Within these areas, there is a focus on low power and low bandwidth versus the high bandwidth with traditional wired power.
  - *Radio:* Most wireless technologies use radio waves. Radio is the technology of using radio waves to carry information, such as sound, by systematically modulating properties of electromagnetic energy waves transmitted through space, such as their amplitude, frequency, phase, or pulse width.
  - *Infrared:* Infrared (IR) wireless is the use of wireless technology in devices or systems conveying data through IR radiation. IR is electromagnetic energy at a wavelength or wavelengths somewhat longer than those of red light.

Impact of Demand Response and Energy Efficiency on Smart Homes and Buildings

Demand response is a process by which electrical providers, distributors, transmitters, and customers (residential, commercial, and industrial) manage their electrical needs, particularly at times of peak usage or in response to market costs, thereby limiting, growing, or eliminating demand for short or extended periods of time.

Demand response has always been identified by practitioners in the industry as the low-hanging fruit of smart grid [14]. Demand response has been around in some form for over 30 years. Early terminology associates names such as demand side management (DSM) and load management (LM) with being focused on reducing grid-wide consumption based on specific system situations, such as peak loads. In the beginning, each utility program was wired to specific customer appliances. During periods of intense system stress with inadequate generation, the utility would send signals and the specific appliances were turned off to reduce system-wide load. While being quite effective in reducing the load when the utility needed it to, it was very expensive to implement. Customer interconnections were wired to specific appliances and required active customer agreement to participate, which was a challenge.

When discussing the concepts of managing demand-side consumption, two terms surface: demand response and energy efficiency (EE). Sometimes these terms are used synonymously. In general terms, demand response is all about reducing the peak, but not necessarily reducing the overall consumption of energy. EE is all about reducing energy consumption. Demand response shaves the peak, but EE brings the overall consumption profile down.

Demand response's value becomes apparent when we can retire inefficient generation used for just a few hours per year with more efficient and cheaper generation. When coupled with distributed renewables by providing some level of local optimization between supply and demand, the value can be enhanced.

Utilities implement programs for demand response, as illustrated in Figure 9.7, where customers sign up to participate. Utilities design programs to achieve specific objectives and can be either dependent on the customer acting to reduce load or on the utility's actions to remotely reduce load.

The advent of smart homes and buildings moves the focus of both demand response and EE from specific connections to specific appliances to a broad connectivity to the home. This change allows either the customer or the utility to control consumption and generation at home as a portfolio of services optimized across the utility or the ISO. This means there are more options available to control consumption, which could now become a request to consume less or generate more. As homes and buildings become smart, the demand reduction signal, instead of being sent to a specific appliance, is now

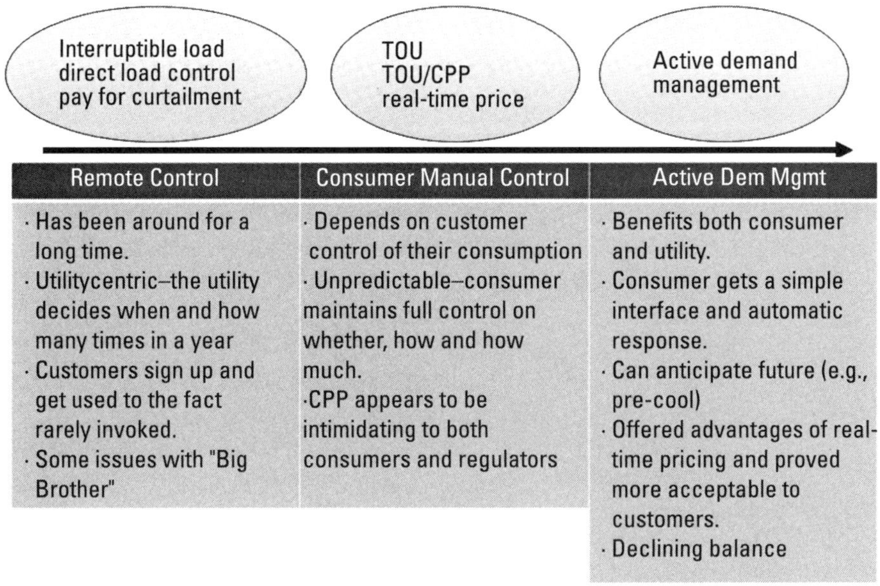

Figure 9.7 Demand response mechanisms in the marketplace. (© Modern Grid Solutions.)

sent to the home gateway where a localized decision is made regarding the action to be taken.

Making the Business Case for Smart Homes and Buildings

The core business case for smart homes and buildings is very straightforward and illustrated by Figure 9.8. The figure shows that for every unit of energy that went into the generating unit, in the form of coal, natural gas, or something else, only about 30% reaches the customer. Every unit of energy saved at the point of consumption is equal to saving roughly three more units of electricity that did not need to get into the network at the source. This also applies to the management of distributed energy.

Regulator Viewpoint

The regulator has two major drivers: the need to cut down ratepayer costs, and the need to implement government policy. Much of the business case for smart homes and buildings comes from providing subsidies and rebates for the utility's customers installing DER technologies, control interfaces, and EVs in their homes. Examples of the regulatory responses impacting smart homes and buildings include:

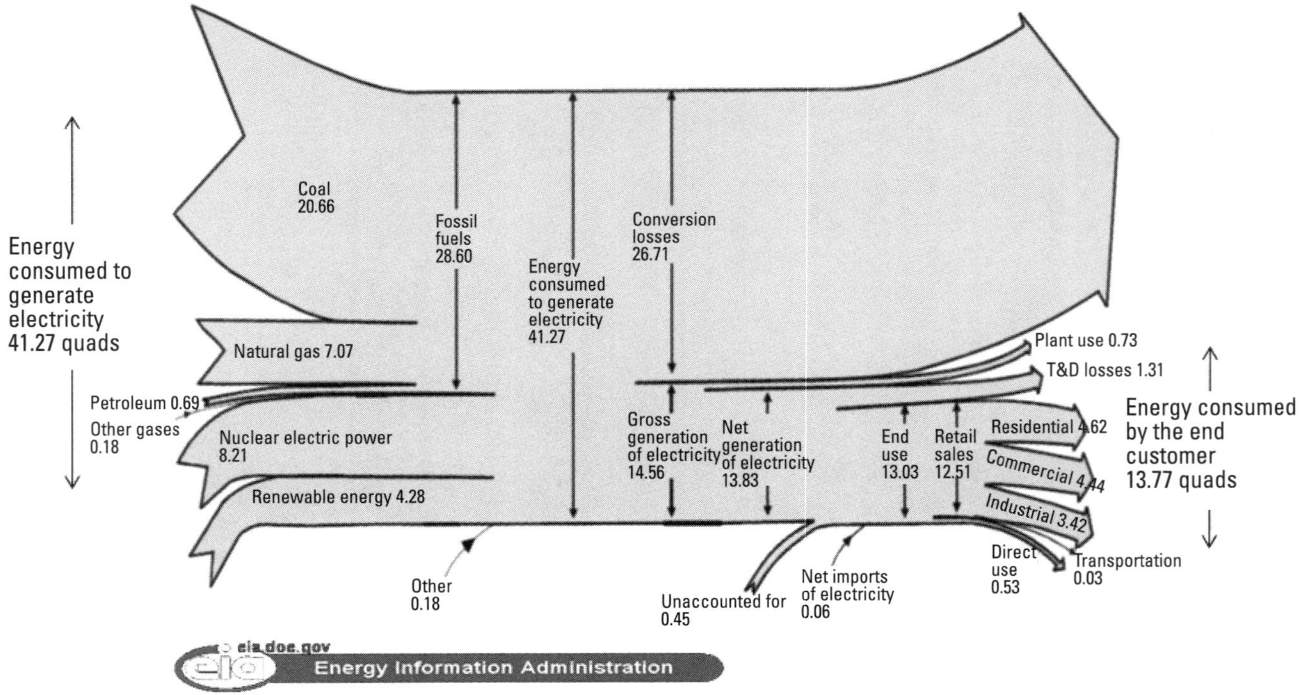

Figure 9.8 Energy conversion analysis from source to consumption. (Source: Energy Information Agency www.eia.gov.)

- *Financial incentives from the federal government and several states:* Including tax credits for lowering the up-front costs of purchasing plug-in electric vehicles, also known as electric cars or EVs. The IRS tax credit is for \$2,500 to \$7,500 per new EV purchased for use in the United States. The size of the tax credit depends on the size of the vehicle and its battery capacity. This tax credit will be available until 200,000 qualified EVs have been sold in the United States by each manufacturer, at which point the credit begins to phase out for that manufacturer.
- *The solar investment tax credit (ITC):* A federal policy mechanism to support the deployment of solar energy in the United States. It is a 30% tax credit for solar power systems on residential and commercial properties under sections 25D and 48. This subsidy is in existence through 2021, providing market certainty for companies to develop long-term investments that drive competition and technological innovation, which, in turn, lowers costs for consumers.
- India's Cabinet Committee on Economic Affairs (CCEA) sanctioned INR 50 billion (\$750 million in U.S. currency) funding for 30% capital subsidy for rooftop solar PV installations. The subsidy will be restricted to residential, government, social, and institutional segments only and the government expects this subsidy to support total rooftop capacity of 4,200 MW until this budget is exhausted.

Utility Viewpoint

The underpinnings of the utility's need to control the consumption of energy at the destination are at the source of the business case for the utility. Figure 9.1 showed the normal daily consumption cycle of energy in a utility and overlaying the curve is the typical generation profile necessary to meet this load profile. This profile is complicated by the inclusion of DERs, which generate more energy at the consumption point.

The utility's business case comes from flattening the load profile, which can result in a combination of reduced costs of generation, reduced greenhouse gas emissions, and a better use of generation resources. Visibility into the home and building supported by some measure of control allows the utility to drive a better solution for all consumers. Reduced costs result in a better payback for the regulated utility and reduced rates to the customer.

Another significant aspect of the utility's business case is to implement nonwire solutions (NWS) to resolve existing capacity problems. The utilities' normal response to problems with increasing load is to enhance the capacity of the infrastructure feeding the increased load. This increase in infrastructure capacity can be through increasing wire size, transformer capacity, and/

or substation capacity. Utilities are turning more to nonwire solutions. Smart homes and buildings are a significant part of that NWS. Some examples of that include:

- *Consolidated Edison:* ConEd has one of the most publicized of these projects in the works for Brooklyn and Queens. The nonwire project will prevent building a $1 billion substation and related infrastructure. The utility plans to solve electrical overload problems by fostering development of $200 million in distributed energy resources [15].
- *Brooklyn and Queens:* Specific neighborhoods are seeing a boom in residents, jobs, and subway riders. Electric use is growing. Rather than making huge investments, ConEd's plan is to meet this growth through energy efficiency, demand management, and distributed generation. The utility is also considering a nonwire solution for a 4-MW system overload expected in the Plymouth Street area of Brooklyn, New York because of growing demand.
- ConEd believes that the final cost of this effort will be about $400 million. This will save a significant amount of rate-payer costs over the traditional cost of $1 billion that would have otherwise been spent on substation enhancements.

Customer Viewpoint

There are two types of customers: the residential customer and the building operator/customer. The business case for the building operator/customer is like that of the investor.

The residential customer's viewpoint focuses on several points:

- *The need to save money:* Programs such as demand response allow the customer to participate in the control of residential load. When customers participate, they can reduce their consumption in response to the utility's request at critical times in the day, month, or year. Based on the program guidelines, the customer gets paid for both participating and for reducing.
- *The need to be more independent of the utility:* Many customers purchase EVs and install a combination of DERs and storage at their residence. This is generally done to increase local generation of electricity and create mechanisms to control consumption, thereby reducing dependence on utility supply. With the installation of these systems, some customers

also allow the utility to send signals to control a combination of generation and load behind the meter.

- *The need to be green:* Customers often install DERs at the home. Examples of PV, storage, EVs, and other devices allow the customer to use their own generation from renewable sources or drive cars running on electricity. For the most part, these technologies are still more expensive than getting their electricity from the grid. The key driver to install them is the need to be green.

Investor Viewpoint

For the building operator or investor, the main business case comes from specific rates governing their utility bill. Commercial buildings tend to be on a commercial rate. Key components of the commercial rate include two primary components: energy charges and demand charges.

- Energy charges are calculated by multiplying the total energy consumed in each time period (e.g., per month) by the energy charge ($/kWh). In some locations, both in the United States and internationally, there are rate structures that combine TOU components and/or layering, which increases the charges based on the amount of energy consumed. These rate structures influence consumption behavior, forcing them to move extended consumption to the off-hours.
- Demand charges are based on the peak electricity usage during a billing period and generally represent the utility's cost of electricity during that period. The demand charge is based on the highest cost of electricity during the year and forms the basis of the demand charge for the following year. Any reduction in the demand charge for 1 year can significantly reduce the building operator or investor's cost of electricity for an entire year.

Building operators and investors install various forms of DER technologies to save on their energy consumption and to reduce the demand charge.

Case Study

Case Study #1: AEP Ohio gridSMART Demonstration Project

AEP Ohio and its partners built a secure, interoperable, and integrated smart grid infrastructure in Ohio [16, 17]. That infrastructure demonstrates the ability to maximize distribution system efficiency and reliability and consumer use of demand response programs to reduce energy consumption, peak demand costs, and fossil fuel emissions. The demonstration area illustrated in Figure 9.9 includes 150 square miles, including parts of Columbus, Bexley, Gahanna, New Albany, Whitehall, Reynoldsburg, Westerville, Blacklick, Johnstown, Alexandria, Minerva Park, and Pataskala. The electric networks covered 70 distribution circuits—fifty-eight 13-kV circuits from 10 distribution stations and twelve 34.5-kV circuits from six distribution stations—and 110,000 consumers in an area selected for its concentration and diversity of distribution infrastructure and consumers. Included in this project are an advanced meter infrastructure, home area networks, community energy storage, sodium sulfur battery storage, and renewable generation sources. These technologies will be combined with two-way consumer communication and information sharing, demand response, dynamic pricing, and consumer products, such as plug-in hybrid vehicles.

A key objective of the gridSMART project was to "[a]ctively attract, educate, enlist, and retain consumers in innovative business models that provided tools and information reducing consumption and peak demand."

It started with the deployment of digital advanced metering infrastructure (AMI) electric meters to the approximately 110,000 premises in the project area to replace their existing analog electric meters. AMI meters feature two-way

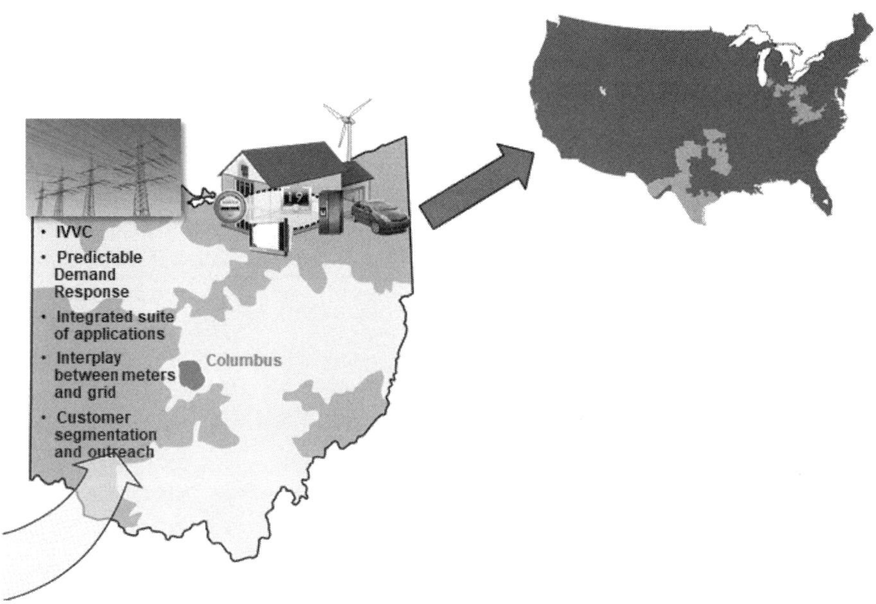

Figure 9.9 AEP-Ohio Energy Project geography and functionality. (Source: Smartgrid.gov https://www.smartgrid.gov/files/AEP_Ohio_DE-OE-0000193_Final_Technical_Report_06-23-2014.pdf.)

communications between the consumer and the utility. This communication provides the ability to measure and record usage in subhourly increments, as well as the ability to integrate supportive technologies such as programmable communicating thermostats, load control devices, and in-home displays into home area networks. Key areas of the project specific to smart homes and buildings included:

- *Tariffs:* Designed to more accurately reflect the underlying variability of the cost of service. Such variable price tariffs ranged from time-of-day rates, where the cost of electricity was lower during off-peak periods and higher during times of peak use, to real-time pricing programs, where rates most nearly aligned with actual wholesale market prices reflected in the locational marginal prices (LMP) of electricity (see Figure 9.10).
- *Cost savings:* Participating consumers could ensure cost savings because of the technology's two-way communication functionality. Consumers could make choices based on the way electric rates varied throughout the course of a day. They had differing ways of interacting with the technology, depending on the program they were enrolled in. These consumer programs included TOU prices, critical peak price events, DLC events, and real-time pricing. Several experimental time-of-day tariffs and DLC riders were tested to determine at what level these tariffs and riders, either directly or indirectly, might reduce a consumer's electricity usage and reduce load for the utility.
- *Monitoring of electric use:* Participants had the opportunity to more closely monitor their electric use and have greater control over their monthly electric costs by shifting usage from higher price periods to

Market name	Program description
eView	Consumer usage feedback device
SMART Shift	Two-tier time-of-day
SMART Shift Plus	Three-tier time-of-day with critical peak pricing events
SMART Cooling	Direct load control thermostat only
SMART Cooling Plus	Direct load control with load control switch
SMART Choice	Real-time pricing with double auction
Standard Residential	Flat tariff with declining block rate, average cost

Figure 9.10 AEP-Ohio gridSMART Project customer tariffs and eView. (Source: Smartgrid.gov https://www.smartgrid.gov/files/AEP_Ohio_DE-OE-0000193_Final_Technical_Report_06-23-2014.pdf.)

lower price periods or by reducing the demand on the electrical system during peak periods.

- *Education and awareness:* Because the rate structures and technologies being introduced were new to most of AEP Ohio's consumers, it was important that AEP Ohio provided consumer education and awareness programs to encourage participation in the programs.

Example portions of the program were the following.

eView

The eView program provided consumers with an in-home device that interacted with the smart meter to provide the consumer with current electrical usage and pricing information, enabling them to make decisions about their energy consumption. Consumers could see the average price of electricity and how much they were using and were encouraged to experiment by turning various household appliances on and off to see the difference in usage and costs. The device held usage and average cost data in memory for 30 days, which helped consumers make comparisons and estimate upcoming bills.

SMART Shift Plus

The SMART Shift Plus program was a three-tiered pricing option that offered the consumer incentives to modify their electric usage patterns during peak load times on weekdays of the summer months. An IHD and optional programmable communicating thermostat (PCT) were installed in the consumer's home to accommodate participation in the program. The PCT gathered and displayed information about how much electricity was being consumed and how much it cost. The IHD displayed the current electricity use and rate and notified consumers when a critical price period occurred.

The tariff for this program permitted AEP Ohio to declare up to 15 critical peak pricing (CPP) events when AEP Ohio experienced unusually high demand. CPP events were not to exceed 5 hours per day during the calendar year. Energy consumed during these events was charged at a substantially higher rate, thus encouraging consumers to reduce their demand for power at times it cost AEP Ohio the most to produce.

As part of the SMART Shift Plus program, the deployment of 33 smart appliances was included in the experiment. The GE smart appliances installed in 20 homes were washer, dryer, range, refrigerator, and electric water heater.

The appliances were equipped with circuitry that communicated with the SMART Shift Plus power display device and allowed the consumer to see in real time how much electricity was used. When the SMART Shift Plus device detected a higher price for power, the appliances responded accordingly. During a price increase or a defined critical peak period when usage costs were higher, the appliances were programmed to respond as follows.

SMART Cooling Plus

This program was an add-on to the SMART Cooling program by installing a load control switch (LCS) in addition to the IHD and PCT devices. The LCS was installed on electric water heaters, pool pumps, or hot tubs as additional power demand that could be managed remotely. These consumers were offered an incentive to reduce demand by allowing the utility to interrupt the devices during DLC events. Consumers with water heaters and hot tubs could experience 15 additional events during the months of October through April. Consumers had the ability to opt out during DLC events.

SMART Choice (Real-Time Pricing with Double Auction)

This program provided consumers the opportunity to participate in real-time pricing based on supply and demand for their specific power circuit. Pricing occurred every 5 minutes for each circuit included in the program. Consumers participated by using an advanced home energy manager (HEM) and the enhanced programmable communicating thermostat (ePCT).

Lessons Learned from the Consumer Programs

Tariff Impacts

Both the SMART Shift and SMART Shift Plus consumers had lower consumption than the standard residential tariff consumers during peak time periods in 2012. In 2013, however, the SMART Shift and SMART Shift Plus consumers had lower consumption during the first hours of the peak period and higher consumption during the last 2 hours and a higher overall peak. After the peak period (approximately 7 P.M. to midnight), the time-of-day consumers' consumption was greater than that of the flat-rate consumers in 2012 and 2013.

In the winter, the SMART Shift consumers had higher overall usage and a similar load shape compared to standard residential consumers. SMART Shift Plus consumers had a lower overnight and mid-day usage with sharper morning and evening peaks compared to standard residential consumers. During the winter months, standard residential consumers were charged a declining block rate. SMART Shift and SMART Shift Plus consumers were charged a flat rate that was lower than the standard residential tariff.

The autumn/spring load profiles appeared similar across all three tariffs, with the TOD consumers exhibiting higher morning and evening peaks.

The TOD behavior may have been driven by PCTs, which allowed consumers to program different set points for specific time periods within a day. A typical PCT program had specific morning and evening time periods, representing the times consumers prepared for work/school and return home, which may have resulted in the higher consumption for the TOD/CPP consumers during the morning and evening time periods.

Consumer Programs Conclusions

The project provided useful information about consumer programs linked to AMI-driven technologies. It demonstrated that programs can be successfully implemented, but significant changes to back-office IT systems and business processes are required. The results indicated that consumers would participate in programs when given adequate information and enabling technologies. Dedicated customer service representatives were essential to handle significant call volume and address concerns as a high priority.

The SMART Shift and SMART Shift Plus programs exhibited lower energy and demand impacts in 2013 than they did in 2012. A milder weather season in 2013, an increased number of participants in 2013, less program communication in 2013, or a combination of those factors may have contributed to the difference in results.

Consumers participating in the SMART Cooling Program significantly reduced their demand during thermostat adjustment events. This reduction resulted in approximately twice the demand reductions achieved by those in SMART Shift Plus Program Equivalent Critical Peak Pricing events. Thermostat adjustment provided the largest demand reduction for approximately the first hour. As temperatures in participants' homes began reaching the new thermostat set-points, HVAC operations resumed. This indicated that proper timing of the events and thermostat adjustment to coincide with AEP Ohio's peak load conditions was a critical component for program success.

Overall, program participants were satisfied with their experience. Customer satisfaction ratings ranged from 67% to 76%, depending on the program. Program participants perceived an average savings of $20 per month on their summer electricity costs.

Next Steps

Based on the success of the project, AEP Ohio has filed a gridSMART Phase 2 project with the Public Utilities Commission of Ohio (PUCO). Phase 2 will extend the benefits demonstrated in the project and deliver additional benefits to a broader set of consumers. Through Phase 2 AMI, AEP Ohio expects to drive significant financial benefits, positively impact customer service and customer satisfaction, improve meter field personnel safety, and reduce environmental impacts. It will also enable demand response and competitive retail electric service (CRES) providers to offer consumer programs.

Case Study #2: Oklahoma Gas & Electric Company—Positive Energy Smart Grid Integration Program

The Oklahoma Gas and Electric Company (OG&E) project involved system-wide deployment of a fully integrated AMI solution and distribution of in-home devices to selected customers. OG&E also deployed a DMS, automated switching, and IVVC. The project included an AMI-enabled study of consumer behavior in response to different forms of dynamic pricing and home area network technologies on an opt-in basis. Additionally, OG&E collaborated with the University of Oklahoma faculty and students to deploy energy saving technologies within 46 buildings on the Norman, Oklahoma, campus and took advantage of opportunities for smart grid education and training.

The project was a partnership with customers aimed at reducing peak loads, overall electricity use, and operations and maintenance costs. Distribution system upgrades were expected to increase operational efficiency, reduce line losses, lower operational costs, and improve service reliability for customers.

Smart grid technologies deployed included:

- *Communications infrastructure:* A secure wireless mesh AMI network enabled two-way meter communications, provided the backbone for energy management programs, and allowed for integration with smart appliances and home area networks. OG&E also installed a WiMAX point-to-multipoint wide area network (WAN) that connected a point-to-point microwave network for backhaul communications.
- *Advanced metering infrastructure:* The project deployed 818,415 smart meters covering OG&E's entire service territory and supporting information technology infrastructure, including a meter data management system (MDMS). The AMI system enabled automated meter reading, fewer estimated bills, enhanced outage response and notification, and improved theft-of-energy detection. More detailed and timely data on peak electricity usage improved load forecasting and capital investment planning.
- *Advanced electricity service options:* The project deployed a variety of customer-focused tools to homes and businesses. These tools included a customer web portal, home area network capability, IHDs, and PCTs, all supported by the AMI system. Customers could view their electricity consumption data at any time through a personalized website, while allowed OG&E to manage, measure, and verify targeted demand reductions during peak periods. Study participants tested IHDs, providing OG&E with information to develop programs to meet consumer needs and have the greatest impact on load reductions.

- *Time-based rate programs:* Customers also participated in studies of variable peak pricing and standard time-of-use pricing, both with peak price elements.
- *Distribution automation deployment:* OG&E deployed reclosers on 50 high-priority circuits integrated with a DMS. The DMS provided monitoring and control of the distribution system. It also supported advanced applications, such as IVVC, which optimized the voltage on automated circuits.
- *Automated Volt/Var control:* OG&E deployed capacitor and load tap changer controllers to facilitate voltage optimization on 100 circuits.

OG&E assessed the relative merits of several time-based pricing options and information treatments on a subset of residential and commercial sector volunteers in Norman, Oklahoma. This was a randomized, statistically based study (see Figures 9.11 and 9.12) involving 4,000 residences and 1,320 small businesses, including the control group. The pricing options included variable peak pricing, a rate plan in which the peak period price changes daily to better reflect exigent system conditions and the real cost of providing power with a critical price element, and a time-of-use rate plan with a critical price element. Smart technologies tested included a personalized website, IHDs, and PCTs. The study explored how consumers responded to the peak price options as well as the relative effects of the various information-providing technologies. The study also estimated impacts on total electricity use, customer bills, and hourly load profiles using information collected through the smart meters.

Benefits realized included:

Recipient information	Input
Total number of customers served by utility/utilities supporting the project	766,886
Total number of residential customers served by utility/utilities supporting the project	657,391
Total number of commercial customers served by utility/utilities supporting the project	84,491
Total number of industrial customers served by utility/utilities supporting the project	9,480
Total number of AMI smart meters installed and operational prior to the SGIG/SGD program	0

AMI smart meters installed and operational	Quantity	Cost
Total	818,415	$153,693,666
Residential	706,501	
Commercial	102,699	
Industrial	9,214	

Figure 9.11 Project customer information. (Source: Smartgrid.gov, https://www.smartgrid.gov/project/oklahoma_gas_electric_positive_energy_smart_grid_integration_program.html.)

Web portal deployed and operational		Quantity	Description
Customers with access to web portal		843,914	
Customers enrolled in web portal		61,097	A web-based customer portal that will provide historical usage information, tools to analyze their...
Customer systems installed and operational	**Quantity**	**Description**	**Cost**
Communications networks and home area networks	N/A	The system that allows smart meters to communicate with programmable communicating themostats (PCT) and in home displays (IHD) utilizing a ZigBee or some other standard protocol.	N/A
In home displays	0		S0
Energy management devices	0		N/A
Direct load control devices	0		S0
Programmable communicating thermostats	28,668	Programmable communicating thermostats (PCT) are capable of accepting commands over the HAN...	$17,265,671
Smart appliances			$0

Figure 9.12 Customer home automation information. (Source: Smartgrid.gov https://www.smartgrid.gov/project/oklahoma_gas_electric_positive_energy_smart_grid_integration_program.html.)

- *Lower electricity bills for customers:* New AMI-enabled energy management tools gave customers the information they needed to make more informed decisions about their peak and overall energy usage on a real-time basis. On average, the demand response participants saved $200 over the course of the summer of 2012, and 99% of the demand response customers saved money through the program.
- *Demand response:* The time-based rate program (SmartHours) and IVVC provided the utility the ability to reduce load by 70 MW and 17 MW, respectively.
- *Reduced operating and maintenance costs:* AMI reduced OG&E's meter reading expenses by approximately $9 million per year.
- *Increased electric service reliability:* The project's distribution automation component facilitated reduction of the frequency and duration of outages through recloser deployment and remote switching via the DMS. Additionally, automated capacitor controls, load tap changer controls, and IVVC application allowed for improved power factor and voltage control, which contributed to overall available demand reduction.

- *Reduced greenhouse gas emissions:* The remote meter readings and the diagnostic and troubleshooting functionalities of the AMI system greatly reduce vehicle fleet fuel requirements associated with manual meter reading and investigative field maintenance [18, 19].

How Smart Homes and Buildings Will Transform the Utility Industry

Most homes are not yet smart. However, buildings are well ahead of single-family homes in terms of smartness because they have a more compelling business case to do so. However, devices and appliances that can enable homes and buildings to become smart are just coming into the market. Experts expect that more of these devices will end up in homes and buildings of the future, making them smarter.

Utility demand in most jurisdictions in North America and Europe is stagnant, but increasing in areas such as Asia and Africa. Homes and buildings have always consumed energy but recently, due to the onset of DERs, are becoming a combination of producers and consumers leading to the emergence of a new term: prosumer. As a result, the load and generation patterns (see Figure 9.13 for generation-load balance equation) are changing dramatically with the advent of newer loads such as EVs and generation such as PVs and other DERs, impacting not just the flow of electricity into the home but also the dynamics of the relationship between customer and utility.

For the most part, the utility's relationship with the home or building owner or operator is reactive and adversarial.

When owners or operators apply for a connection permit, the utility sees them as customers and registers them and interacts with them as outsiders. It takes money from them when a bill is paid and penalizes them for nonconformance or noncompliance or nonpayment. In many places, it sees them as

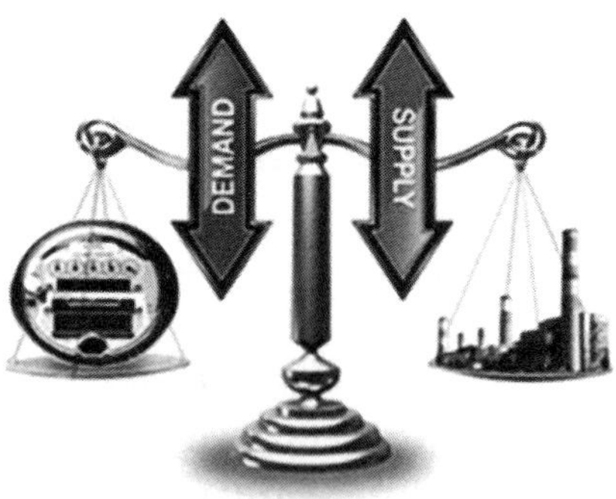

Figure 9.13 Energy balance equation. (Source: www.Smartgrid.gov.)

competitors and, in many jurisdictions, pushes back on regulations that encourage homes and buildings to increase their installations of DERs, such as the implementation of microgrids.

As a result, the owners or operators are turning to third-party technology providers, aggregators, and even other service providers to meet their needs and aspirations.

Utilities are in a unique position. They have a 100-year relationship with customers, which is greater than any other service industry. They also serve a critical need to homes and buildings and deliver a product that is critical to the building's existence. The utilities also have multiple gateways and deliver multiple products and services to the house such as electricity, gas in some places, and water/waste water in some places. The utility has an opportunity to work with homeowners and building owners or operators. However, this cannot be a technological solution alone. These technology implementations need to be supported and supplemented by changes to the utility's core business operating model. These are very fundamental changes and could take any one of several different paths:

- As homes and buildings become smart and become prosumers, utilities will not see them as competitors, but as a part of the energy equation.
- Offer services and options to install more DERs (including demand response and energy storage) into their premises and define ways and means to interact. This will allow utilities to find advantages for the utility and bring the prosumer into the planning process.
- Change how the customer is perceived. No longer a customer but a collaborator and a part of the utility's service model. For example, bring the customers into the generation discussion and assist them with the decision-making process with one or more of the following supports: DER/generation, DR, and/or PV integration, storage, or V2G.
- Providing new services and products to the customer. Examples of this could include servicing the DER installations at the house and providing prosumers with a smart inverter that can be controlled from the utility to provide KW/KVAR support.

In effect, move them from a customer to a collaborator.

A key part of this equation is the regulator. The regulator cannot sit on the sidelines transforming policy objectives into regulations that impact everyone else, but keep the utility-customer contract intact as it has been for the last several decades. The regulator must change and allow the utility to do more.

The utility, customer, and other service providers need to work together, not independent of each other, with the assistance of the regulator.

Conclusions

Why is this important? Smart homes and buildings signify an important sea change in the electrical industry. This is where many of the technologies discussed earlier in this book come together.

- *Distributed energy resources:* Many of the DER technologies being released in the marketplace are targeted to smart homes and buildings. Specific examples such as the solar roof (solar PV built into the roofing shingle) and others are being installed at a rapid pace in many parts of the world.
- *Smart transportation:* With every major automobile manufacturer planning to deliver EVs over the next few years, it is safe to assume that these will be parked at the homes and garages of many smart homes and buildings.
- *Energy storage:* As the ultimate disruptor of the energy equation, energy storage is moving from niche applications to the mainstream. The Tesla Power Wall—storage designed for the home—is an example of technology that is becoming mainstream and could, for the first time, allow the homeowner or building operator to break the instantaneous connection with the grid to one of a transactional connection.
- *Microgrids and nanogrids:* Like how the utility brought homes, buildings, commercial properties, and industries together into a single energy community, microgrids and nanogrids are now enabling homes and buildings to band together into their own energy communities.

In addition, the utility is making technological changes to the grid to handle these and other technology implementations that are impacting grid operations by focusing on areas such as distribution automation and data analytics.

The utility business is all about the customer and will likely stay that way. Unless it transforms its operating model, however, and redefines the relationship with the customers, there is another option, one in which the utility may become irrelevant.

It is time to change this relationship and bring the customer in as a collaborator.

References

[1] U.S. Department of Energy, "Staff Report to the Secretary on Electricity Markets and Reliability," 2017, https://energy.gov/sites/prod/files/2017/08/f36/Staff%20Report%20on%20Electricity%20Markets%20and%20Reliability_0.pdf.

[2] TechTarget, "Smart Home or Building," 2017, http://internetofthingsagenda.techtarget.com/definition/smart-home-or-building.

[3] Safewise, "What Is Home Automation and How Does it Work?" 2017, https://www.safewise.com/home-security-faq/how-does-home-automation-work.

[4] Khoi, N. M., et al., "IReHMo: An Efficient IoT-Based Remote Health Monitoring System for Smart Regions," LTU, http://ltu.diva-portal.org/smash/get/diva2:1005647/FULLTEXT01.pdf.

[5] BlueSpeed, "The 7 Greatest Advantages of Smart Home Automation," June 14, 2016, http://bluespeedav.com/blog/item/7-greatest-advantages-of-smart-home-automation.

[6] Hurst, D., "Utilities and Home Automation," NextEnergy, February 4, 2016, https://nextenergy.org/utilities-and-home-automation/.

[7] Safewise, "What Makes Your House a Smart Home? How Home Automation Protocols Work," 2017, https://www.safewise.com/blog/makes-house-smart-home-home-automation-protocols/.

[8] U.S. Department of Energy, "What Is the Smart Grid?" SmartGrid.gov, 2017, https://www.smartgrid.gov/the_smart_grid/smart_home.html.

[9] Microchip Atmel, "In-Home Display Units," 2016, http://www.atmel.com/applications/smart_energy/in-home-display-units/default.aspx.

[10] Huq, M. Z., "Home Area Network Technology Assessment for Demand Response in Smart Grid Environment," Semantics Scholar, https://pdfs.semanticscholar.org/236e/0ecdd3c26f0b36a6550e156749c3b3621be3.pdf.

[11] Trilliant, "The Home Area Network: Architectural Considerations for Rapid Innovation," 2010, http://trilliantinc.com/library-files/white-papers/HAN_white-paper.pdf.

[12] Mistral Solutions, "Building Intelligence with Smart Building Automation System (SBAS)," January 2014, http://www.mistralsolutions.com/newsletter/Jan14/Application-Home_Automation.pdf.

[13] U.S. Department of Energy, "Home Area Networks and the Smart Grid. Pacific Northwest National Laboratory," April 2011, http://www.pnnl.gov/main/publications/external/technical_reports/pnnl-20374.pdf.

[14] Vadari, M., "Challenges and Success Factors for Demand Response," *Grid Insights*, Intel at Energy Central, December 8, 2014.

[15] Wood, E., "Utilities Embracing Disruptive Energy, Non-Wires Alternatives," *Microgrid Knowledge*, May 19, 2015, https://microgridknowledge.com/utilities-embracing-disruptive-energy-non-wires-alternatives/.

[16] U.S. Department of Energy. "AEP Ohio gridSMART Demonstration Project: Fact Sheet," September 2015, https://www.smartgrid.gov/files/OE0000193_AEP_FactSheet.pdf.

[17] U.S. Department of Energy, "AEP OHIO gridSMART Demonstration Project: Final Technical Report," SmartGrid.gov, June 23, 2014, https://www.smartgrid.gov/files/AEP_Ohio_DE-OE-0000193_Final_Technical_Report_06-23-2014.pdf.

[18] U.S. Department of Energy, "Oklahoma Gas & Electric Company-Positive Energy® Smart Grid Integration Program," U.S. Department of Energy, Office of Electricity Delivery and Energy Reliability, September 2014, https://www.smartgrid.gov/files/OGE_Project_Description.pdf.

[19] U.S. Department of Energy, "Oklahoma Gas & Electric (Positive Energy Smart Grid Integration Program)," SmartGrid.gov, 2017, https://www.smartgrid.gov/project/oklahoma_gas_electric_positive_energy_smart_grid_integration_program/latest_data.html.

10

Electric Utility Transformation

The crisis takes a much longer time coming than you think, and then it happens much faster than you would have thought.

—Rüdiger "Rudi" Dornbusch, economist, U.S. Congress and Senate, Committee on the Budget (2012) Concurrent Resolution on the Budget Fiscal Year 2013. p. 95

There are only patterns, patterns on top of patterns, patterns that affect other patterns. Patterns hidden by patterns. Patterns within patterns. If you watch close, history does nothing but repeat itself. What we call chaos is just patterns we haven't recognized.

—Chuck Palahniuk American novelist, freelance journalist, and author of *Fight Club*

Utilities must continue to learn from other industries, both regulated and unregulated industries especially, about how to manage through change in the business environment. What happened to them can also happen to the electric utility.

Introduction

The electric utility is in the throes of a major change, but change is not new to this industry.

While the Public Utilities Regulatory Policies Act[1] (PURPA) is credited as being one of the first major changes to hit the modern electric utility in the United States, real meaningful change came in 1996 with the advent of FERC orders 888/889 [1–3], which played a major role in opening the U.S. electricity markets to competition. Order 888 required two major changes, the unbundling of electricity services and separation of all the marketing functions for these services. These changes required utilities to provide open access to tariffs by making services available on the open market. Order 889 took it one step further and specified the information that needed to be placed on the open market. This set the stage for a new system called Open Access Same-Time Information System (OASIS) [4, 5]. Immediately following this, existing power pools such as Pennsylvania Jersey Maryland (PJM), New York Power Pool (NYPP), and New England Power Pool (NEPool) formed independent system operators. Independent system operators became regional transmission operators (RTOs) with the release of FERC Order 2000, which administered the transmission grid on a regional basis across the United States and Canada.

As RTOs were settling into wholesale markets, a new change appeared. The smart grid[2] started a new revolution, with the focus on the distribution grid [6–10]. Key technologies such as smart meters, home energy managers (HEM), distribution automation (DA), and wireless communication (RF mesh, cellular and others), enabled the smart grid by becoming available at scale. They were also more cost-effective and capable of performing multiple actions.

The next major change in the electric industry was legislation called the American Recovery and Reinvestment Act (ARRA) in 2009, a stimulus package that included the following [11]:

- Approximately $8 billion under a program called Smart Grid Investment Grant (SGIG). There were 99 projects involving more than 200 participating electric utilities and other organizations to modernize the electric grid, strengthen cybersecurity, improve interoperability, and collect an unprecedented level of data on smart grid operations and benefits [12].
- Approximately $1.6 billion under two programs with 16 projects each, Smart Grid Regional Demonstrations and Energy Storage Demonstrations [13].

1. PURPA was designed to promote energy conservation (reduce demand) and promote greater use of domestic energy and renewable energy (increase supply). The law was created in response to the 1973 energy crisis.
2. The first official definition of smart grid was provided by the Energy Independence and Security Act of 2007 (EISA-2007), signed to law in December 2007. Title XIII of this bill provides a description, with 10 characteristics, that can be considered a definition for smart grid.

Although the utility industry and its collaborative efforts made progress in moving smart grid along the road to maturity, they were generally stuck in an eternal proof-of-concept mode. This was the first time the U.S. federal government spent this much money on improvements to the grid in such a short period of time. ARRA was an acceleration event bringing with it pilots that were better planned, demonstrations, and shovel-ready technologies. ARRA investments allowed several new and innovative technologies to be tested, but, some of them failed, due to a lack of viable performance. Many others succeeded as pilots and were put into full production, capable of standing on their own.

Between 2009 and the present, other technologies also came to the forefront. Energy storage, microgrids, electric transportation, and smart homes and buildings came into their own along with other technologies. In addition, as utilities collected more and more data, they tried to develop analytics to extract intelligence from this data to improve operations.

Challenges Faced by Utilities

Figure 10.1 presents a high-level view of the challenges faced by today's utilities worldwide.

- *Workers retiring and fewer people joining the workforce:* In 2006, a leading utility industry magazine stated, "assuming only nominal growth, by 2010 the industry in the U.S. will need to hire some 10,000 new skilled workers each year" [14]. This statement referred to both utilities and their vendors.

  The actual loss of skilled employees was substantially slowed by the financial meltdown of 2008. As the economy continues to improve, the trend is reversing and number of retirees has begun to increase. The situation is exacerbated by the fact that utilities did not hire people for the last few decades, creating a skill gap between personnel with deep expertise and no one ready to take on the mantle after their retirement. To replace this loss of expertise, utilities and their vendors are beginning to actively look to hire new workers. A host of social and market factors constraining the supply of skilled workers complicates this process.

  As the industry looks to the future workforce, a few difficult questions come to mind. What skill sets do these newer personnel need? What impacts does this have on the universities that are generating the next generation of the utility employee or power system engineer.

Figure 10.1 Utility challenges. (© Modern Grid Solutions.)

In addition, utilities are outsourcing more work because it is economical and allows them to develop a workforce that is competent and conversant with technology.

- *Declining revenues:* Houses and buildings are becoming more energy-efficient, so they consume less energy. In addition, the introduction of DERs into the marketplace are creating the first real alternate supplier of electricity allowing customers to generate power by themselves and, in turn, buy less energy from utilities. If not managed properly, these technologies have the potential to relegate the utility to a standby asset which could further impact future revenues.
- *More automation:* Utilities are adopting new technologies that are automating more of the work that is done, thereby requiring fewer workers. This move also includes moving away from paper-based data processes toward digital and electronic data processes.

- *New technologies coming at a pace previously unseen in the utility industry:* The old paradigm had regulators and utilities that controlled the pace of innovation through collaborative research. This is no longer working. Some of these technologies are being implemented by the utilities themselves. Examples of these are DA and other advanced systems such as advanced distribution management system (ADMS) and distributed energy resource management system (DERMS). There are also other technologies being implemented by the customers, and require a response by the utility. Examples of these are solar panels with smart inverters and EVs.
- *New players challenging utility's status quo:* For a long time, the paradigm was the utility that generated the power and invested in and maintained the infrastructure to deliver reliable power to the customer. This paradigm is changing, however, through the introduction of newer technical constructs such as microgrids and business models such as aggregators. These constructs challenge the utility's right to deliver power to its customers.
- *Retail markets and choice:* Until the introduction of DERs and storage, there was no fundamentally different way of creating electricity as a product and/or delivering it to the customer. As a result, there was some limited switching in several states that had retail choice. However, fundamental technological alternatives such as the DERs and storage supported by strong value propositions are threatening the legacy of the utility's monopoly of centralized generation. Various states are beginning to discuss the potential to bring in retail-level markets followed by the creation of distribution system operators (DSO). This will change the dynamics between the utility and its customers along with relationships within its own internal departments.
- *Shifting regulatory climate:* The existing regulatory framework is outdated and is still mostly based on a one-size-fits-all across the entire customer base. It also depended on an endless cycle of network expansion, rate increases, and capital requests, creating a backlash from customers and regulators. This dissatisfaction created headroom for disruptors to come in and change the game by providing services directly to the customer.

The normal utility response to rate pressures was to cut the budgets (resulting in dissatisfied customers) or decoupling of energy delivery from infrastructure costs. However, this was creating a situation that was not enabling the customer's needs or aspirations and also not posturing what an efficient delivery mechanism should look like.

State and federal policy makers and regulators are now fostering direct, not-for-profit competition in existing utility service territories and distributed renewable expansion and emission reductions.

The Case for a Transformed Utility of the Future

Let's revisit the hypothetical case study[3] from Chapter 2, with a slight modification.

"I just returned from Houston and my friend got a message on her cell phone that the power was out at their house, but that it would be back on in two hours, so we kept playing tennis. When she checked the app, she also showed me her car was only charged 80% and because of solar PV on her roof, the storage at her house was fully charged and providing power to her refrigerator and the critical appliances in her house. She smiled and said she sold $75 worth of power last month back to her retailer and it paid for lunch today. She said her electricity bill now only includes a connection charge unless she does her clothes washing and baking on the same day. I am calling my retail energy provider to see what they can provide."

An analysis of this case study-based on the knowledge gained from reviewing Chapters 3 through 9 makes a case for utility transformation.

New Technologies and Technical Constructs

New technologies such as distributed renewable generation, energy storage, electric vehicles, smart building systems, and data analytics are allowing customers to contemplate disconnecting from the grid. While most customer-generation capabilities are not yet fully renewable, or cost-competitive, the trend is moving towards reduced costs and greater availability of power from renewable sources. The tipping point has been achieved in some conducive markets (such as California) and is imminent in others in the not-too-distant future.

Deploying these hardware and software platforms to homes and buildings in a geographical area allow the formation of microgrids or community electrification options[4] that will represent a new opportunities and challenge to existing utility operations.

3. Case credit to Charles Filewych, CEO, Smart Grid Interconnect. Used here with permission.

4. Community Choice Aggregation (CCA) is a system adopted into law in Massachusetts, New York, Ohio, California, New Jersey, Rhode Island, and Illinois. It allows cities, counties, and some special districts to aggregate the buying power of individual customers within a defined jurisdiction to secure alternative energy supply contracts on a community-wide basis, but allowing consumers not wishing to participate to opt out. CCAs now serve nearly 5% of Americans in over 1,300 municipalities.

Today, this kind of living is attractive to a unique population base. This population generally lives off the land and in remote locations, either individually or in communities [15]. In the case of microgrids, these capabilities are attractive to college campuses which often use microgrids more to study the impacts of becoming a microgrid, rather than a desire to disconnect from the grid. These technologies are going mainstream, however, and could attract a larger segment of the population, which may want to disconnect from the grid or have no grid to which to connect.

Making the case for electric utility transformation: customers starting to live off-grid[5] may be a trickle today, but may in the future become a flood once the tipping point of technology availability at an attractive price-point is reached. If that happens, it may be difficult to stem the tide and could lead to utilities losing customers at an unacceptable pace.

Utilities would do well to learn lessons from other once-successful companies such as AT&T, Kodak, and Xerox to ensure that they do not make the same mistakes. Utilities should act now to anticipate actions their customers may take in the future and work towards embracing it into their business model. This may be their only hope to retain their customer connection and their business. The regulator is an important part of this equation, since they determine that kinds of activities and services that the utility can offer to their customers.

New Business Constructs and Models

Retail energy providers (REPs) have been in place since the days of ERCOT[6] and its grand introduction of competition at both the wholesale and retail level in Texas[7]. Since then, retail-level competition (retail choice) has been established in in other states such as Pennsylvania and in several countries.

The advent of new technologies such as DERs, storage, and microgrids has given rise to the emergence of a new business construct under the New York REV: the aggregator. The aggregator is more a generic name for someone who can combine generation assets and load responsibility of a broad range of residential, commercial, and other customers and use this clout to interact at the retail and wholesale level, releasing more value from the same assets.

Aggregators do not have the burden of legacy systems owned by incumbent electric utilities. They can jump into this new world by taking advantage of newer technologies. By combining technologies such as IoT, distribution automation (DA), wireless communications, cloud computing, and smartphone

5. A not-so-new term, but one gaining in popularity. It means completely disconnecting oneself from the electric grid.
6. The Electric Reliability Council of Texas (ERCOT) is also the operator of its wholesale and retail markets.
7. Even before ERCOT, California had also established retail markets but it did not last.

apps, they can deliver an enhanced set of services to their customers. Taking lessons from their counterparts in Texas, these aggregators are able to bundle in other services, such as home security and internet access, combined into one bill with a common customer service location. All of this leads to improved customer satisfaction.

The aggregator is still a new construct and the business models in the marketplace is equally nascent. The introduction of the microgrid and the more granular nanogrid as a technical construct and the controls associated with them have created new tools for the aggregator, which can only get more sophisticated over time.

Making the case for electric utility transformation: how or even when these changes to the business models will appear or become prevalent are unknown. However, business model changes are coming, and they will disrupt the utility of the future from multiple directions. These disruptions may include customers looking at other options, businesses coming into this arena to provide services previously provided by the utility, and severe changes in the regulatory mandate that either changes the utility's role or splits them up into regulated and unregulated components of the same whole.

Retail Markets

A full-scale retail market did not evolve into other states besides Texas, until now. New York has started the discussion of a retail electricity market under the REV umbrella as filed in the DSIP guidance [16]. The distribution system platform provider will be responsible for three main tasks: planning, grid operations, and market operations. California has also been talking about implementing markets in the near to mid term.

While the exact form the market would take remains unclear, North America and other parts of the world are talking about the concept of the distribution system operator (DSO), which is intended to be an organization with functionality parallel to that of the RTO/ISO but focused at the distribution level.

There are many discussions at the U.S. DOE and at industry conferences about the nature of these markets, with some expecting a top-down approach, in which the RTO markets will expand in scope and expand into the distribution grid to the extent that it is still networked. Another school of thought looks at a bottom-up approach focused on the principles of transactive energy [17]. The concept of transactive energy at a retail level is gaining traction as spurred by rising number of prosumers, which required a retail market equivalent to the power exchange/market operator at wholesale level.

Making the case for electric utility transformation: if the wording in New York is any indication, it is possible the distribution utility may also see changes like that of transmission through FERC orders 888/889 and a separation of the monopoly infrastructure owning portion of the distribution utility from the rest of the utility. Utilities have gone through these changes once before during the unbundling required for the transition to wholesale markets. At that time, transmission operations had to separate from generation and trading operations. The utility industry expects that when retail markets come to bear, distribution portions of the utilities may need to undergo a similar transition.

Characteristics of the Transformed Utility of the Future

Characteristics that define the transformed utility of the future are listed below.

Have a Flexible Operating Model

The utility environment is amid major change: new problems, new systems, new competitors, new constructs (business and technical), new regulations and regulators, and new policy requirements. The list keeps growing, appearing to be never-ending. In addition, a new threat is also on the horizon. Right now, 100% of customer load is delivered by the customer. The introduction of DERs and storage is bringing in new possibilities that will change this paradigm; their share of the percentage of customer load can only go down.

To meet these new threats and changes, the utility operating model will need to shift/evolve if they are to make up for the decline of the traditional market share. Each utility must decide its long-term business cope and possibly extend to newer areas, possibly through partnerships, acquisitions, or other alliances to preserve shareholder value.

The utility's operating model needs to become flexible, allowing it to move quickly as conditions change. An operating model requires the right mix of flexibility in systems, organizational culture, and processes, all functioning seamlessly under a forward-looking strategy and a new way of working.

Wires, Pipes, and Service-Centric, Not Energy-Centric

Utilities recognize the complexity of delivering power from disparate sources with differing generating characteristics to a broad set of customers of different classifications: residential, commercial, and industrial. The smart grid has not changed the foundational physics responsible for delivering this power. Physical equipment such as substations, circuit breakers, and transformers still exist and, for the most part, are still necessary for the delivery system to work. The commodity being transferred over the delivery infrastructure may become almost irrelevant and, in fact, could be one or more of a combination of electricity, natural gas, or even hydrogen.

The transformed utility of the future needs to first focus on the wires and pipes and then focus on delivering the most reliable and resilient service to their connected customers. These wires and pipes need to be bidirectional, open-access, and transactive.

Focus on the Customer and Their Desires

The customer of the future is changing. Many of them are installing PV on their roofs, buying electric cars, smart homes with battery packs, and exercising green power choices using a variety of mechanisms including demand response. Some of them are generating their own power and selling it back to the grid, while others are joining up in a microgrid or developing a relationship with an aggregator. Yet others may interact with an energy market and sell energy or ancillary services.

The utility of today may not be able to predict the needs of the customer of tomorrow, but one thing is certain. It will look very different from today's customer. Customer needs and expectations are changing, as influenced by other industries and utilities need to fulfill to those needs, or someone else will.

Manage DERs: They Are Coming, Like It or Not

DER-based technologies are getting more mature, lower in cost, and better in performance. They are also becoming more prevalent in the marketplace and when combined with each other (e.g., solar PV and storage) are also dispatchable. In addition, supporting their installations are a host of federal and state mandates and incentives such as rebates for electric car purchases and residential solar PV installations, which have resulted in a tremendous growth in the generation of energy from DERs.

Much of the compensation for DERs today is based on tariffs set by the PUCs. However, if the New York REV effort has any say in this, it is foreseeable that the transformed utility of tomorrow needs to be able to use market mechanisms to capture DER benefits. Utilities can use approaches such as transactive pricing, locational pricing, and even some demand response.

Redefine Planning and Asset Management

The advent of DERs, storage, and microgrids, along with other new and innovative technologies on the horizon, has changed the utilities' approach to power system planning. Planning is no longer solely validating different options to deliver power from centralized generating stations to the load through the transmission and distribution infrastructure. Most options result in either the addition of new infrastructure components or increasing the capacity of existing components. While this process is complicated due to the need to assess different scenarios to deliver to the increasing load, the advent of DERs increases the complexity of this analysis exponentially.

Newer terms like probabilistic hosting capacity[8] and nonwire alternatives[9] are being identified and will begin to impact the planning processes at utilities as they use their original set of scenarios and then add DER penetration rates and other options such as storage. The transformed utility of the future will need to perform this analysis on the uncertain future of trying to perform their power system planning tasks while taking into consideration customer and other-owned DERs.

Data and Digital Insights Driven

Customer load patterns are changing. Distribution grid flow patterns are changing. Smart homes and buildings are changing how the customer consumes energy, control the consumption in real time, and interact with utility demand response programs. Transmission and distribution automation have spawned a plethora of sensors and controls using legacy and new approaches such as the IoT, all leading to a dramatic increase in the amount of data collected by the utility and stored in diverse databases within different systems.

The transformed utility of the future should take full advantage of all the data that it gathers to make decisions that impact and enhance its day-to-day operations and planning processes. The data gathered will also be able to support the transformed utility of the future as it exploits the insights to continuously change the operational model in response to changes in the marketplace.

Embrace Change and Innovate to Turn Threats into Opportunities

The transformed utility of the future needs to be in a constant state of innovation at every level: technology, business processes, organizational structure, cultural, and even strategic. As indicated earlier in this section, a host of new technologies, with new competitors along every step of the value chain, and a rapidly changing regulatory/policy landscape require innovation. Trying to stay the course will result in someone else defining the future, leading to unpredictable outcomes.

A transformed utility needs to exploit new value opportunities. Once you have the right insights, it is time to join forces with key stakeholders, including

8. The increased connection of distributed energy resources (DERs) has changed distribution networks from being primarily a passive (consuming energy) into active (consuming/producing energy). The hosting capacity is a probabilistic amount that determines the maximum amount of DERs that can be connected to a feeder.

9. Nonwire alternative solutions are the use of nontraditional investments such as transmission and/or distribution substation infrastructure enhancements and by exploiting localized technologies such as DERs, storage, and microgrids.

prosumers, cities, and aggregators. Actively building an innovative ecosystem with participants from across the energy value chain to optimize investments by exploring and capturing the value opportunities the energy transition provides [18].

A Path Forward to the Utility of the Future

Utilities worldwide are largely regulated organizations with solid iron-on-the-ground assets and a large legacy workforce required to deliver reliable and resilient electric and sometimes gas and water services to its customers over large geographically diverse territories. They cannot change or transform overnight. They need to define a future state destination that reflects the new priorities of customers, regulators, and shareholders and that provides a place for motivated employees. They need a roadmap to the vision which stages the transformation journey over time. Some utilities are already on this journey. The key takeaway here is not about which step on the roadmap the utilities are on now, but what step(s) to take next.

The end game needs to be the use of smart grid and associated technologies to fundamentally reposition the utility to become a full partner across industry verticals, in partnership with public entities, as opposed to just having customers. Utilities need to move from self-interest towards customer-centricity delivered through high performance. If this happens, the transformed utility will be able to demonstrate customer and societal and environmental impact and, in turn, change their image and their relationship with customers, regulators, and government.

In this section, the author presents a roadmap for change with two separate and equally important realizations, these are not in sequence and many utilities worldwide are already in various stages on this roadmap.

Reinvent

This is the first step. A large percentage of utilities are currently in this step and involves a reinvention of the business model built around a shared vision with the notion that stakeholders will support it.

Utilities are focused on protecting their existing revenue streams by placing restrictions on other organizations from delivering energy related services to their customers. Examples of the utility industry's action towards DERs and microgrids are reminiscent of similar actions taken by Yellow Cab against ride-sharing companies such as Uber, Lyft, and others.

Utilities are also looking to upgrade/enhance existing resources to enable a higher revenue stream and to better fit into their new role as the backup energy supplier.

Focus and Prioritize

The next step along this roadmap is to focus on activities that produce quality financial outcomes. Some of the steps that fall into this stage include:

- Consolidation, which is necessary and will occur with those companies having the financial strength acquiring those who do not. Since much of this consolidation will occur outside the regulated footprint, the ability of states or other jurisdictions to control who and how this occurs will be significantly reduced. These acquisitions will deliver the benefits of consolidating back-office departments such as human resources, corporate governance, and purchase strength. Examples of consolidation occurring in North America include Exelon's purchase of Pepco Holdings, Hydro One's acquisition of Avista, and Sempra Energy's anticipated acquisition of Oncor.
- Consideration of new business areas, such as utility funded roof top solar, and storage. Initially, these activities will be pilots, but will be expanded if they prove fruitful and are allowed by the regulator.
- Investment in innovative companies, such as technology start-ups or other service companies that may appear attractive.
- Wire expansion. New investment will focus on the wire space while maintaining the generation assets that fit into the future. The rest of the generation assets, such as the older fossil, will be mothballed and written down. Some of those writedowns will be allowed by the regulators and many will be on the back of the shareholders.

This is more of a response to financial pressures to the existing business model as new entrants begin to make a difference.

Realign and Pivot

The next step along the roadmap is the realization that change is coming and utilities need to get their core systems standardized and future-facing. They must prepare for more consolidations/acquisitions or get ready for the next set of challenges or competition. This requires standardizing people, processes, and technologies across the company. Examples of these rationalizations include implementing a consolidated company-wide common CIS, GIS, ADMS, or EMS and other such core systems supported by a standardization of the organization and processes.

Rationalization sets the stage for further consolidations, which will be a necessary next step in the transformed utility roadmap. Some of the more advanced utilities are already here and looking to the next set of steps.

Diversify and Expand

The definition of a utility has and will continue to change, but the rate of change will be much faster in the future. Utilities will look more different than the same by 2030, with each exploiting or choosing the future they can realize, with more of them looking very different from each other as well as from how they look today.

- Service will be a commonality, but that is where the commonality ends. The type of service will vary with geography, economic status of the service territory, competitive landscape (specifically what do the aggregators bring and how competitive are they with the utility), and what each utility believes to be their strengths. This could include everything from backup generation for commercial buildings to grid services to retail solutions such as facilities management.
- Diversification will also foster less focus on regulated activities and a greater focus on nonregulated activities with the utilities trying to attain a better return using profit from the regulated activities.
- Become the energy services and solutions marketplace for their customers, which, in turn, moves the customer relationship from an adversarial one to that of a collaborator. This needs to be supported by a demonstrated customer, societal, and environmental impact, thereby changing the image and relationship with customers, regulators, and government.

With wholesale deregulation, the industry has already seen the disaggregation of vertical utilities. This final stage of transformation will split the existing vertical activities into more granular businesses, both to remove them from the regulated activities and to broaden the reach of those services across utilities and geographies.

Case Studies

Case Study #1: ARRA and Its Impact on the Smart Grid

The American Recovery and Reinvestment Act of 2009 (ARRA) included several measures that were necessary to modernize the U.S. energy and communications infrastructure and enhance its energy independence[10]. ARRA provided the DOE with $4.5 billion to modernize the electric power grid. There were two main programs under this act that impacted the electricity industry:

- *Smart Grid Investment Grant (SGIG):* In this program, the DOE and the electricity industry jointly invested $8 billion in 99 cost-shared projects involving more than 200 participating electric utilities and other organizations to modernize the electric grid, strengthen cybersecurity, improve interoperability, and collect an unprecedented level of data on smart grid operations and benefits [19]. Figure 10.2 shows the extent of the investments that were made in the SGIGs on specific technologies. A total of almost $8 billion was spent.
- *Smart Grid Demonstration Projects (SGDP):* This program was intended to demonstrate a suite of existing and emerging smart grid technologies and concepts to prove technical, operational, and business-model feasibility. Two types of smart grid projects were selected for the SGDP. The first group included regional smart grid demonstrations, which were intended to test the viability of these technologies as a set and validate the new business models at scales that could be replicated across the country. The other group focused on energy storage technologies such as batteries and flywheels, for their ability to deliver services such as load shifting, frequency regulation services, and integration of renewable re-

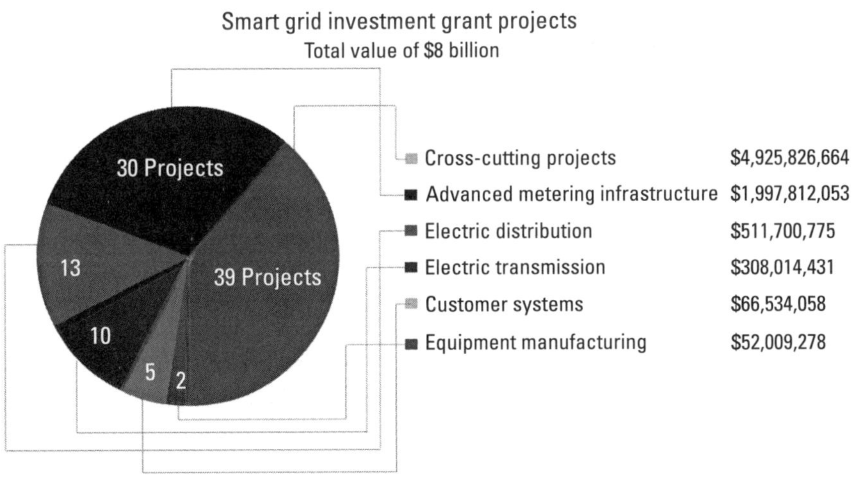

Figure 10.2 SGIG investments on specific technologies. (Source: Energy.gov [20].)

10. It is important to note that this was a U.S. law supported with budgets and projects. At this same time, almost every country in the world had similar programs focusing on their own electricity infrastructure, some of the programs continuing till now and beyond.

sources. The program had 32 projects split between the two areas for a total budget of about $1.6 billion with the federal share at about $600 million [13].

Benefits Delivered

These investments were intended to serve a dual mission. The first was to deliver an economic stimulus and the second was to focus on the modernization of the U.S. electricity grid. Both missions generated economic benefits as well as operational, customer, and reliability benefits from the deployment of smart grid technology [13]. An analysis from the DOE, "Economic Impact of Recovery Act Investment in the Smart Grid," reported that smart grid projects funded through these two programs created nearly $7 billion total economic output, approximately 50,000 jobs, with over $1 billion in government tax revenue. This represented a nearly 2-to-1 return on investment for the U.S. government's investment. In addition, the DOE estimates ARRA smart grid investments boosted overall gross domestic product (GDP) by $4.18 billion. The smart grid GDP multiplier is higher than many other forms of government investment. For every $1 million of direct spending, the GDP increased by $2.5 to $2.6 million [21]. These investments also helped utilities acquire and deploy the technologies that enable a more intelligent electricity delivery system, such as 15 million smart meters, 20,000 substation monitors, 1,000 new synchrophasors, and over 492 electric vehicle charging stations.

Figure 10.3 presents a specific example of the benefits delivered at the pricing pilot implemented at Oklahoma Gas and Electric (OGE), which delivered a savings of 1.3 KW of capacity reduction per customer and about $150,000 customers who enrolled in its DR pilot.

In addition to the equipment installed and specific improvements such as the number of truck rolls avoided, these two programs had an immeasurable impact on the future operations of the electric power industry by providing the largest ever one-time investment in upgrading the U.S. electric infrastructure. It also helped utilities take the first steps towards this bold new world, share what they learned with others so the industry could be better prepared to meet the needs of a growing digital economy, enabled greater levels of clean energy deployment, and strengthened the electric grid to be more resilient to natural disasters and cyberattacks.

In addressing these and other technology, policy, and market challenges, the DOE continues to be an important contributor to grid modernization through research, development, demonstration, analysis, and technology transfer activities. SGIG showed what can be achieved in grid modernization through public-private partnerships involving DOE and the electric power industry.

Pricing Pilot at Oklahoma Gas & Electric

OGE deployed TOU-CP and VPP-CP programs in Summer 2011, VPP-CP is highlighted here.

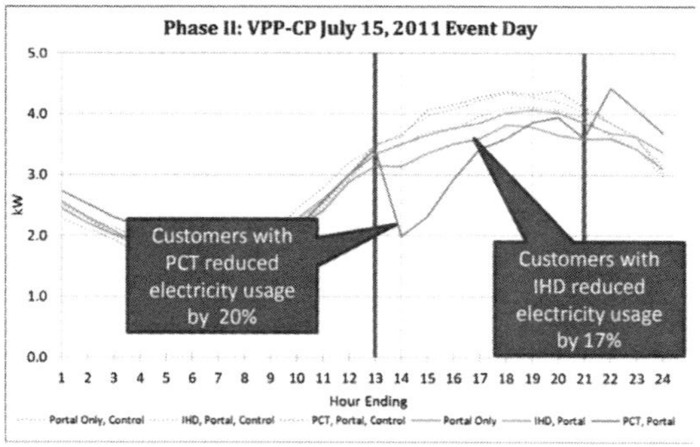

Price Level	Residential VPP-CP Price	Number of days in summer 2011 at each price level
Low and off-peak	4.5¢ per kWh	63
Standard	11.3¢ per kWh	25
High	23.0¢ per kWh	28
Critical	46.0¢ per kWh	6
Critical Event	46.0¢ per kWh	7 (included in the above)

Potentially Avoid Future Generation:

- Study results show a 1.3 kW reduction per customer is possible (max reduction: 1.97 KW)
- Hoping for 20% participation by Dec 2014
- Targets: Enroll ~ 40K customers in 2012 with 72MW peak reduction; 150K customers by Dec 2014 with 210 MW peak reduction (offsets a natural-gas fired peaking plant)
- Discontinue roll out of IHD in 2012

Figure 10.3 Example project: operational benefits demonstrated. (Source: Energy.gov [22].)

New technologies drive changes on multiple fronts and the need continues for strong national efforts to modernize the grid [23].

Lessons Learned

In addition to the hits, there were several lessons learned as well. Some of those lessons are listed here [23, 24]:

- Utilities have untapped opportunities to maximize the capabilities of new smart grid technologies. For example, several utilities installed smart meters, but have not yet used the embedded capabilities to monitor customer voltage levels or have combined voltage monitoring with automated controls for voltage and reactive power management. Planned follow-on activities include expanding deployments to larger portions of service territories, expanding communications networks, and integrating various information management systems to realize untapped automation capabilities.
- Vast amounts of new data require the development of new data exchange and management capabilities, including new models and analysis tools to unlock the full value of smart grid technologies. Smart meters, PMUs, and other devices provide timely and granular data at large volumes that require investments in high bandwidth communications networks and advanced data analytics to better automate controls and inform operator decisions. These advanced capabilities are necessary to integrate large amounts of distributed and renewable generation, reduce susceptibility of the system to destabilizing events, and bring together utility functions for generation, transmission, distribution, and demand-side programs.
- Cybersecurity systems, both processes and personnel, continue to be a critical component of utility operations. Smart grid technologies provide many benefits, but are also open opportunities for adversaries to attack critical infrastructures. Generation and utility operators need to continue efforts to identify and deploy protections against ever-evolving cybersecurity threats.
- The acceptance and ultimately the degree to which customers will engage with smart grid technologies needs continued attention. Reliance on manual actions by customers as opposed to "set it and forget it" constructs will limit consumer participation. Also, the segment of the customer population participating with smart grid technologies should be increased, as many of them are technologically savvy.
- The SGDP projects were expected to quantify smart grid costs and benefits; however, this was not performed in a consistent manner. The

lack of complete cost/benefit data prevented any meaningful financial analysis from being performed [24]. Adequate financial analyses of such implementations are necessary to determine if they represent positive business cases and/or are competitive with other least-cost options.

- Too much emphasis was put on AMI implementations. The emphasis on the shovel-ready aspect of the ARRA law resulted in more than $2 billion (about 25%) being spent on these implementations without adequate thought given to realize their full potential or even taking advantage of the increased amounts of new data available at the utility because of these installations.
- Substantial opportunity may exist in leveraging smart grid technologies by participating as a market participant with RTO and ISOs. Most of the ARRA implementations did not investigate the potential for new sources of benefits from these implementations, which could have been released by interacting with the RTOs and ISOs in their jurisdictions.

For the most part, the projects focused on technology implementations without enough emphasis put on customer and operational value creation and the need for business models to change to deliver business transformation. As this chapter and others before it have stated, utilities need to explore new business models that focus on grid modernization and the integration of distributed energy resources. While the technological implementations took the industry deep into understanding the benefits of their implementations, their cost reductions have unleashed a plethora of increases in DER installations such as PV, EVs, and microgrids.

These are transformational changes that require new approaches to resource planning, economic and environmental regulations, and market development to sustain reliability and boost resilience while involving consumers and third parties in electricity management and generation to a much greater extent. The states of New York and California are taking the lead in bringing policymakers, regulators, consumer advocates, utilities, and other service providers together to continue working to ensure grid capabilities and keep pace with changing requirements for DER integration, reliability, security, and regulations [24].

Case Study #2: Lessons Learned from Other Industries

Transformation is not unique to the electric power industry. In fact, it has been happening in many industries with both winners and losers. Some of these were in the regulated industry space, while others were in the unregulated space. The utility industry can learn from those who succeeded in their transformations and those who did not[11].

AT&T and Ma Bell

The history of AT&T dates to the invention of the telephone itself. The Bell Telephone Company was established in 1879 by Alexander Graham Bell, the inventor of the telephone. Bell also established American Telephone and Telegraph Company in 1885, which acquired the Bell Telephone Company and became the primary phone company in the United States. Because of a combination of regulatory actions by government and actions by AT&T, the firm eventually gained what most regard as a monopoly status.

This company maintained a monopoly on telephone service in the United States until antitrust regulators split the company in January 8, 1982, the date of settlement of *United States v. AT&T*, a 1974 United States Department of Justice antitrust suit against AT&T [25]. Under the settlement, AT&T's local operations were split into seven independent Regional Bell Operating Companies known as Baby Bells. The AT&T Corporation was eventually purchased by one of its Baby Bells, the former Southwestern Bell, in 2005 and the combined company became known as AT&T, Inc.

While the breakup of the telephone monopoly was a significant event, there were three other events happening in parallel, which were even more significant. The internet, cellular service, and cable all underwent major changes and drove the communications world in new directions [26].

1. *Open access:* The breakup helped other long-distance companies, such as Sprint and MCI—that did not own local lines—compete with the AT&T network on an equal footing. At the same time, other networks from outside the traditional telecom world were just beginning to start up, bringing with them the ability to challenge the local/long-distance strength of AT&T split but also transform communications in more fundamental ways.
2. New technologies:
    - *Internet:* About a year before the breakup, Arpanet, the precursor to the internet, adopted TCP/IP as its operating protocol, which, in turn, became the foundation for general-purpose data networks, taking over most every network: voice, local, long-distance, and others.

11. The analysis presented in Case Study #2 is entirely my own and backed up by some facts and anecdotal information. This case study is not intended to be authoritative in respects to what happened inside those companies. Rather, the cases are presented here only to present the lessons learned as they can be applied to the utility industry.

- *Cellular:* In October 1983, just months before the AT&T split, Ameritech launched its cellular wireless network. With nationwide sign-ups under 100,000, very few people could understand the impact of mobile phones on the economy. Here again, a revolutionary stage was being set by technologies and networks that were originally not as ubiquitous, reliable, or even as cheap as traditional telecom, and certainly not directly competitive.

3. *An outside industry, cable TV:* Cable TV, at that time, was a young, flourishing industry and still considered a complement and competitor to broadcast TV, but not competitor with phones, mobile, or the internet. Cable has built a broadband platform that has adopted the digital internet and is now challenging the telecom industry in nearly every way.

The key takeaway for the utility of tomorrow is that AT&T's monopoly was a creation of government regulation that forbade and discouraged competition and innovation in numerous ways. Its primary competition did not come from other telephone companies, managed competition designed by the government, or the creation of similar firms that competed with AT&T's core businesses. The main competition and the end to its monopoly communications business came from innovative new-wave companies, which brought new technologies, new platforms, and new business models from both outside and inside the telecom world.

Utilities cannot expect to stay the course as they have over the last 100 years. They need to assess the competition that will try to enter this marketplace with competing technologies and new/innovative business models and chip away at the utilities' customer base. Utilities must transform to stay relevant.

Cable Industry

Consumer behavior and the way people consume content is changing. Viewers no longer want a passive experience with a TV program in one location in one room. They want to watch catch-up TV on a laptop, download a phone app on the move, engage with content on a tablet, or talk about their experiences on social media sites. They want different experiences on each platform, different flavors of content, and they want it all on their personal schedules [27]. This "access anywhere, any time, and on any platform" consumer impacts the entire value chain of broadcasters, content owners, and cable networks that provide access at the home.

Included next are some key facts:

- Cable subscriptions are going down, replaced by people with internet connections getting their information through a computer.

- Sales of TV sets are going down [28]. The millennials watch much of their TV on the PC monitor, laptop, or smartphone.
- Cable broadband superiority is under threat from telecom operators' widespread deployment of fiber-to-cabinet in their access networks. At the same time, the internet is emerging as a viable platform for broadcast quality content delivery and content providers are using the internet as a complement to their traditional distribution partners.
- Intense competition between the players has triggered an increasing degree of consolidation, as horizontal and vertical integration can yield significant economies of scope and scale. Companies that integrate across the value chain can combine different infrastructure and technologies to serve consumers better and offer better options.
- Technology innovations are rapidly bridging the last mile between the digital enterprise and the physical world. Broadcast, fixed, mobile, and Wi-Fi networks are converging and becoming tightly integrated to provide a seamless, anytime/anywhere user experience for voice, data, and video services [29].

All this combined demonstrates that digital delivery, which represents the future, is coming far faster than previously expected. Streamed pay-television services are already going mainstream with services from companies such as DirecTV, Hulu, Amazon, and YouTube. This, combined with individual cable networks losing subscribers at a much faster rate, is causing a crisis in the TV industry, which will finally drive it to embrace new business models and broader distribution or fail [30].

The key takeaway for the utility of tomorrow is that cable TV providers, which until now held the franchise monopoly with the cable connection to the home, are moving to content and other related connectivity-based services, such as telephone, home security, and mobile smartphone. Cable providers are doing this because competition from other service providers outside their normal operating realm are now providing services to their customer that are indistinguishable, at least in terms of access to content.

Utilities need to assess their competition that will try to enter this marketplace with competing technologies and new/innovative business models and get access to the home or business directly bypassing the utility-connected wire into the home.

Electric Cars

The electric transportation industry is also in the throes of their biggest shakeup since the days of Henry Ford. For the past 100 years, especially since the move away from the steam engine, this industry has been based on the gasoline-powered engine [31–33].

Included next are some key facts:

- *Electric cars:* The global automobile industry has been impacted by electric cars, starting with the emergence of hybrid cars that came to the forefront of the burgeoning industry in the last 10 years to the latest model of the all-electric cars now emerging. The sales of these cars are increasing at a rapid pace, resulting in a simultaneous reduction of the number of gasoline-power cars. The entry and acceleration of EVs in the marketplace will also impact the gasoline distribution industry through reductions in the usage of gasoline and the introduction of the EV-charging network, which will require more installations and servicing.
- *Self-driving cars:* Another new technology platform appearing on the horizon is self-driving technology. Companies such as Google, Tesla, and Nissan are working on this and related technologies necessary for cars and other means of transport to become self-driving. The entry of, and eventual acceptance of, the self-driving car will disrupt a dramatic number of people in various businesses including trucking, limo/taxi, and package delivery.

The key takeaway for the utility of tomorrow is that the scenario that will unfold here is the same as that faced by the horse and carriage industry when gasoline-powered automobiles arrived on the scene. The electric car industry will cause disruption, not just for the automobile industry, but several other industries as well, as it moves through its period of innovation.

Let us look at some of the reactions to this disruption, from some of the industries most impacted by this change.

- Volvo has announced that all cars they manufactured will be electric by 2019.
- Consumers reserved more than 500,000 cars for Tesla's model 3, an amazing number for a car that had not yet been designed. With its purchase of SolarCity, Tesla is now moving outside its mainstream automobile business into solar roofs, home storage, and other devices sold directly to the customer.
- China, the world's largest auto market, is working on a plan to ban the production and sale of vehicles powered solely by fossil fuels. Regulators have not yet decided when the Chinese ban would take effect, but work has begun on a timetable.

- India, the world's fourth-largest market for automobiles, announced that all cars sold in the country after 2030 would be based on electric. Other countries that are announcing similarly aggressive goals are France/Britain (2040) and Norway (2025); China has announced a shift in 2019, and the state of California is also announcing a ban on cars with internal combustion engines.
- Most of the major automakers have announced an all-of-the-above strategy focusing on both electric cars and gasoline-powered cars at least into the near-to-mid-term future.

While this book is not about the analysis of the impacts on other industries, a transformed electric utility should not be looking at electric cars as more than just another load. New and innovative technologies such as V2G could be used to enable EVs as a mobile source of energy and/or ancillary services instead of being in a single location.

Conclusions

This chapter makes a case for the utility industry to transform. Transformation is not new, nor is it unique to the electric utility industry. It is not even an option. Examples from other industries have been provided showing both successes and failures. One only needs to look at historical companies such as Kodak, Xerox, and AT&T/Ma Bell, all of which did not transform in time, which ultimately led to either their demise or loss of market share. Utilities were immune to these changes for a long time, but if the events of the past 20 years and the forecasted changes over the next couple of decades are any indication, it is all going to change.

Utilities need to learn from companies that have survived and thrived despite intense competition, in order to compete with the new and often much nimbler companies just appearing on the horizon. Utilities also need to learn from the mistakes of companies that did not survive to ensure they do not make the same mistakes.

Lastly, Chapter 11 will present, a key step in this transformational journey is for a utility to move beyond the scope of what it does today and become a partner with other entities and take an important position on the roadmap to a smart city. This will only happen if the utility positions itself within the future landscape in a confident and mature way. This transformed utility will then drive value and be operationally sophisticated, ensuring that cities open their doors to make utilities full-time partners in the complex and integrated landscape of the future.

References

[1] United States of America76 FERC 61,009, Federal Energy Regulatory Commission, Order Clarifying Order Nos. 888 and 889 Compliance Matters, July 2, 1996, https://www.ferc.gov/legal/maj-ord-reg/land-docs/rm95-8-0aj.txt.

[2] Luke, M., "Why We Need Bigger Electricity Markets," *World Economic Forum,* February 27, 2015, https://www.weforum.org/agenda/2015/02/why-we-need-bigger-electricity-markets/.

[3] Bock, E. P., "The Electric Utility Restructuring Debate: A Primer," *Fifth Annual North American Waste to Energy Conference, Panel on Electric Utility Restructuring,* Research Triangle Park, NC, April 1997, http://www.seas.columbia.edu/earth/wtert/sofos/nawtec/nawtec05/nawtec05-05.pdf.

[4] "Open Access Same-Time Information System," November 9, 2017, https://en.wikipedia.org/wiki/Open_Access_Same-Time_Information_System.

[5] Federal Energy Regulatory Commission, "History of OATT Reform," June 28, 2010, https://www.ferc.gov/industries/electric/indus-act/oatt-reform/history.asp.

[6] Litos Strategic Communication, "The Smart Grid: An Introduction," https://energy.gov/sites/prod/files/oeprod/DocumentsandMedia/DOE_SG_Book_Single_Pages%281%29.pdf.

[7] SmartGrid.gov, "What Is the Smart Grid," https://www.smartgrid.gov/the_smart_grid/smart_grid.html.

[8] Office of Electricity Delivery & Energy Reliability, "Grid Modernization and the Smart Grid," https://energy.gov/oe/activities/technology-development/grid-modernization-and-smart-grid.

[9] Carvallo, A., and J. Cooper, *The Advanced Smart Grid: Edge Power Driving Sustainability,* 2nd ed., Norwood, MA: Artech House, 2017.

[10] Fox-Penner, P., *Smart Power: Climate Change, the Smart Grid, and the Future of Electric Utilities,* 2nd ed., Island Press, 2014.

[11] Wikipedia, "American Recovery and Reinvestment Act of 2009," November 10, 2017, https://en.wikipedia.org/wiki/American_Recovery_and_Reinvestment_Act_of_2009.

[12] Office of Electricity Delivery & Energy Reliability, "Recovery Act: Smart Grid Investment Grant (SGIG) Program," https://energy.gov/oe/information-center/recovery-act-smart-grid-investment-grant-sgig-program.

[13] SmartGrid.gov, "Smart Grid Demonstration Program," https://www.smartgrid.gov/recovery_act/overview/smart_grid_demonstration_program.html.

[14] Burr, M., "Baby Boom Blues," *Public Utilities Fortnightly,* Vol. 144, July 2006, pp. 28–30.

[15] Pocock, C., "Is Off-Grid Living in Washington State Possible?" *Financier Global,* July 2017, https://financierglobal.com/off-grid-living-in-washington-state-possible/.

[16] State of New York, Department of Public Service, "Staff Proposal, Distributed System Implementation Plan Guidance, Case 14-M-0101: Proceeding on Motion of the Commission in Regard to Reforming the Energy Vision," October 2015, http://www3.

dps.ny.gov/W/PSCWeb.nsf/All/C12C0A18F55877E785257E6F005D533E?OpenDocument.

[17] GridWise Architecture Council, "Transactive Energy: A Primer," October 2017, http://www.gridwiseac.org/about/transactive_energy.aspx.

[18] van Ginkel, S. J. M. Mazurek, and S. Vardy, "Surviving Shocks to the System: Becoming a Grid Optimizer in the New Age of Energy," Accenture report, 2017.

[19] U.S. Department of Energy, "Economic Impact of Recovery Act Investments," April 2013, https://www.smartgrid.gov/files/Smart_Grid_Economic_Impact_Report.pdf.

[20] U.S. Department of Energy Report, "Jumpstarting a Modern Grid: The American Recovery and Reinvestment Act, Smart Grid Highlights," October 2014, https://energy.gov/sites/prod/files/2014/12/f19/SGIG-SGDP-Highlights-October2014.pdf.

[21] U.S. Department of Energy, "Economic Impact of Recovery Act Investments in the Smart Grid," April 2013, https://www.smartgrid.gov/files/Smart_Grid_Economic_Impact_Report.pdf.

[22] Paladino, J., "The Impact of Smart Grid Projects Funded by the Recovery Act of 2009," *U.S. Department of Energy, Electricity Advisory Committee Meeting*, June 2012, https://energy.gov/sites/prod/files/Presentation%20to%20the%20EAC%20-%20Impact%20of%20Smart%20Grid%20Projects%20Funded%20by%20ARRA%20-%20Joe%20Paladino.pdf.

[23] U.S. Department of Energy, "SG Investment Grant – Final Report," December 2016, https://www.smartgrid.gov/files/Final_SGIG_Report_20161220.pdf.

[24] U.S. Department of Energy, "2015 Progress Report for OE ARRA Smart Grid Demonstration Program Aggregation of RDSI, SGDP, and SGIG Results," May 2015, https://energy.gov/sites/prod/files/2016/12/f34/Activity%206%20Report_Public_Version_051415%20FINAL.pdf.

[25] Wikipedia, "History of AT&T," November 10, 2017, https://en.wikipedia.org/wiki/History_of_AT%26T.

[26] Swanson, B., "Lessons from the AT&T Breakup, 30 Years Later," *AEIdeas*, January 2014, http://www.aei.org/publication/lessons-att-break-30-years-later/.

[27] "What Are the Most Important Opportunities and Challenges Facing the TV Industry Today?" *Kit::plus*, January 2017, https://www.kitplus.com/articles/What_are_the_most_important_opportunities_and_challenges_facing_the_TV_industry_today/391.html.

[28] Molla, R., and P. Kafka, "TV Sets Are Starting to Disappear from American Homes," *Recode*, July 2017, https://www.recode.net/2017/7/30/16035706/tv-sets-american-home-decline.

[29] Tieri, S., et al., "Sustaining Success in the Digital Era: The New Challenges for Cable," *Accenture*, 2015, https://www.accenture.com/t20151210T113807Z__w__/us-en/_acnmedia/Accenture/Conversion-Assets/DotCom/Documents/Global/PDF/Dualpub_10/Accenture-New-Challenges-Cable-Operators.pdfla=en#zoom=50.

[30] Dawson, J., "The TV Industry Is Facing a Crisis and Needs to Immediately Embrace New Business Models," *Recode*, January 2017, https://www.recode.net/2017/1/5/14180704/ces-tv-television-content-rights-streaming-video-mohu-sling.

[31] Dow, J., "Mazda Exec Presents Bold 'Let's Pretend the EV Future Will Never Come' Strategy," *Electrek*, August 1, 2017, https://electrek.co/2017/08/01/mazda-exec-presents-bold-lets-pretend-the-ev-future-will-never-come-strategy/.

[32] Pham, S., "Volvo: Gas-Only Cars Are History After 2019," *CNN*, July 5, 2017, http://money.cnn.com/2017/07/05/autos/volvo-electric-cars-internal-combustion-engine/index.html.

[33] Agerholm, H., "India to Make Every Single Car Electric by 2030 in Bid to Tackle Pollution That Kills Millions," *Independent*, May 1, 2017, http://www.independent.co.uk/news/world/asia/india-electric-cars-2030-fossil-fuel-air-pollution-piyush-goyal-climate-change-a7711381.html.

11

Transformed Utility: Springboard to a Smart City

A smart city is a confluence of existing, new, and emerging technologies, which span a city's infrastructure and services such as energy, water, transportation, administration, education, and safety, making it more intelligent, interconnected, and efficient.

Introduction

Several chapters have focused on the changes occurring in electric utilities, starting with individual pieces and how they contribute to that transformation. Chapter 10 defined the transformed utility as a provider of clean energy. The transformed utility also has a flexible operating model, driven by data and digital insights. The new utility also has a redefined planning model, with the expectation of increased DER penetration, which is both customer service and wires/pipes centric. They are also on a concurrent path with their cities with the grand objective of achieving greenhouse gas reductions in line with the needs of their citizens.

If the electric utility has transformed, what comes next? One answer is the transformed utility is now ready to be a springboard for the future smart city.

Just as is the case with the utility's transformation, a smart city is not just an objective, it is also a journey. Cities worldwide face new challenges. With over 5 million people moving out of rural areas every month, over 50% of the world's population now lives in cities. As a result, cities account for over 70% of the global energy consumption, resulting in growing pollution and increasing congestion from the increase in carbon emissions. These same cities are also rehabilitating aging infrastructure and decreasing budgets leading to a need to do more with less [1–6].

In parallel with these challenges, the cities' needs are increasing. Citizens in this new world expect everything from better parking to efficient lighting, improved traffic flow, and smarter security to improved waste management and disaster planning. Cities already provide services, but people expect improvements and additions to those services, and they want them now.

The present trajectory of these global cities is not sustainable on several levels. One solution is for the cities to get smart at several levels: smart at using budgets, at the use of common infrastructure components, and at the planning level. Planning can become smarter by working in an integrated manner with other neighboring cities or with county, state, or federal governments. This approach is at the core of the smart city. Let's look at a few examples:

- *India:* Urbanization in India is expected to reach 40% with over 600 million people living in urban areas contributing to 70% of GDP by 2030. The 100 Smart Cities Mission is an urban renewal and retrofitting program instituted by the government of India. The mission is to develop 100 cities into citizen-friendly and sustainable places to live. Prime Minister Narendra Modi launched the mission in June 2015. A total of $15 billion was approved to cover the cost of the development of the 100 smart cities and the rejuvenation of 500 others [7]. The primary focus of the mission is on core infrastructure services such as clean water, greater power supply, robust information technology (IT) connectivity, governance, e-governance and citizen participation, safety and security of citizens, health, education, and sustainable urban environment. In addition, the objective of building 100 smart cities is to promote the adoption of resources, assets, and infrastructure, to enhance the quality of urban life, and to promote a clean and sustainable environment.
- *Barcelona:* This has been named the smartest city in the world for 2015 by Juniper Research [8] The list ranks cities across a combination of smart capabilities, emphasizing smart grids, smart traffic, smart lighting, and other aspects of social cohesion and technological capabilities. Barcelona is dedicated to becoming smarter with technological advancements and fostering a collaborative environment that spurs economic development as businesses seek to succeed in smart city innovations [9]. The city has over 22 ongoing programs of public-private collaboration, many of which utilize small enterprises, most notably in the strategic areas of traffic reduction and waste collection. To foster an environment of innovation from the private sector, the city has made several power-

ful tools available to the public such as financial support and open data initiatives.

- *Rio de Janeiro:* Rio de Janeiro's development is not unusual in its history. Rapid urban growth often coincided with periods where a lack of public policy and governmental oversight produced overpopulation in several areas of the city. This overpopulation stressed the infrastructure, overloading the system. The challenge for the city manager was to plan for meeting the basic demands of the population in the future and simultaneously respond to emergencies in the city. In addition, Rio was selected to host a string of extremely prestigious events ranging from the Pan American Games in 2007, the Rio + 20 International Conference in 2012, the XXVIII World Youth Day in 2013, the FIFA World Cup in 2014, and the Olympic Games in 2016. Since 2009, when it was chosen to host the Olympic Games, Rio needed to plan the greatest urban transformation in its history with more than 200 structural works and initiatives executed simultaneously in the areas of infrastructure, mobility, accessibility, the environment, social integration, and connectivity. These initiatives contributed to building a new Rio de Janeiro. To meet these challenges, the local government had a strategic plan for the period with 56 goals and 58 initiatives, which included responses to the issues of urban mobility, housing, urbanization, and sanitation.

Bringing it all together, energy, telecommunications, and mobility/transportation are the three main pervasive dimensions or platforms of every smart cities' vision. Cities cannot function without energy. Energy fuels our cars, subways, and trains. It cools, heats, and lights our homes and businesses. It pumps our water and processes the food that we eat, and it powers the technologies that are the foundation of a smart city. Energy profoundly impacts livability, workability, and sustainability and includes the entire range from small-scale power plants that generate energy close to where it is used to large centralized power stations. It also includes advanced technologies that help keep the lights on during power outages.

Given the critical role energy plays in a city, smart energy and the smart grid should be a priority. The success of a smart city instituting many of its clean energy, environmental, and transport objectives relies on creating and supporting a smart energy system. A smart energy system analyzes in real time where a transformer has blown and automatically reroutes power to keep the lights on in homes and businesses. It is a system that collects and manipulates data from sensors and smart devices to give operators a complete view of the energy infrastructure. For example, a smart energy system would monitor how much power solar installations are generating at any given time or when they need to signal a demand response call to help balance the load on the electric and gas grid.

> When Indian Power Minister Piyush Goyal announced that "by 2030, not a single petrol or diesel car should be sold in the country," he created a new bond between utilities and the smart city—one in which additions of new electric load is supported by policy.

To ensure a smart energy future, cities and utilities must work together, regardless of whether the utility is part of local government or a private investor-owned utility. This chapter will delve into how the transformed utility can become an invaluable partner to cities as they strive to realize their smart city vision.

This chapter starts with the definition of a smart city focusing on the key, relevant attributes. It then moves to defining the relationship between the smart city and the transformed utility starting with the attributes that drive areas of commonality. The next section looks at how commonalities enable a transformed utility to become the springboard supporting the aspirations of a smart city. This is followed by an assessment of the impediments that could stop utilities and cities from working together. Two case studies are included, one focusing on a utility and a city working together and the second focusing on a specific implementation, smart streetlights, which required collaboration across different city departments.

Smart City Defined

What is a smart city? Even the people who build them do not agree. If you ask 10 people to define a smart city, you will get 10 different answers. This is not because the answers are wrong. This is because each person starts from a different point. The answer may also come because people have different pain points and often emphasize different goals.

In its *Readiness Guide* [1], the Smart Cities Council begins with definitions.

- *City:* Real-world smart city examples are rarely a city in the strictest term. Many are more than a single city, such as a metropolitan region, a cluster of cities, counties and groups of counties, a collection of nearby towns, or a regional coalition. Other examples are less than a full-scale city, such as districts, neighborhoods, townships, villages, campuses, and military bases. Indeed, many municipalities are taking a neighborhood-by-neighborhood approach to modernization.

The use of the word city in this chapter also follows the same definition.

- *Smart city:* There are three directional goals that drive transition to a smart city. Those goals are enhanced livability, enhanced workability, and enhanced sustainability.
  - Enhanced livability provides a better quality of life for city residents through access to a comfortable, clean, engaged, healthy, and safe lifestyle. This can include access to inexpensive and clean energy, convenient mass transit, good schools, faster emergency responses, clean water and air, low crime, and diverse entertainment and cultural options.
  - Enhanced workability accelerates economic development through access to more and better jobs and increased local GDP. In a smart city, people may have greater access to the foundations of prosperity—the fundamental infrastructure services that let individuals compete in the world economy. Those services include broadband connectivity; clean, reliable, inexpensive, reliable and resilient energy; educational opportunities; affordable housing and commercial space; and efficient transportation.
  - Enhanced sustainability improves access to the resources people need without compromising the environment for future generations. To accomplish this, the city and its citizens need to focus every part of their ecosystem on greenhouse gas reductions. Examples include homes that consume less energy, increased use of public transportation to lower fuel consumption, aggressive recycling, increased generation of energy from renewables, and planting more trees [10]. Smart cities enable the efficient use of natural, human, and economic resources, while promoting cost savings in times of austerity. Smart cities are typically careful stewards of taxpayer dollars. Their goal is not to invest huge sums of money into new infrastructure; their goal is to make infrastructure do more and last longer for less.

Life can be better in a smart city—better for people and better for businesses. Smart cities can alleviate the pollution, gridlock, and even civil unrest brought on by an exploding world population.

Let's now look at how the smart dimension comes to life.

A smart city uses information and communications technology (ICT) to enhance its livability, workability, and sustainability. In simplest terms, there are three parts to ICT: collecting, communicating, and crunching. First, a smart city collects information through sensors, other devices, and existing systems.

Next, it communicates that data using wired or wireless networks. Third, it crunches that data to understand what is happening now and what is likely to happen next.

Table 11.1 presents a comparison of how typical city operations will change into the smart city future.

Figure 11.1 provides a framework as presented by the Smart Cities Council [1, 3]. This framework aligns the smart city responsibilities and enablers. The vertical responsibilities denote essential services cities provide to their citizens. The horizontal enablers are technology capabilities that improve those responsibilities. Cities that do not coordinate their various departments at the technology planning level often end up with redundant investments in technologies, training, and personnel.

However, there is an even deeper connection between smart city responsibilities that should not be overlooked: the matter of dependencies. As many city

Table 11.1
Comparison of Traditional Cities versus Smart Cities

	Traditional City	Smart City
Planning	Ad hoc and decentralized, cost savings are not realized, and limited potential for investment scalability	Coordinated and holistic, resources are shared across city responsibilities, cost savings are fully realized, investments are scalable, and improved city planning and forecasting
Infrastructure	Runs inefficiently and costs more money and resources to run	Optimized with cutting-edge technology, saves money and resources, and improved service-level agreements
Distribution System Operations	Guess at infrastructure conditions, react to problems, and difficulty in deploying resources efficiently to address problems	Enjoy real-time reporting on infrastructure conditions, predict problems, automate maintenance, deploy resources more efficiently, and save money
ICT Investments	Piecemeal and siloed, deliver suboptimal benefit, and do not realize economies of scale	Centrally planned, deployed across city departments and projects, deliver optimal benefit, provide maximum value and savings
Citizen Engagement	Limited, scattered online connection to citizens and citizens cannot make optimal use of city services—or easily find them	Complete and singular online presence, citizens can easily find and use services, citizens can participate in smart city initiatives, two-way communications between government and people, specialized services focused on the individual citizen, and citizens can both contribute to and access real-time intelligent city data
Sharing Data	Departments and functions are siloed and departments rarely share data or collaborate on initiatives	Departments and functions are integrated and/or shared, data is shared between departments and better correlated with other data services, results are improved, and costs are cut

Source: Smart Cities Readiness Guide [1].

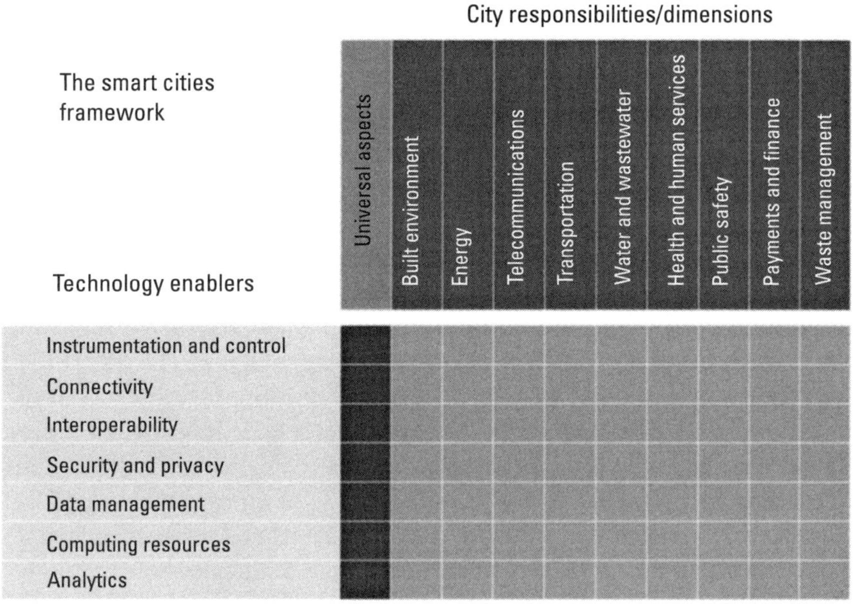

Figure 11.1 The smart cities framework. (Source: Smart Cities Council Readiness Guide [1].)

systems, services, and infrastructures are connected, becoming smart in one area is often dependent on progress being made in another. The following situations illustrate the concept of dependencies.

1. A healthy population requires high water quality. Water systems rely heavily on energy systems to pump and move water through a city's infrastructure. As projects are planned to improve water infrastructure, an assessment of the electrical systems and the distribution grid should also take place. This holistic thinking can avert a major system change or unanticipated course correction during further smart city planning.
2. Improvements in a city's energy infrastructure such as deploying a smart grid, rarely occurs without an understanding of the dependencies between all of a city's systems and services. The distribution lines and underground cables of the energy grid often follow the layout of city streets, which is part of the built environment. This creates dependencies between utility services and the various transportation systems that also rely on streets. The built environment is also a major consumer of electricity and natural gas. The built environment also has the potential to produce electricity. As distributed generation evolves and building owners adopt solar, fuel cell, and related technologies, utilities and city governments should form even closer alliances.

Figure 11.2 takes the smart energy and city services interdependencies one step further, as illustrated by how and where a utility can impact attainment of smart city outcomes through the city's responsibilities. The Harvey Ball scale captures the relative contribution to typical smart city goals from each responsibility or dimension area, as exemplified by Portland, Oregon's Climate Action Plan [10]. The green colored balls are directly addressable by utility participation in the smart city's endeavors.

The key takeaway from Figure 11.2 is that utilities have a naturally aligned interest in outcomes and can materially advance multiple dimensions of a smart city. The last consideration is an assessment of degree to which a transformed utility is positioned to do so. The placement of the full green Harvey Ball in the intersection of greenhouse gas reduction and transportation is an example of maximum impact. The City of Seattle example (see Case Study #1) will present the potential for the incumbent utility to work with the city in its ambitions to reduce greenhouse gas emissions. It will do so by ensuring adequate charging infrastructure. That will include not just commercial areas, but also in residential portions of the city, so that apartment dwellers and low-income housing residents also have access.

The Relationship Between a Transformed Utility and a Smart City

The key takeaways from Table 11.1 are compared with the characteristics of a transformed utility in Table 11.2.

Table 11.2 identifies multiple intersections between smart cities and utilities as outlined here.

- *Assets:* Both manage a large number of a diverse set of assets. The costs of these assets can vary widely and require a capital budget for their periodic replacement. The assets also require maintenance on a periodic basis and the failure of some of the assets can result in a tremendous disruption.

  Both need capabilities (organization, processes, and technology). Examples are the asset management system (AMS) to manage the assets, geospatial information system (GIS) to manage their locations and connectivity, the work and resource management (WRM) system to manage the work, and the field dispatch system (FDS) to manage and dispatch the field crews.

  While most utilities deliver one or two services such as electricity and/or gas that require assets, cities deliver many more services such as water or wastewater, solid waste disposal, built environment, and transportation, all of which require asset management capabilities as described

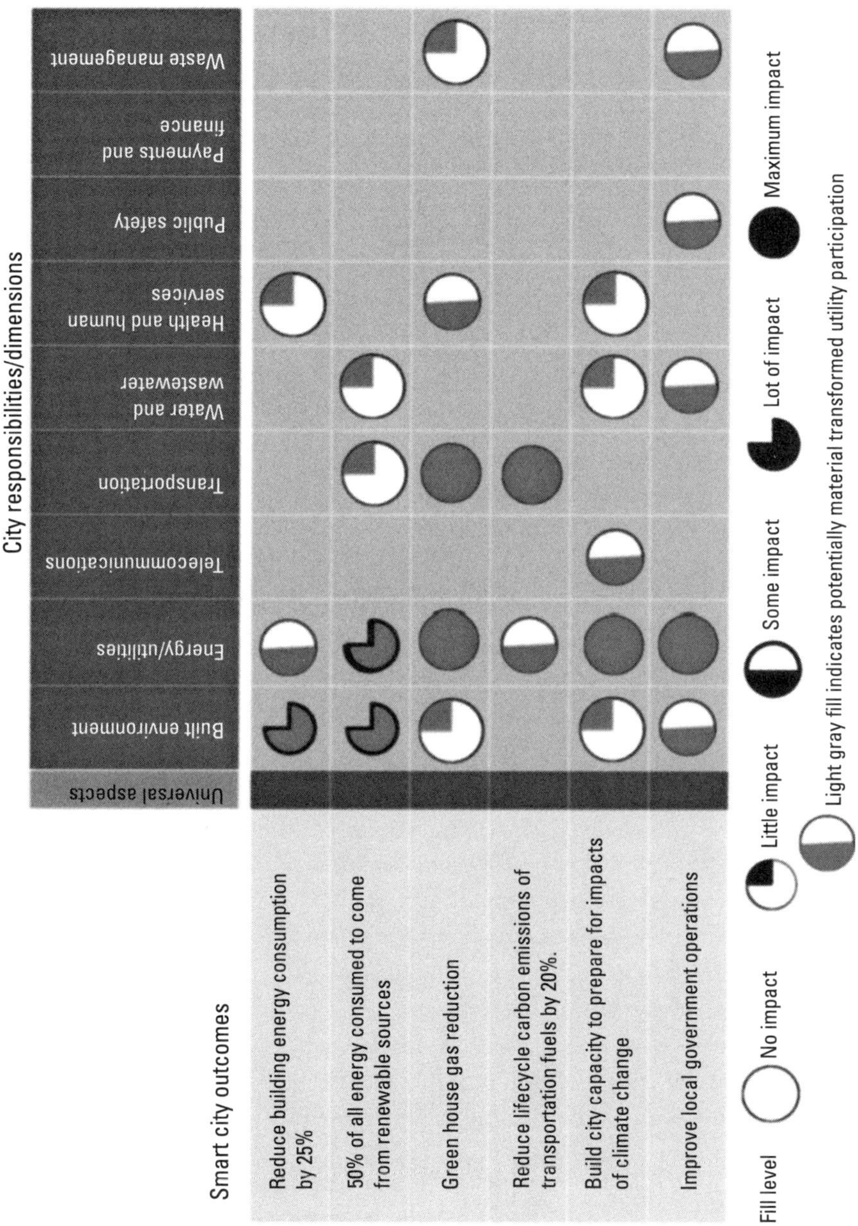

Figure 11.2 Impact assessment of smart cities' goals on dimensions. (Source: Modern Grid Solutions.)

Table 11.2
Comparison Between Smart City and Transformed Utility Characteristics

Key Takeaways From a Smart City	Chapter 10 Definition of a Transformed Utility
The energy or grid is smart, delivering real-time reporting on infrastructure conditions, predicting problems, and anticipating the use of resources, which are environmentally sustainable. Planning is coordinated and holistic, resources are shared and investments are scalable; ICT investments are centrally planned, deployed across city departments, and provide maximum value and savings; departments and functions are integrated and/or shared; and data is shared between departments and better correlated with other data services.	Wires, pipes, and service-centric, not energy-centric; focus on the customer and their desires; manage DERs; redefine planning and asset management; driven data and digital insights; and embrace change and innovate to turn threats into opportunities.

above. Equally important is that many of the assets of the utility and the city are located adjacent to each other.

- *Services to customers:* Both deliver services to a large number of customers and, due to overlapping jurisdictions, often the same customers. In fact, for every utility customer in an overlapping jurisdiction, the city may have two to four customer information systems (CIS) for its services.

  Both the smart city and the utility need capabilities to manage customers, provide services, bill and collect, and interact with assets when repairs or other updates are needed. Both have the need for one or more CISs to be used by customer service representatives to deliver these services.

  The situation in the United States and other countries often leads to multiple entities, some public and some private, delivering services to citizens. All of them have overlapping capabilities using similar systems.
- *Greenhouse gas goals and mitigation:* The city, state, or federal agency typically defines the goals for reaching greenhouse gas emission reductions, while the utility is one of the prime sources of greenhouse gas emissions.

  Utilities take action to reduce their customers' energy footprint through programs such as energy efficiency home improvements, lightbulb replacement, and appliance replacement, but the city tends to be one of the agencies responsible to set the stage for the programs and pay for them.
- *Deal with DERs:* Cities generally take the first step towards implementing DERs in public buildings, setting up electric vehicle charging networks in public places, and act such as converting waste to energy.

While there are a host of capabilities within cities that must interface with these changes, their outcome always impacts the incumbent utility, which must either to deliver energy or ingest energy from new locations.

Every capability described above will come into play to interact and deliver on DER-related services. Other capabilities, such as distribution operations and even energy markets may also become involved. Many of these capabilities tend to be parallel, with a lot of duplication between the city, the electric utility, and other entities that provide these services.

- *Citizen quality of life:* The city government's primary role and mandate is to help with improving its citizens' quality of life. They do this by providing for tangible things such as better transportation, housing, parks, recreation facilities, and access to better jobs. They also focus on less tangible things such as reduced pollution and traffic congestion. Given the anticipated population growth for many cities, a do-nothing approach will lead to poor quality of life and a loss of confidence in city leadership.

  Utilities contribute to the quality of life of individuals by providing reliable and clean power to its customers. This, in turn, allows business to get established there and offer better jobs for all. Replacing dirty power from coal-fired plants with energy from DERs along with a reduced consumption of overall power can also lead to reduced pollution and a higher quality of life.

The electric utility is one of the most valuable teammates a smart city can have. The utility's impact goes beyond electricity. Utilities make significant investments into the efficiency and resiliency of their infrastructure. Cities have a tremendous opportunity to partner with utilities on those investments so that both can increase their returns. Approximately 5% of the U.S. GDP comes directly from the electric utilities, which, in turn, harden the systems, making them more efficient in responding after natural disasters and storms. For example, within a week after Hurricane Irma struck in 2017, 95% of the people and businesses in Florida had their power back on, which was four times faster than after Hurricane Wilma in 2005. These investments provide cities with a valuable opportunity to work with utilities, allowing for a greater outcome for both. When the electric grid is reliable and resilient, the business community often responds by investing in the community, thereby improving the livability, workability, and sustainability of that city [11].

The intersections between an electric utility and a city are important to study. These two entities overlap is almost every aspect of the services provided to citizens.

A Hypothetical City Case Study

Let's look at a hypothetical case study of a series of events, starting with a car hitting an electric utility pole. This hypothetical situation illustrates a common set of occurrences.

A car hits an electric utility pole. The driver is injured. The pole also falls on the road, causing fire, blocking traffic, and splitting the wires, which causes a localized blackout. The blackout is in an area of the city with a hospital. The hospital loses electric utility grid-fed power, which forces it to go to emergency backup power. In addition, the same pole was also carrying communication wires. The neighborhood therefore loses telephone, cable, and internet connectivity.

The problems due to this string of events are not restricted to the electric utility, police, fire, telecom, or medical services area. It is a comprehensive problem. This is also a city problem. The response to this event will typically involve the majority of a city's departments as identified here:

- Fire and police will typically be the first responders. They will be directed to the location by their own control centers, which received the 911 calls or other using specific communications that were implemented just for them. They will potentially need to negotiate traffic jams to get to the accident location.
- Ambulances will be dispatched simultaneously, navigating to the same location to take the injured driver to the hospital. They will be directed by the 911 systems and the associated communications networks to get in and out of the congested area.
- Multiple phone calls will be made to multiple locations, including calls to 911 about the accident, to the utility customer service center about the outages, and to the customer service centers about the outages to power, cable, internet, and phones.
- The multiple phone calls lead to multiple field crews being sent to the same location to fix the different problems on the same pole, each trying to fix the problem so that their critical clients are restored first.
- Once the problem is resolved by each agency, each one of them closes out their respective incidents or outages and the reports get filled in within their own reporting system.

The key takeaways from this hypothetical case are:

- Multiple telecommunications systems performed all the communications related to a single event. The communications included updating key emergency response personnel, and the meters and sensors that send data and control information from the same event to different control centers managed by different agencies.
- Each of these agencies uses its own systems and processes such as CIS, AMS, WRM, GIS, and telecommunications systems. Regardless of each city's specific breakdown of systems, any event such as the one set out hypothetically would require several parallel capabilities.
- Each involved agency has their own organization with people performing similar roles to solve similar problems.

Many cities, large and small, struggle with managing these capabilities while continuing to provide basic and new services during an economic time of shrinking budgets.

Making the Case: Transformed Utility Springboarding a Smart City

The assessment of the response to the hypothetical case study in the previous section put focus on multiple intersecting capabilities in different city agencies functioning in a siloed environment (illustrated in Figure 11.3). As defined earlier, the smart city of the future cannot exist in a sustained manner when each service provider functions independently of the others. This is true regardless of whether the service(s) are provided by the city itself or by another entity or a private or public sector.

There is another aspect to be considered. In most cities, the maturity level of these capabilities is less than optimal. Many of these capabilities have been built on shoestring budgets and with minimal IT support. Some larger cities may be exceptions, but this is the norm with most cities worldwide.

In contrast, utilities tend to have more mature capabilities supported by advanced technology platforms such as AMS, GIS, WRM, FDS, and CIS. This maturity comes from a much longer history of delivering services to their customers. In general, utilities are also held to a higher standard for service delivery, stemming from the critical nature of the product they deliver to their customers—electricity.

The following list provides a set of considerations for smart cities:

- Implementation of a citywide communications infrastructure, designed with a holistic view of the city's needs. This infrastructure would focus on overall usage rather than the agency or individual user.

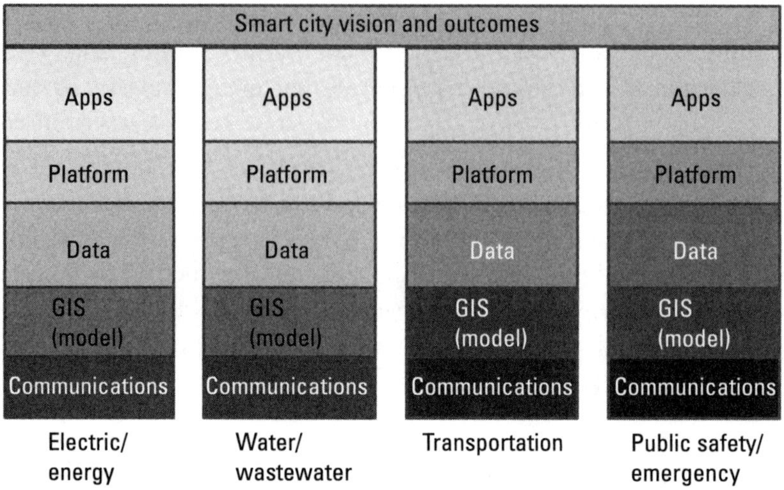

Figure 11.3 Siloed smart city components. (Source: Smart Cities Council Readiness Guide [1] and adapted by Modern Grid Solutions.)

This infrastructure could be city-owned, privately owned, or a combination. An example would be the AMI communications infrastructure implemented by a local utility. The city would be able to use the same network for its own sensors, paying the utility for the service. This change would allow the use of a single infrastructure component for multiple utility services, supporting a smart city vision as described in Figure 11.4.

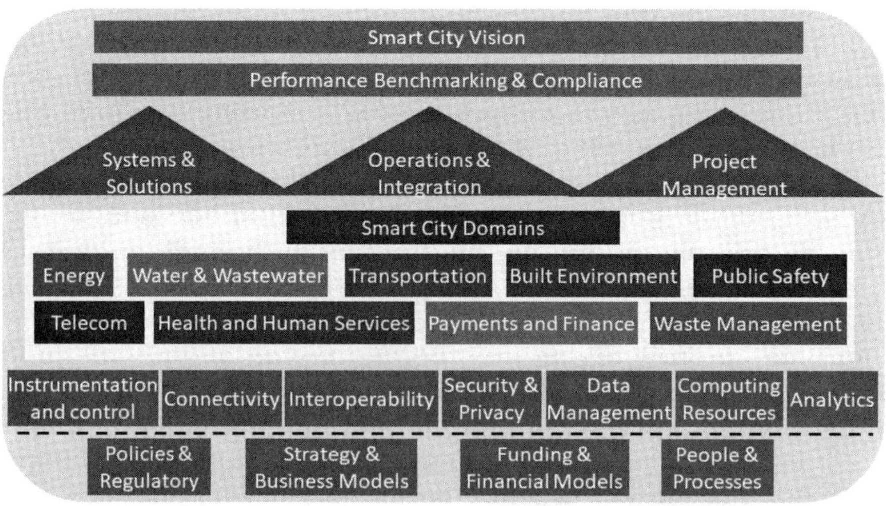

Figure 11.4 Integrated smart city components. (© Modern Grid Solutions.)

This concept already shows up in certain locations. Some utilities and the cities within their jurisdictions already use and possibly share third-party systems. Extending the concept to private networks owned by the utility and used by the city benefits both parties.

- Utility systems such as AMS, GIS, WRM, and FDS that focus on assets, their locations, and work to be done on those assets (planned or unplanned outage) could be leveraged to also include city's assets. Each service providing department could still follow their own processes but getting a consistent set of data from common applications.

Rio Operations Center in Cidade Nova, Rio de Janeiro

Rio Operations integrates 30 municipal and state agencies and utilities to monitor and optimize the operation of the city. In addition, Rio Operations also minimizes occurrences of problems and anticipates solutions. This unit will alert the responsible sectors about the risks and urgent measures that must be taken in cases of emergencies such as heavy rains, traffic accidents, and landslides. The center will run 24 hours a day, 7 days a week, interconnecting information from various city systems for real-time visualization, monitoring, analysis, and performance. This center integrates all stages of crisis management: from anticipation, mitigation, and preparation, to immediate response to events and system feedback with new information that can be used in future cases.

One evening, a 20-story office building next to the municipal theater downtown collapsed, taking two other buildings down with it. The operations center sprang into action. At the operations center, employees alerted the fire and civil defense departments and requested the gas and electric companies to shut down service around the scene. Communications also temporarily closed the subway underneath the site, blocked off the street, dispatched ambulances, alerted hospitals, sent in heavy equipment to remove the rubble, and alerted civil guards to evacuate nearby buildings and secure the accident site. The operations center's Twitter feed alerted followers about blocked streets and alternate routes [12].

Looking back at the hypothetical case presented above, if the information about the assets was in one place and the organizations and processes were integrated, the response to the incident would be well coordinated. The agencies, city or utility and public or private, could work together to solve the problem, resolving the incident faster and more efficiently.

Combining all asset-related activities into one integrated capability would allow cities to deliver a sophisticated set of services at a lower cost.

- The utility's CIS and associated capabilities could be extended to provide similar services to other city agencies such as water and wastewater and solid waste disposal and allow citizens to report other city-related problems such as potholes on roads, accidents, or streetlight issues.

  As we can see in the hypothetical case, if all calls related to the incident were routed to one call center, the response could be more efficient and be internally coordinated among the various agencies to ensure the most optimal response with the best use of scarce city resources. The Rio Operations Center (see inset) is an excellent example of a coordinated control center implementation. Although not a utility control center, lessons learned from that example are very appropriate for this chapter.

For the smart city, it exposes it to one more network and a set of capabilities that has already been installed and works. For the utility, it presents a new revenue stream for an existing asset. Both benefit the ultimate stakeholders, the customers, who are also citizens of the same city.

Impediments to a Transformed Utility Springboarding a Smart City

Using a transformed utility's capabilities to springboard a smart city is not easy. There are several impediments, some on the part of the utility, some from the regulator, and some from the city, which need to be resolved before moving forward.

Financial

The systems and infrastructure components are expensive to install, operate, and maintain. Extending them from utility use to utility and smart city use will add to this cost. The added expense may still be less than if the city agencies implemented these capabilities on their own. Extending the utility's systems will get more expensive if there are multiple cities within the jurisdiction of a single utility.

An agreement between the city and the utility about the sharing of the costs and associated revenues would address this problem. This agreement would need to be approved by the regulator(s).

The business case for this move requires thoughtful consideration of disparate use cases.

Control of Assets

The primary issue of asset control resides in the availability of systems use when required. Each agency and/or utility will have high and low activity periods. Coordinating these peaks and valleys and planning the design and architecture appropriately is very important.

However, a critical aspect that can resolve the control issue would be putting the right service level agreements (SLAs) in place between the utility, as the owner of the systems, and the city, as one of a set of users of the system. Again, consideration of all the use cases prior to instituting an SLA would be paramount.

Communication providers and contractors frequently use SLAs. The utility industry and city governments should become equally familiar with their use.

Data

Both the utility and the city collect a lot of data. This data comes from sensors spread across networks. The use of common systems means that all of the data will be held in common databases for systems owned by the utility and used by the city under SLAs.

The city could consider that data as a part of the transparency compact with its citizens. In fact, smart cities often provide more data to their citizens and many encourage the start-up/app culture to develop smartphone and other online applications to provide advanced visualization and benefits to individuals. However, utilities consider data as a strategic and competitive advantage and may be reluctant to share with anyone external to the organization.

This impediment must be addressed in advance with both the SLA and the architecture to ensure mutually beneficial outcomes.

Technological

The identified systems are complex. Extending their reach outside the utility will require architecture that supports users delivering services, managing assets, and taking actions, while simultaneously delivering different services. Key issues such as cybersecurity and privacy also come into play since city policies may be different from utility regulations.

This is also not new for utilities, many of whom already provide electricity, gas, and even water services to their customers many times through a single set of systems.

Regulatory

Cities are not regulated worldwide, but utilities are. Depending on the jurisdiction, utilities are either regulated at the state or province level or the federal level. This means that any use of the assets outside of the normally stated purpose of that asset needs to be approved by the regulator to ensure that the expenditure was in the best interests of the customers. The customers in this case are the citizens of the city. If the utility spans multiple cities and/or states, multiple sets of regulators need to agree and approve the changes, complicating this impediment.

This should not be a significant issue, as regulators are typically focused on protecting the interests of their constituents.

Experience dictates that the biggest impediment to this movement of a transformed utility springboarding a smart city is the need for utilities and cities to advance from today's typical provider and customer relationship to a partnering relationship based on shared vision, shared objectives, and commitments and ultimately underpinned by trust to do the right thing by one another.

The smart grid allows the traditional utility to become more efficient. Traditional utilities had the monopolistic franchise, which provided a degree of certainty, around investments and earnings, and therefore returns. Over a period of decades, however, the downside was that regulators and customers have been wary of the utility putting its self-interest over everything else. If the utility continues along this path, it will not be allowed to come out of this self-interest box to partner with the city.

However, if the utility can transform and streamline its operations and become more customer-friendly, then its value in the customer's eyes could improve. The customer may perceive the utility as driving value and being operationally sophisticated. As the utility moves on its path to transformation, it must develop an inclusive message and become transformed into a services and solutions marketplace around energy. With this, the utility shows itself to be fully equipped, responsive, and ready to work together with the city.

This, in turn, opens the door to a fundamental repositioning of the utility in the future landscape that will enable it to become a full partner in the smart cities evolution across industry verticals and in partnership with public entities as opposed to just having them as customers. The utility can partner with the city and bring that vision to life. When this state is achieved, the customer and the city will open the doors to the utility to become a partner on the smart city landscape.

The first step in this movement toward a smart city and transformed utility relationship is the creation of a shared vision. A vision that galvanizes the joint commitment, supports specific roles and actions, and moves that vision toward reality.

Case Studies

Case Study #1: Seattle City Light/Seattle Public Utilities

Seattle MetroLab [13] is a part of the national MetroLab Network City-University pairs from 20 major metropolitan areas in the United States. The Seattle entity will engage on multiple smart cities projects over an 18-month period on topics that span technology, data sharing, and infrastructure. The University of Washington's Urban@UW and eScience Institute will work collaboratively with the city on this effort.

The lab will work on algorithms and social bias and emphasizing the "responsible smart city."

Seattle is already a sustainability superstar [14]. Sustainability Tools for Assessing and Rating (STAR) Communities named Seattle the nation's most sustainable city, with the highest score to date in its program that evaluates the livability and sustainability of U.S. communities. STAR partnered with local governments to develop its national rating program to enable communities to evaluate themselves in eight sustainability goal areas: built environment, climate and energy, economy and jobs, education, arts and community, equity and empowerment, health and safety, and natural systems.

According to STAR Communities, Seattle scored well overall, and stands out in several efforts:

- *Commitment to carbon neutrality:* Seattle City Light was the first electric utility in the United States to become carbon-neutral. Additionally, Seattle became one of the first cities in the nation to set a goal of carbon neutrality by 2050.
- *Leading-edge energy efficiency programs:* Community Power Works, an energy efficiency and economic development program, has led to home energy upgrades for over 3,000 families while helping grow local home contractors' businesses. Seattle's Energy Benchmarking program, which requires commercial property owners to track and report energy usage, has one of the highest compliance rates in the country.
- *The Green Seattle Partnership (GSP):* A public-private partnership restores and maintains Seattle's forested parklands. GSP has engaged thousands of volunteers since 2005, resulting in nearly 650,000 hours spent planting over 150,000 trees and restoring over 1,000 acres of parkland throughout the city. GSP also partners with Seattle Goodwill Industries targeting at-risk young adults and providing them with on-the-job training in restoration and trail maintenance.

- *Transportation choices:* Seattle is one of only five cities in United States where less than half of commuters are driving alone to work. Since 2011, Seattle has seen a 59% increase in bicycling and a 27% increase in pedestrian traffic.

Seattle is not doing this to gain kudos. The city and its leadership believe that it is an economic and sustainability necessity. Seattle has grown by 70,000 people in the past 5 years and will grow by 120,000 more people by 2035. This 31% population increase cannot be met with a similar increase in city government costs to provide services to the ever-increasing needs of the city and its citizens.

To this end, the city's departments of transportation, fire, and public utilities all have plans under way to implement smart city projects in the coming years, projects that will streamline traffic flow, foster economic development, and enhance aspects of the city's livability, workability, and sustainability.

Coming Back to the Topic of the Transformed Utility Becoming a Springboard

Seattle City Light is wholly owned by the City of Seattle, which, in addition to electricity, also delivers water, wastewater, sewer, and garbage (including recycling, food, and yard waste) services to its customers through its other public utility, Seattle Public Utilities (SPU).

> To get this right we need to have a shared vision of what a smart city can really be for our city and how we can best move forward collectively to make it a reality.
>
> —*Michael Mattmiller, Seattle CTO*

Seattle City Light is by no means a transformed utility. However, as it moves along the road map to transformation, it has a lot going for it, starting with high reliability and some of the lowest rates in the country. Its owner, the City of Seattle, also wants to be a thought leader in the smart utility and smart city space. The City of Seattle is constantly looking to get ahead of change and define what is possible, and strives to be an example for others. Key examples of actions taken at the city and utility level in support of this case study include the following.

- *IT consolidation:* The city consolidated all its IT functions into a new department, Seattle Information Technology Department (Seattle IT), in April 2016 with the transition to complete by the end of 2018. Some of the objectives behind this move were to:

- Create capacity to deliver on the most important technology projects in the city within current levels of staffing;
- Make IT a strategic business partner, enabling the delivery of IT solutions for a safe, affordable, vibrant, and innovative city;
- Establish consistent standards and priorities for IT investments.

The key takeaway for the city is that while this move was not without its critics, it changed the city's approach towards implementing IT-related systems from the legacy siloed approach to a city-enterprise-wide approach to key systems such as CIS, AMI, and GIS.

- *Permitting process consolidation* (this consolidation effort is still in progress and not completed)*:* The City of Seattle is developing a common portal for residential and commercial building construction in which all the permits required can be requested from one location. This allows for common information such as premises or ownership information to be entered once, and then every city agency including Seattle City Light and SPU will get the same information to perform their assessments and necessary approvals. This singular portal also allows for any changes to be propagated to all the agencies also.

  The key takeaway for the city is that a clear understanding of the process and transparency on where customer requests stand is a key goal of this effort. When completed, this common portal would allow customers to interact from one location to get all their permitting needs met. Happy customer, happy citizen.

- *Interdepartmental cooperation for customer energy solutions:* For the longest period, energy efficiency services were offered under the city's purview, but the impact was only mildly felt by the engineering and distribution planning departments within City Light. Engineering, focused on system peak and energy efficiency, reduced the energy consumption, which, in turn, impacts the peak. Now, with the advent of DERs and EVs, this impact is no longer mild; it is significant. Power is moving in all directions, fault current could come from the customer side of the feeder (if photovoltaic [PV] is installed at the customer), and new loads from EVs are felt in locations much different than before.

  The key takeaway for the city is that under normal circumstances, City Light would have responded to the needs with an extensive engineering study that mostly resulted in the increased capacity in locations when there was more load. Now the two departments work together to assess other nonwire alternatives (such as local storage, demand response) to deliver to the same needs in a way that is transparent to the customer.

- *NeighborGrid (powered by Seattle City Light):* Seattle City Light's Technology Innovation group, in coordination with Customer Energy Solutions, will partner with the city's Office of Housing low-income developers to create mixed-use buildings that will incorporate smart devices, direct current distribution, and integrated electric vehicle charging all powered by solar energy and battery storage for resiliency during outages (as illustrated in Figure 11.5).

  The building will be designed to maximize energy efficiency and comfort for the residents. Additionally, smart devices such as thermostats, smart light controls, and DC distribution systems will be implemented.

  The key takeaway for the city is that NeighborGrids can offer an opportunity for Seattle's low-income housing to become hyper-efficient and resilient communities. These benefits, including lower electric bills, electricity during outages, and healthy living spaces, are usually only available to wealthy homeowners.

  In addition, the benefits to City Light include:

  - *Customer service resiliency:* Batteries and controller technologies to allow the utility to charge or discharge energy storage devices either individually or collectively and provide emergency backup for unplanned power outages.

Figure 11.5 NeighborGrid example. (Source: Seattle City Light.)

- *Grid services and net-metering:* Solar panels will generate power for the building. These panels will be equipped with a smart inverter to provide additional grid benefits.
- *Load control:* Explore schedulable electric vehicle charging and load management equipment. Technologies include programmable thermostats, load response or curtailment, water heater switches, and other related controls.

- *Electrification of transportation (City of Seattle Initiative):* In any city, electricity accounts for about 30% of the emissions. However, since Seattle's electric grid is carbon neutral, 65% of the carbon emissions are from transportation. With Seattle becoming carbon-neutral, focus needed to go to electric vehicle (EV) infrastructure, a necessary requirement to enable citizens to make the switch away from fossil fuel.

  This is an initiative supported by Seattle City Light, Seattle Department of Transportation, and the Seattle Office of Sustainability and Environment. Analysis identified a strong business case with a net benefit for vehicle charging (illustrated in Figure 11.6) because the distribution system can largely handle the increase in transportation load. In addition, the city found a very strong customer demand, particularly electrification of the shared transportation sector.

  The key takeaway for the city is that a perfect example of joint action by the utility in working towards the city's goals to become carbon-neutral, resulting in a positive business case for the utility.

Case Study #2: Intelligent Streetlights—The Smart App for the Smart City

Streetlights are one of the most ubiquitous pieces of equipment in any city. Most streetlights are powered by fluorescent bulbs or high-pressure sodium bulbs. They light downtowns, parks, and public gathering places. They light

Figure 11.6 Public fast charging program. (Source: Seattle City Light.)

industrial centers and residential neighborhoods. In fact, streetlights appear in most places in a city where people live or congregate regularly.

That is changing, albeit slowly. Cities and utilities are becoming aware of the benefits of LED technology, especially the operational cost savings from LED street lights. Worldwide, there is a movement to replace other bulbs with LED bulbs because they generally result in reduced energy consumption. This reduction can range from 50% to 70% depending on the starting point. A recent study at a U.S. city municipal utility demonstrated the conversion to LED would have reduced the streetlight account from the third highest user of energy to the 29th highest, creating a substantial savings.

However, LEDs and simple control systems that allow dimming merely scratch the surface of potential advantages that cities can extract from their lighting assets. Some pioneering cities have recognized that potential and are running pilot programs that test various IoT scenarios. They are installing components enabling streetlights to serve as Wi-Fi hotspots [15].

The following list shows potential improvements for a typical city through changes in their streetlights.

1. Converting the existing lights to LED would provide energy savings upwards of 50% to 60%.
2. Adding timer-based local controls to reduce intensity in the middle of the night or even turn every other light off late at night. Add a motion detector to perform the same action as above but, if motion is detected, would turn all lights in an area to their full intensity provides an additional savings of 10% to 15%.
3. Adding a communications infrastructure would allow for streetlights to be monitored and controlled remotely. This communications infrastructure already used in many cities worldwide, is the same as the one for the AMI system. With this infrastructure, streetlights could interface with the utility to do the following:
    a. AMI system feeding into the CIS to calculate energy consumption and invoice the city or other responsible party;
    b. Outage management system (OMS) to detect outages;
    c. AMS and GIS to track the streetlight assets and locations;
    d. WRM and FDS to manage the work and resources associated with replacing and supporting the streetlight assets in the field;
    e. CIS to address customer calls on streetlight outages.
4. With power and communications available at each streetlight, streetlights could become Wi-Fi hotspots thereby offering Wi-Fi to every citizen. These same Wi-Fi hotspots can be used to deliver announce-

ments and host digital signage advertising nearby events or provide directions.

5. With power and Wi-Fi available at each streetlight, other services become possible, such as:
   a. Putting a camera at locations and providing services such as crime surveillance and/or support to traffic monitoring;
   b. Sensors installed to collect data about air quality, mosquito detection, available parking spaces, and the number of people on the street. One city was even considering gunshot detection on their streetlights in some parts of the city.

Let's look at a few case studies of intelligent streetlights.

Commonwealth Edison (ComEd)

The communities of Bensenville and Lombard in Chicago, Illinois, are getting smart LED streetlights. The installation is part of two proof-of-concept projects being rolled out by ComEd [16]. These projects are designed to demonstrate, in a real-world environment, the array of benefits that can result when cities switch to digital, energy-saving technologies.

At its core, the LED "guts" of a smart streetlight deliver greater energy efficiency and cost savings by consuming as little as one-third of the energy compared to the high-pressure sodium and mercury vapor streetlights they replace. They also last up to 1.5 times as long and offer better quality of light.

However, the benefits of smart streetlights go beyond just that of the LED energy efficiency by leveraging digital technologies. Like smart meters, the streetlights tap into a wireless network, which allows for two-way communication with grid operators. This talking ability enables Bensenville and Lombard communities to work with ComEd to easily adjust the daily scheduling of lights. It also enables the streetlights to take the initiative and call in to notify operators when they need maintenance or replacement.

So far, approximately 750 ComEd-owned fixtures have been replaced with smart streetlights, which will leverage the advanced communications network already being installed by ComEd as part of its ongoing installation of about 4 million smart meters throughout northern Illinois. ComEd owns and maintains approximately 175,000 streetlights in its service territory.

The Chicago streetlights become part of the data collection infrastructure. The new streetlights in Chicago help city officials understand traffic patterns and environmental conditions using sensors. The project will give the city unprecedented insights into air quality, temperature, sound levels, and other environmental data. In addition, the sensors on streetlights will be able to track traffic by counting the mobile devices in use around them. By being able to detect how many people are in an area and where they are going in real time, the city gains data that it can use to alleviate traffic congestion [15]. To offset privacy concerns about the data collection effort, the Chicago project makes its data available for review to citizens through the city's data portal.

Jamaica Public Service

Jamaica Public Service (JPS) [17] is planning to implement a countrywide smart streetlight control program to connect 110,000 smart LED streetlights in population centers around the island, including Kingston, Spanish Town, Negril, and Falmouth, serving more than 600,000 customers.

The streetlights will be centrally managed and controlled, thereby helping to drive additional energy efficiency, which is crucial for high-cost, petroleum-based island energy grids like those found in Jamaica. Smart streetlight controls can help increase citizen safety by alerting a utility to light failures and allowing for speedier repair. The smart streetlight system can detect and report light failure per location, report maintenance and repairs of lights, and measure and report energy usage by each lamp. They can also help increase safety and energy efficiency through adaptive dimming and brightening capabilities based on pedestrian and vehicular traffic. In addition, smart streetlights can facilitate smart technology to support crime fighting and using image sensing, including use of closed-circuit TV.

Conclusions

Cities and utilities are in this new future together: the new future of smart cities amid more technologically complex future.

- *Smart cities:* These struggle as all cities worldwide struggle with the complications of exponential population growth and climate change. This population explosion increases the urgency for cities to react to climate change. These changes require better, cleaner, and sustainable use of resources and new or additional physical infrastructure. Lastly, an aging population requires an ever-greater need for better access to healthcare, public safety, and other government services. This requires more money in an age of shrinking budgets. The smart city must address new conditions and requirements by its citizenry. In addition to going across the service siloes, it also needs to partner with external entities, core to a city's success. If approached properly, the transformed utility could be that partner.
- *Transformed utilities:* These are also struggling with reduced revenues from lower customer energy sales due to a combination of energy ef-

ficient homes and buildings along with the introduction of DERs. The DERs allow customers to reduce their utility bills. Declining sales volumes will lead to a rate escalation firestorm, an unacceptable alternative. Progressive utility commissions from states such as New York and California are reopening the utilities' distribution operating model to enable multiple stakeholders. To survive and thrive, utilities must co-opt private entity investment, which could either be a major threat to asset-based driven earnings. It could also present new earnings opportunities. Utilities can also directly drive more than 50% of state or smart city greenhouse gas reduction or renewables goals and can accelerate smart energy penetration in open retail markets, the electrification of transportation, and smart building energy efficiency or demand response. Utilities should be full partners with the public sector in achieving smart city outcomes: economic vitality, environmental sustainability, and quality of life.

The jointly led smart city and smart energy is an idea whose time has come. Transformed utilities that partner with smart cities can raise the trajectory to greenhouse gas targets by driving smart energy, electrifying transportation, amplifying smart homes and buildings, boosting smaller scale renewables, such as solar and microwaste to energy, by extending their large annual capital expenditure to help cities, who also invest over the long term. In return, the smart city public sector "what's in it for me" (WIFM) approach encourages money or energy asset-based services from the utility to help scale, share cost of larger IT/operations technology (OT) enabling platforms, accelerate outcome attainment, and acquire energy expertise or help to manage the vendor ecosystem. In parallel, the utility WIFM approach of either new asset investment ownership or managed service will help secure their earnings into the future, with the state or city as a powerful advocate to secure PUC approval of asset extensions and/or new managed services to the city, businesses, and citizens.

For this to work, the value proposition of the partnership must be positioned as a smart city and transformed utility win-win. The core utility stakeholders could be swayed, could be led, and could be energized by painting the big vision and outcome trajectory and establish the case that it is worth investing time and effort to come together. There may be a need to amplify the urgency for the utility by indicating that accelerating revenue decline from DER penetration requires they move quickly and embrace the smart city.

The following points may assist to implement the partnership between utility and city:

1. Start with a definition of what a deep partnership looks like. This includes a shared vision with features such as intersecting road map(s), combined governance, shared investment, and mutual resource commitment to run the program. For this vision to work, the city and the utility need to think big. This is a vision that needs to be sold to the citizens and the regulators.
2. Agree on a stakeholder alignment game plan that includes a large group such as state legislators and government, utility commissions, county, city departments, citizens or businesses, and also the vendor ecosystem.
3. Define the approach, the path, and the roles required to mobilize the partnership and phase the work in stages: (1) envision and architect, (2) plan or design, (3) execute pilots in parallel, and (4) move over time to a full implementation.

The vision also needs to have clarity, benchmarks, and steps for the interim states of the journey. This will give a clear picture of what success looks like for the city and the utility in year 1, year 3, year 5, year 10, and so on into the future.

References

[1] Smart Cities Council, "Smart Cities Readiness Guide," http://readinessguide.smartcitiescouncil.com/article/how-use-readiness-guide.

[2] Rouse, M., et al., "Smart City," *TechTarget*, http://internetofthingsagenda.techtarget.com/definition/smart-city.

[3] Smart Cities Council, "Smart Cities India Readiness Guide," 2017.

[4] Saiidi, U., "How Smart Cities Are Building the Future," *CNBC*, February 9, 2017, https://www.cnbc.com/2017/02/09/how-smart-cities-are-building-the-future.html.

[5] Hultin, J., and F. van Beuningen, "Smart Cities: Acceleration, Technology, Cases and Evolutions in the Smart City," https://www.i-scoop.eu/smart-cities-smart-city/#The_essence_of_a_smart_city_and_community.

[6] Totty, M., "The Rise of the Smart City," *Wall Street Journal*, April 2017, https://www.wsj.com/articles/the-rise-of-the-smart-city-1492395120.

[7] Wikipedia, "Smart Cities Mission," http://smartcities.gov.in/content/ and https://en.wikipedia.org/wiki/Smart_Cities_Mission.

[8] Juniper Research, "Barcelona Named 'Global Smart City – 2015,'" 2015, https://www.juniperresearch.com/press/press-releases/barcelona-named-global-smart-city-2015

[9] Sheu, T., "Barcelona's Quest to Become the Smartest City of Them All," April 2015, http://www.barcinno.com/smart-city-barcelona/.

[10] City of Portland and Multnomah County, "Local Strategies to Address Climate Change," June 2015, www.portlandoregon.gov/bps/climate, www.multco.us/sustainability.

[11] Ebi, K., "Cities That Partner with Utilities Have a Powerful Advantage," Smart Cities Council Global," http://na.smartcitiescouncil.com/article/cities-partner-utilities-have-powerful-advantage.

[12] Singer, N., "Mission Control, Built for Cities – I.B.M. Takes 'Smarter Cities' Concept to Rio de Janeiro," *New York Times*, March 4, 2012, http://www.nytimes.com/2012/03/04/business/ibm-takes-smarter-cities-concept-to-rio-de-janeiro.html.

[13] Seattle MetroLab, http://metrolab.uw.edu/.

[14] Enbysk, L., "Why Seattle Is a Sustainability Superstar," Smart Cities Council, September 2014, http://smartcitiescouncil.com/article/why-seattle-sustainability-superstar.

[15] Smart Cities Council, "Smart Street Lighting 101 – How Advanced Street Lighting Systems Can Transform Cities in Remarkable Ways," Foreword, J. Berst.

[16] ComEd, "In Chicago's Western Suburbs, the Street Lights Are Smart Enough to Phone Home," August 2016, https://poweringlives.comed.com/in-chicagos-western-suburbs-the-street-lights-are-smart-enough-to-phone-home/.

[17] "Smart Street Lights Installed in Jamaica," https://smartcitiesworld.net/news/news/smart-street-lights-installed-in-jamaica-1814.

Afterword

As I reflect on my 30+-year utilities industry journey, from operations roles in two large IOUs situated in progressive states, to UK industry privatization, to leading industry restructuring, strategy, and operating model transformation engagements, my grand takeaway is that the advent of pervasive fundamental change is now finally at hand. Where regulators have successively opened up wholesale generation and transmission markets over the years to bring greater solution/fuel source diversity, promote alternative/cleaner technologies, and lift industry efficiency to the benefit of consumers, it is the more recent shift of power from suppliers directly to energy consumers that will lead to the greatest change of all. Technology-enabled disrupters have emerged from all quarters to deliver ever more tailored solutions to customers as their needs are readily captured and expediently targeted. While utilities and regulators cannot control this rapid state of innovation, they can architect a smart energy playing field that will require considerable change for incumbents and ample opportunity for rapidly emerging entities that directly or indirectly impact the industry. These technologies are already altering the fundamental physical landscape of where and how power is generated, stored, delivered, and efficiently utilized for traditional (i.e., light, heat, electronics, commercial and industrial processes) and emerging needs/applications (e.g., mobility/transportation). As true broad-scale competition takes flight at the customer gateway (and beyond), regulators and incumbents must reset downstream regulatory and operating models to promote solution innovation while providing a secure highway through which these solutions will be delivered; in short: *a smart grid with equitable access coupled with a vibrant retail marketplace.*

While those two interwoven concepts are easily said, they are challenging to align, integrate, and operationalize across every dimension (e.g., regulatory policy, utility charter and future capabilities, standards definition and adherence, reliability, security, privacy).

Into this perfect storm of change, threat, and opportunity, Dr. Mani Vadari builds on his first book, *Electric System Operations: Evolving to the Modern Grid*, by extending and crystallizing his vision for how the future may constructively unfold in this compelling sequel, *Smart Grid Redefined: Transformation of the Electric Utility*. While there are many books on smart grid or smart energy, they typically focus on specific technologies or technical architectures, which are essential elements no doubt. However, it is the more recent exposure of the last decade to large scale transformation and downstream industry restructuring that now provide the multidimensional perspective to present an integrated view of how a future state may be organized to run effectively while incorporating disruptive solutions at scale. By calling out the category of technology (e.g., distributed energy resources) that is inducing fundamental shifts on how the customer may be served (disrupting/bypassing the traditional supply, energy delivery, customer service operating model), *Smart Grid Redefined: Transformation of the Electric Utility* creates a provocative forward view that promises far more customer impact for entities who ride (versus try to thwart) the smart energy change wave. In this spirit, Mani reimagines the role the industry may play on the broader stage of smart cities, where societal benefits are further enhanced.

Smart Grid Redefined: Transformation of the Electric Utility brings a pragmatic view to life by setting the change manifesto (it is "when and where," not "if"), defining the integrated nature of future landscape, and how disruptive technologies/advanced capabilities can provide customer and operational value at scale. It is now up to stakeholders to engage and tailor the picture to their circumstances, then decide whether they want to the institutional and personal bets to proactively ride the smart grid change wave or be left in its wake.

Michael Harrison
Retired Executive Partner, Accenture
Founder, President, Realize Visions LLC

Acronyms and Abbreviations

ABB	Asea Brown Boveri
AC	alternating current
ACCC	aluminum conductor composite core (cables)
ACCR	aluminum conductor composite-reinforced
ADMS	advanced distribution management system
AEP	American electric power
AFV	alternative-fuel vehicle
AI	artificial intelligence
AMI	advanced metering infrastructure
AMR	automated meter reading
AMS	asset management system
API	application programming interface
ARRA	American Recovery and Reinvestment Act of 2009
BAS	building automation system
BEMS	building energy management system
BEV	battery electric vehicle
BI	business intelligence
CAES	compressed air energy storage

CAIDI	Customer Average Interruption Duration Index
CCA	community choice aggregation
CDC	Control Data Corporation (RTU Protocol)
CEO	chief executive officer
CERTS	Consortium for Electric Reliability Technology Solutions (established 1998)
CES	community energy storage
CEZ	Ceske Energeticke Zavody (Czechoslovakia)
CHP	combined heat and power
CIM	common information model
CIO	chief information officer
CIS	customer information system
CPP	critical peak pricing
CRM	customer relation management
CSP	concentrating solar power
CT	current transformer
CTO	chief technology officer
CVR	conservation voltage reduction
DA	distribution automation
DARPA	Defense Advanced Research Projects Agency (U.S. Department of Defense)
DC	direct current
DER	distributed energy resources
DERMS	distributed energy resources management system
DEMS	distributed energy management system
DESS	distributed energy storage system
DEV	Drive Electric Vermont
DG	distributed generation
DGA	dissolved gas analysis
DLR	dynamic line rating

DMS	distributed management system
DOE	U.S. Department of Energy
DOD	U.S. Department of Defense
DPFC	distributed power flow controller
DR	demand response
D-SCADA	distribution – supervisory control and data acquisition
DSIP	distribution systems implementation plan
DSM	demand side management
DSO	distribution system operator
DSP	distribution systems platform (provider)
DTE	Detroit Edison
DVAR	dynamic volt-VAR controller
EAM	enterprise asset management
EE	energy efficiency
EERE	energy efficiency and renewable energy
EES	electric energy storage
EIA	Energy Information Agency (DOE-EIA)
EMS	energy management system
EMU	electrical multiple unit (electrically powered train)
ENIAC	Electronic Numerical Integrator and Computer (first computer)
EPA	Environmental Protection Agency
EPRI	Electric Power Research Institute
ERCOT	Electric Reliability Council of Texas
ERDF	European Regional Development Fund
EREV	extended range electric vehicle
ESA	energy storage association
ETR	estimated time of restoration
EU	European Union
EV	electric vehicle

EVSE	electric vehicle supply equipment
FACTS	flexible AC transmission system
FCA	Fiat Chrysler Automobile
FCEV	fuel-cell electric vehicle
FCV	fuel-cell vehicle
FDS	field dispatch system
FERC	Federal Energy Regulatory Council
FIFA	Fédération Internationale de Football Association
FLISR	fault location isolation and system restoration
GDP	gross domestic product
GFAC	Grid Friendly Appliance Controller
GHG	greenhouse gases
GIS	geospatial information system
GPS	global positioning system
GSE	ground service equipment
GTM	GreenTech Media
HA	home automation
HAN	home automation network
HDFS	Hadoop Distributed File System
HEM	home energy manager
HEMS	home energy management system
HEV	hybrid electric vehicle
HTS	high-temperature superconductor
HVAC	heating, ventilation, and air conditioning
HVDC	high-voltage direct current
IC	internal combustion
ICE	internal combustion engine
ICT	information and communication technologies
IED	intelligent electronic device

IEEE	Institute of Electrical and Electronic Engineers
IHD	in-home display
IOU	investor-owned utility
IP	Internet Protocol
ISO	independent system operator
ISO-NE	Independent System Operator – New England
IT	information technology
IVR	interactive voice response
IVVC	integrated volt-VAR control
LCC	line commutated converter
LCD	liquid crystal display
LED	light-emitting diode
LIRR	Long Island Railroad
LV	low voltage
MAIFI	Momentary Average Interruption Frequency Index (power generation)
MDMS	meter data management system
MDPT	market design platform technologies
MIT	Massachusetts Institute of Technology
MOU	memorandum of understanding
MPP	message processing program
MV	medium voltage
MVAR	megavolt-ampere reactive
MW	megawatts
MWFM	Mobile Workforce Management
NAS	sodium sulfur
NEMA	National Electrical Manufacturers Association
NETL	National Energy Technology Laboratory (U.S. Department of Energy)
NIST	U.S. National Institute of Standards and Technology

NISTIR	U.S. National Institute of Standards and Technology Report
NREL	National Renewable Energy Laboratory
NWA	nonwire alternatives
NWS	nonwire solutions
NYISO	New York Independent System Operator
OASIS	Open-Access Same-Time Information System
OGE	Oklahoa Gas and Electric
OMS	outage management system
OT	operations technology
OTS	Operational Training Simulator
OV	overvoltage
PD	partial discharge
PDA	personal digital assistant
PEPCO	Potomac Electric Power Company (Washington, D.C.)
PEV	plugged-in electric vehicle
PHEV	plugged-in hybrid electric vehicle
PHI	PEPCO Holding Inc.
PJM	Pennsylvania Jersey Maryland
PLC	power line communications
PMU	phasor measurement unit
PNNL	Pacific Northwest National Laboratories
PQ	power quality
PSEG	Public Service Electric and Gas
PT	potential transformer
PUC	Public Utility Commission
PURPA	Public Utility Regulatory Policy Act of 1978
PV	photovoltaic
RDSI	renewable and distributed systems integration
REV	Reform the Energy Vision

RF	radio frequency
ROI	return on investment
ROW	right of way
RRPFC	real and reactive power flow control
RTO	regional transmission operator
RTU	remote terminal unit
RWE	Rheinisch-Westfälische Elektrizitätswerke (German power supplier)
SAIDI	System Average Interruption Duration Index
SAIFI	System Average Interruption Frequency Index
SCADA	supervisory control and data acquisition
SCE	Southern California Edison
SE	State estimator
SEP	smart energy profile
SGIG	Smart Grid Infrastructure Grant
SGDP	Smart Grid Demonstration Program
SIR	Standardized Interconnection Requirements (State of New York)
SLA	service-level agreement
SMES	superconducting magnetic energy storage
SOA	service-oriented architecture
SQL	Structured Query Language (database query language)
SST	solid-state transformer
SVC	static VAR compensator
TA	transmission automation
TOU	time of use (rates)
UPB	universal powerline bus
UPFC	unified power flow controller
UPS	uninterruptible power supply
VAR	volt-ampere reactive

VEIC	Vermont Energy Investment Corporation
VSC	voltage source converter
VVO	volt-VAR optimization
WAMS	wide-area monitoring system
WMS	work-management system
WRM	work and resource management

About the Author

An electricity industry visionary and leader, Dr. Subramanian (Mani) Vadari delivers strategic services to a global set of smart grid companies (utilities and vendors) seeking deep subject matter expertise in setting the business and technical direction to develop the next-generation electric/energy system.

Dr. Vadari is the founder and president of Modern Grid Solutions. In addition, he is also an affiliate professor at the University of Washington. Prior to founding Modern Grid Solutions, Dr. Vadari was the vice-president at Battelle, where he led the development of an industry-leading demand management product. Earlier, Dr. Vadari was a partner at Accenture, where he founded and led the global system operations and smart grid practice. His efforts led to establishing Accenture as a major presence in smart grid, winning several foundational projects in North America, Europe, the United Kingdom, and China. Dr. Vadari has also served as a lead engineer at ESCA (now GE Grid Solutions). At ESCA, Dr. Vadari developed their Transient Stability application and Dispatcher Training Simulator product creating a successful niche business and helped launch their market dominance in energy markets.

Dr. Vadari has been delivering solutions focusing on transmission, distribution, and generation operations, energy markets, and smart grid for over 25 years. His roles have ranged from business architect to lead solution delivery. A smart grid and smart cities expert, he offers sought-after perspectives on the entire electric value chain.

A frequent keynoter at industry events in United States and abroad, he also provides advice to several companies and industry groups. Dr. Vadari's book, *Electric System Operations: Evolving to the Modern Grid*, is being used as a textbook at several universities and is trending to 5-star reviews on Amazon. Dr. Vadari has also published more than 50 articles and over 40 blogs on a broad range of subjects.

Index

Recent Artech House Titles in Power Engineering

Andres Carvallo, Series Editor

Advanced Technology for Smart Buildings, James Sinopoli

The Advanced Smart Grid: Edge Power Driving Sustainability, Second Edition, Andres Carvallo and John Cooper

Battery Management Systems, Volume I: Battery Modelings, Gregory L. Plett

Battery Management Systems for Large Lithium Ion Battery Packs, Davide Andrea

Battery Power Management for Portable Devices, Yevgen Barsukov and Jinrong Qian

Big Data Analytics for Connected Vehicles and Smart Cities, Bob McQueen

Design and Analysis of Large Lithium-Ion Battery Systems, Shriram Santhanagopalan, Kandler Smith, Jeremy Neubauer, Gi-Heon Kim, Matthew Keyser, and Ahmad Pesaran

Designing Control Loops for Linear and Switching Power Supplies: A Tutorial Guide, Christophe Basso

Electric Power System Fundamentals, Salvador Acha Daza

Electric Systems Operations: Evolving to the Modern Grid, Mani Vadari

Energy Harvesting for Autonomous Systems, Stephen Beeby and Neil White

GIS for Enhanced Electric Utility Performance, Bill Meehan

Introduction to Power Electronics, Paul H. Chappell

Introduction to Power Utility Communications, Dr. Harvey Lehpamer

IoT Technical Challenges and Solutions, Arpan Pal and Balamuralidhar Purushothaman

Plug-in Electric Vehicle Grid Integration, Islam Safak Bayram and Ali Tajer

Power Grid Resiliency for Adverse Conditions, Nicholas Abi-Samra

Power Line Communications in Practice, Xavier Carcelle

Power System State Estimation, Mukhtar Ahmad

A Systems Approach to Lithium-Ion Battery Management, Phil Weicker

Signal Processing for RF Circuit Impairment Mitigation in Wireless Communications, Xinping Huang, Zhiwen Zhu, and Henry Leung

The Smart Grid as An Application Development Platform, George Koutitas and Stan McClellan

Smart Grid Redefined: Transformation of the Electric Utility, Mani Vadari

Synergies for Sustainable Energy, Elvin Yüzügüllü

Telecommunication Networks for the Smart Grid, Alberto Sendin, Miguel A. Sanchez-Fornie, Iñigo Berganza, Javier Simon, and Iker Urrutia

For further information on these and other Artech House titles, including previously considered out-of-print books now available through our In-Print-Forever® (IPF®) program, contact:

Artech House
685 Canton Street
Norwood, MA 02062
Phone: 781-769-9750
Fax: 781-769-6334
e-mail: artech@artechhouse.com

Artech House
16 Sussex Street
London SW1V 4RW UK
Phone: +44 (0)20 7596-8750
Fax: +44 (0)20 7630-0166
e-mail: artech-uk@artechhouse.com

Find us on the World Wide Web at: www.artechhouse.com